Cellular Dialogues in the Holobiont

Evolutionary Cell Biology

Series Editors

Brian K. Hall – Dalhousie University, Halifax, Nova Scotia, Canada

Sally A. Moody – George Washington University, Washington DC, USA

Editorial Board

Michael G. Hadfield – University of Hawaii, Honolulu, USA

Kim Cooper – University of California, San Diego, USA

Mark Martindale – University of Florida, Gainesville, USA

David M. Gardiner – University of California, Irvine, USA

Shigeru Kuratani – Kobe University, Japan

Nori Satoh – Okinawa Institute of Science and Technology, Japan

Sally Leys – University of Alberta, Canada

Science publisher

Charles R. Crumly – CRC Press/Taylor & Francis Group

Published Titles

Cells in Evolutionary Biology: Translating Genotypes into Phenotypes – Past, Present, Future
Edited by Brian K. Hall and Sally Moody

Deferred Development: Setting Aside Cells for Future Use in Development in Evolution
Edited by Cory Douglas Bishop and Brian K. Hall

Cellular Processes in Segmentation
Edited by Ariel Chipman

Cellular Dialogues in the Holobiont
Edited by Thomas C. G. Bosch and Michael G. Hadfield

Cellular Dialogues in the Holobiont

Edited by
Thomas C. G. Bosch
Michael G. Hadfield

CRC Press
Taylor & Francis Group
Boca Raton London New York

CRC Press is an imprint of the
Taylor & Francis Group, an **informa** business

First edition published 2021
by CRC Press
6000 Broken Sound Parkway NW, Suite 300, Boca Raton, FL 33487-2742

and by CRC Press
2 Park Square, Milton Park, Abingdon, Oxon, OX14 4RN

First issued in paperback 2022

© 2021 Taylor & Francis Group, LLC
CRC Press is an imprint of Taylor & Francis Group, an Informa business

No claim to original U.S. Government works

**Visit the Taylor & Francis Web site at
http://www.taylorandfrancis.com**

**and the CRC Press Web site at
http://www.crcpress.com**

ISBN: 978-0-367-51375-7 (pbk)
ISBN: 978-0-367-22881-1 (hbk)
ISBN: 978-0-429-27737-5 (ebk)

DOI: 10.1201/9780429277375

Typeset in TimesLTStd
by Nova Techset Private Limited, Bengaluru & Chennai, India

Contents

Series Preface (Evolutionary Cell Biology)

In recent decades, the central and integrating role of evolution in all of biology was reinforced as the principles of evolutionary biology were integrated into other biological disciplines, such as developmental biology, ecology, and genetics. Major new fields emerged, chief among which are Evolutionary Developmental Biology (or Evo-Devo) and Ecological Developmental Biology (or Eco-Devo). Evo-Devo, inspired by the integration of knowledge of change over single life spans (ontogenetic history) and change over evolutionary time (phylogenetic history), produced a unification of developmental and evolutionary biology that is generating many unanticipated synergies. Molecular biologists routinely employ computational and conceptual tools generated by developmental biologists (who study and compare the development of individuals) and by systematists (who study the evolution of life).

Evolutionary biologists routinely use detailed analysis of molecules in experimental systems and in the systematic comparison of organisms. These integrations have shifted paradigms and answered many questions once thought intractable. Although slower to embrace evolution, physiology is increasingly being pursued in an evolutionary context. So too, is cell biology, a rich field in biology with a long history. Technology and instrumentation have provided cell biologists the opportunity to make ever more detailed observations of the structure of cells and the processes that occur within and between cells of similar and dissimilar types. In recent years, cell biologists have increasingly asked questions whose answers require insights from evolutionary history. As just one example: How many cell types are there and how did these different cell types evolve? Integrating evolutionary and cellular biology has the potential to generate new theories of cellular function and to create a new field, which we term *Evolutionary Cell Biology*.

A major impetus in the development of modern Evo-Devo was a comparison of the evolutionary behavior of cells, evidenced in Stephen J. Gould's 1979 proposal of changes in the timing of the activity of cells in development (heterochrony) as a major force in evolutionary change and in Brian Hall's 1984 elaboration of the relatively small number of mechanisms used by cells in development and in evolution. Given this conceptual basis and the advances in genetic analysis and visualization of cells and their organelles, cell biology is poised to be transformed by embracing the approaches of Evo-Devo as a means of organizing and explaining diverse empirical observations and testing fundamental hypotheses about the cellular basis of life.

Importantly, cells provide the link between the genotype and the phenotype, both during development and in evolution. No books that capture this cell focus exists.

Hence the proposal for a series of books under the general theme of "Evolutionary Cell Biology" (ECB) to document, demonstrate, and establish a long-sought level in evolutionary biology: namely, the central role played by cellular mechanisms in translating genotypes into phenotypes in all forms of life.

Brian K. Hall
Sally A. Moody

Preface

A dialogue that matters: microbe–host interactions in protists, plants, and animals.

Animal evolution appears intimately linked to the presence of microbes. A continuously increasing number of studies demonstrate that individuals from sponges to humans are not solitary, isolated entities, but consist of complex communities of many species that likely coevolved during a billion years of coexistence (McFall-Ngai et al., 2013). This progress is due in large part to the application of "metagenomic" methods: a series of experimental and computational approaches that allows a microbial community's composition to be defined by DNA sequencing without having to culture its members. This work has yielded catalogues of microbial species, many previously unknown and belonging to all three domains of life, as well as lists of millions of microbial genes collectively known as a host's microbiome. Research on host–microbe interactions has become an emerging cross-disciplinary field.

Contrary to the classical view that microbes are primarily pathogenic and disease-causing, there is now a multitude of studies indicating that a host-specific microbiome provides functions related to metabolism, immunity, development, and environmental adaptation to its animal, plant, or fungal host. Similarly, microbes have been documented as important for environmental sensing, inducing colony formation and sexual reproduction in choanoflagellates, and contributing to developmental transitions and life history traits such as development pace and longevity. Similarly, the microbiome of plants impacts the phenotype and fitness of the plant host. It has become increasingly clear that animals, plants, and fungi evolved in a microbial world and that multicellular organisms rely on their associated microbes to function. Symbiosis appears as a general principle in eukaryotic evolution (Douglas, 2014).

Why this book?

This is a book about interactions between protists, plants, and animals as hosts and their closely associated microbes. There are at least three major reasons why those interactions deserve our attention. First, research on microbial communities inhabiting protists, animals, and plants has progressed at a spectacular rate in the last decade, and interactions between microbes and their hosts have been found to be ecologically and evolutionarily transformative across the tree of life. Second, even though a broader appreciation of the importance of microbes has emerged, little has been written about how widespread and conserved the mechanisms underlying symbiotic interactions really are. Third, defining the individual microbe–host conversations, that is, the cellular dialogues in the holobiont, is a challenging but necessary step on the path to understanding the function of the associations as a whole and across the global biota. We asked, is there a common "language" employed in host–microbe interactions and, if so, does that language provide evidence of common evolutionary threads in the establishment of symbioses? These host–microbe symbiont interactions, in a number of plant and invertebrate hosts, are at the heart of this book.

This book examines how the growing knowledge of the huge range of protist–, animal–, and plant–bacterial interactions, whether in shared ecosystems or intimate symbioses, is fundamentally altering our understanding of biology. The establishment and maintenance of these symbioses and their contributions to the health and survival of both/multiple partners rely on continuous cell-to-cell communication. This dialogue matters because it may be concerned with cellular regulation, nutrition, or provision of signals for physiological homeostasis and development of the host. The book includes 16 chapters devoted to exploring, explaining, and exposing these dialogues across a broad spectrum of protist, plant, and animal eukaryotes to a broad field of biologists. Recent technological advances have greatly accelerated the ability to generate genetic and genomic tools to develop practically any eukaryotic species into an accessible and convenient research object for understanding the interactions between symbionts and their hosts. Many chapters in this volume include descriptions of organisms that demonstrate the potential for "non-model" organisms (sponges, corals, Hydra, sap-sucking insects, weevil beetles, and squid) to provide important insights into the fundamental processes of symbiosis. This comparative approach leverages the power of cross-species analyses and promises a new understanding of the fundamental drivers controlling a strictly microbe/symbiont-dependent lifestyle and its evolutionary consequences. It may also impact how we approach complex environmental diseases such as coral bleaching. *"Enhanced understanding of the interactions between marine invertebrates and their microbial communities is urgently required as coral reefs face unprecedented local and global pressures and as active restoration approaches, including manipulation of the microbiome, are proposed to improve the health and tolerance of reef species"* (O'Brien et al., 2019). We are convinced, therefore, that assembling chapters describing symbiotic interactions across species in this book will significantly contribute to new understandings of the ways multiple organisms interact and affect each other.

Thomas C. G. Bosch
Michael G. Hadfield

References

Douglas, A.E. 2014. Symbiosis as a general principle in eukaryotic evolution. *Cold Spring Harb Perspect Biol.* 6(2): pii: a016113. doi: 10.1101/cshperspect.a016113.

McFall-Ngai, M., Hadfield, M.G., Bosch, T.C.G., Carey, H.V. et al. 2013. Animals in a bacterial world, a new imperative for the life sciences. *Proc. Natl. Acad. Sci.* 110: 3229–36.

O'Brien, P.A., Webster, N.S., Miller, D.J., Bourne, D.G. 2019. Host-microbe coevolution: applying evidence from model systems to complex marine invertebrate holobionts. *mBio.* 10(1): doi: 10.1128/mBio.02241-18.

Contributors

Hisashi Anbutsu
Computational Bio Big-Data Open
    Innovation Laboratory
National Institute of Advanced Industrial
    Science and Technology
Tokyo, Japan

and

Bioproduction Research Institute
National Institute of Advanced
    Industrial Science and Technology
Tsukuba, Japan

John M. Archibald
Department of Biochemistry and
    Molecular Biology
Dalhousie University
Halifax, Nova Scotia, Canada

Jay Bathia
Zoological Institute
Kiel University
Kiel, Germany

Gordon Bennett
Department of Life and Environmental
    Sciences
University of California, Merced
Merced, California

Aileen Berasategui
Department of Biology
Emory University
Atlanta, Georgia

Thomas C. G. Bosch
Zoological Institute
Kiel University
Kiel, Germany

Anny Cardenas
Department of Biology
University of Konstanz
Konstanz, Germany

Rebecca Choi
Department of Integrative Biology
University of California, Berkeley
Berkeley, California

Morgan J. Colp
Department of Biochemistry and
    Molecular Biology
Dalhousie University
Halifax, Nova Scotia, Canada

Sebastian Fraune
Heinrich-Heine-Universität Düsseldorf

Marnie Freckelton
Kewalo Marine Laboratory
Pacific Biosciences Research Center
University of Hawai'i at Mānoa
Honolulu, Hawaii

Takema Fukatsu
Bioproduction Research Institute
National Institute of Advanced
    Industrial Science and Technology
Tsukuba, Japan

and

Department of Biological Sciences
Graduate School of Science
University of Tokyo
Tokyo, Japan

and

Graduate School of Life and
    Environmental Sciences
University of Tsukuba
Tsukuba, Japan

Karen J. Guillemin
Humans and the Microbiome Program
CIFAR
Toronto, Ontario, Canada

Michael G. Hadfield
Kewalo Marine Laboratory
Pacific Biosciences Research Center
University of Hawai'i at Mānoa
Honolulu, Hawaii

Janine Haueisen
Environmental Genomics Group
Christian-Albrechts University
Kiel, Germany

and

Environmental Genomics Group
Max Planck Institute for Evolutionary
    Biology
Plön, Germany

Ute Hentschel
GEOMAR Helmholtz Centre for
    Ocean Research
and
Christian-Albrechts-Universität
Kiel, Germany

Martin T. Jahn
GEOMAR Helmholtz Centre for Ocean
    Research
Kiel, Germany

Dan Kim
Department of Integrative Biology
University of California, Berkeley,
Berkeley, California

Stacy Li
Department of Integrative Biology
University of California, Berkeley
Berkeley, California

Colin R. Lickwar
Department of Molecular Genetics and
    Microbiology
Duke Microbiome Center
Duke University School of Medicine
Durham

Nicole B. Lopanik
Affiliations School of Earth and
    Atmospheric Sciences
School of Biological Sciences
Georgia Institute of Technology
Atlanta, Georgia

Cecile Lorrain
Environmental Genomics Group
Christian-Albrechts University
Kiel, Germany

and

Environmental Genomics Group
Max Planck Institute for Evolutionary
    Biology
Plön, Germany

Michelle S. Massaquoi
Institute of Molecular Biology
University of Oregon
Eugene, Oregon

Meril Massot
Department of Integrative Biology
University of California, Berkeley
Berkeley, California

Margaret J. McFall-Ngai
Kewalo Marine Laboratory
Pacific Biosciences Research Center
University of Hawai'i at Mānoa
Honolulu, Hawaii

Timo Minten-Lange
Heinrich-Heine-Universität Düsseldorf

Vivek Narayan
Department of Integrative Biology
University of California, Berkeley
Berkeley, California

Brian T. Nedved
Kewalo Marine Laboratory
Pacific Biosciences Research Center
University of Hawai'i at Mānoa
Honolulu, Hawaii

Lucía Pita
GEOMAR Helmholtz Centre for Ocean
    Research
Kiel, Germany

Claudia Pogoreutz
Department of Biology
University of Konstanz
Konstanz, Germany

Nils Rädecker
Department of Biology
University of Konstanz
Konstanz, Germany

Jean-Baptiste Raina
Climate Change Cluster
University of Technology Sydney
Sydney, Australia

John F. Rawls
Department of Molecular Genetics and
    Microbiology
Duke Microbiome Center
Duke University School of Medicine
Durham

Hassan Salem
Mutualisms Research Group
Max Planck Institute for Developmental
    Biology
Tübingen, Germany

Lara Schmittmann
GEOMAR Helmholtz Centre for Ocean
    Research
Kiel, Germany

Hinrich Schulenburg
Zoological Institute
University of Kiel
Kiel, Germany

Michael Shapira
Department of Integrative Biology
University of California, Berkeley
Berkeley, California

Samuel Slowinski
Department of Integrative Biology
University of California, Berkeley
Berkeley, California

Eva H. Stukenbrock
Environmental Genomics Group
Christian-Albrechts University
Kiel, Germany

and

Environmental Genomics Group
Max Planck Institute for Evolutionary
    Biology
Plön, Germany

Christian R. Voolstra
Department of Biology
University of Konstanz
Konstanz, Germany

Virginia Weis
Department of Integrative Biology
Oregon State University
Corvallis, Oregon

1 When does symbiosis begin? Bacterial cues necessary for metamorphosis in the marine polychaete *Hydroides elegans*

Marnie Freckelton and Brian T. Nedved

Contents

1.1   The symbiosis space

Symbiosis, the living together of unlike organisms (de Bary 1879), has become one of the most important concepts in biology today. This has been especially true since the discovery of the prevalence of microbiomes (Woese 2004; Moran and Dunbar 2006) and the role of the holobiont (Margulis 1971) in shaping both individual health and development, the evolution of biological complexity, and ecosystem-wide processes (Bordenstein and Theis 2015). It is through this cooperation of individuals that new habitats and energy sources can be accessed and utilized (Moran and Dunbar 2006; Fisher et al. 2017). In the marine environment, cooperative prokaryotic-eukaryotic

1

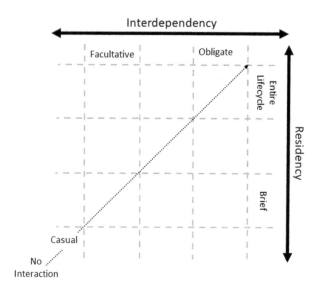

FIGURE 1.1   Defining the symbiosis space. X-axis represents the level of interdependency between host and symbiont from casual to facultative to obligate. Y-axis represents the length of the interaction from brief to whole lifecycle. Dotted arrow represents a possible trajectory from free-living to endosymbiotic state.

interactions have drastically increased the rate at which organisms adapt to utilize new environments and energy sources (Apprill 2020).

Symbiosis, however, is not a discrete concept but rather extends across a spectrum of interactions defined by three axes: impact on host fitness (positive to negative), the dependency of the relationship, and the residency of the symbiotic relationship. In prokaryotic–eukaryotic symbioses, the evolutionary relationship is considered to start with complex, nonessential interactions, progress to facultative and then finally to obligate symbiotic relationships (Moya et al. 2008). At the terminus of this progression, we have obligate endosymbiotic relationships (mutualisms) where neither individual can survive independently, for instance, the pea aphid *Acyrthosiphon pisum* and its endosymbiont *Buchnera aphidicola* (Scarborough et al. 2005; Shigenobu and Wilson 2011), whereas legumes and rhizobial bacteria, a facultative commensalism, sit earlier in the continuum. Current models of symbiosis are heavily skewed towards the terminus of this evolutionary progression, obligate mutualistic endosymbiosis. However, many symbioses in the marine environment fall early in the continuum, and, while some of these relationships progress to obligate interactions, many are maintained at the casual or facultative level (Fisher et al. 2017). Thus, important questions arise: how do symbioses evolve from casual to facultative to dependent relationships? When does symbiosis begin (Figure 1.1)?

1.2   Chemical cues mediate symbiotic interactions

The successful establishment of many classic models of horizontally acquired symbiosis (*Euprymna-Vibrio*; legumes-Rhizobia/mycorrhizal fungi) depend upon successful and specific signal exchanges between partners. Biologists have begun to decipher the chemical languages and cellular and molecular events involved in these exchanges revealing them to be multistage events of some complexity. The complexity of these exchanges strongly suggests that (in their current form) they almost certainly postdate the evolutionary start of the relationship, and likely have limited value in discovering how symbioses develop initially. Furthermore, in the marine environment, most host organisms first interact with multispecies bacterial biofilms as larvae (Lema et al. 2019). How are the "right" signals heard amongst the noise? We argue that examination of pre-symbiotic bacteria–eukaryotic interactions in non-model organisms will greatly enhance our understanding of the interactions and ecological drivers that establish and maintain symbioses.

1.3   How do specific symbiotic interactions begin? Examples from the pre-symbiosis space

Inter-kingdom associations that exist early in the symbiosis continuum (Figure 1.1) can inform our understanding of how symbioses develop. Bacteria can interact with eukaryotes in a number of ways that are not classically defined as symbiosis but result in modifications to the eukaryote's physiology or behavior. While not strictly symbioses, these interactions can be viewed as existing on the edge of the symbiotic continuum space defined above. For example, rosette (colony) formation and sexual reproduction in choanoflagellates are induced and inhibited by different bacterial metabolites (Levin and King 2013). Rosette formation in *Salpingoeca rosetta* is induced by a combination of lipid molecules from the prey bacterium *Algoriphagus machipongonensis* (Alegado et al. 2012; Beemelmanns et al. 2014; Woznica and King 2018). In this example, one form of eukaryotic development is dependent on the presence of appropriate prey microbes; however, the microbes themselves experience only the negative selection pressure of being eaten. Although not a symbiosis, the choanoflagellate example demonstrates how interactions with bacteria can not only drive eukaryotic development but also reveal potential drivers for multicellularity.

Marine sponges and corals provide another valuable space in which to examine not only the interplay between host eukaryotes and their associated bacteria, but also how symbionts can be found amongst the "noise" of diverse bacterial populations. Marine sponges contain diverse consortia of bacteria: currently at least 39 microbial phyla have been demonstrated to associate with sponges (Pita et al. 2018). Additionally, the density of those bacteria can reach 10^9 microbial cells/cm^3 of sponge (Hentschel et al. 2006). The coral microbiome is also a complex consortium: up to 69 phyla have been identified in association with stony corals (Pollock et al. 2018; Huggett and Apprill 2019). In contrast to sponges and corals, most other animals interact with only 3–5 bacterial phyla (Kostic et al. 2013). The natural variability of both microbes and symbioses is reflected in the variability of interactions, from transients to true symbionts, and positive to negative (Figure 1.1). While unravelling the complex mechanisms that allow hosts to identify friend from foe amongst such complex communities remains experimentally challenging (Pita et al. 2016), a number of commonalities are emerging, for example, core groups have been identified that associate with taxonomic consistency (Rohwer et al. 2002; Erwin et al. 2012; Hester et al. 2015; Ainsworth and Gates 2016; Bourne et al. 2016; Thomas et al. 2016; Ainsworth et al. 2017). Many of the identified core microbiomes of these groups have also been demonstrated to possess genes valuable for vitamin synthesis (Thomas et al. 2010; Fan et al. 2012; Fiore et al. 2015; Lackner et al. 2017), nutrient cycling (Rohwer et al. 2002), and are valuable contributors to the highly effective chemical defenses of many host species (Ritchie 2006; Wilson et al. 2014; Lackner et al. 2017). Importantly, the bacterial communities of these animals can vary considerably across their life histories (Apprill et al. 2009; Sharp et al. 2010), providing an important avenue of approach to begin understanding how cues and signals might be heard within this overwhelming diversity.

1.4   Bacterially induced metamorphosis of marine invertebrate animals

Perhaps, the most diverse examples of bacteria-regulated eukaryotic development are seen in the regulation of metamorphosis in marine invertebrates. For more than 100 years, marine biologists have sought to understand how the minute larvae of marine invertebrate animals, spawned into the global ocean, find and establish themselves in the right ecological settings for survival, growth, and reproduction. During this transition, larvae contact submerged surfaces and interact with dense assemblages of bacteria on these surfaces. Cues produced by either specific species, strains, or assemblages of bacteria residing in these microbial films are sensed by larvae, and the sensation of these cues triggers both behavioral and developmental cascades that culminate in an ecological shift from a planktonic to a benthic existence.

The interaction between bacterial biofilms and marine invertebrate larvae appears to be the norm rather the exception among diverse invertebrate taxa (reviewed by Hadfield 2011, 2014). Bacterially induced metamorphosis has been observed for nearly every major marine phylum, from sponges (Woollacott and Hadfield 1996; Wahab et al. 2014; Whalan and Webster 2015), cnidarians (Negri et al. 2001; Seipp et al. 2007; Tran and Hadfield 2011; Sneed et al. 2014; Klassen et al. 2015; Sharp et al. 2015; Tebben et al. 2015; La Marca et al. 2018), polychaete worms (Hamer et al. 2001; Huang and Hadfield 2003; Shimeta et al. 2012; Sebesvari et al. 2013; Freckelton et al. 2017; Vijayan and Hadfield 2020), molluscs (Bao et al. 2007; Chiu et al. 2007; Alfaro et al. 2011; Campbell et al. 2011; Pachu et al. 2012; Yang et al. 2013; Liang et al. 2020), arthropods (Lau et al. 2005), echinoderms (Cameron and Hinegardner 1974; Huggett et al. 2006; Nielsen et al. 2015), and chordates (Roberts et al. 2007). This breadth of examples strongly suggests that the establishment and maintenance of most benthic marine populations, including those on rocky shores, coral reefs, mudflats, and subtidal regions, depends largely on specific bacterial stimulation for recruitment. Outside of the polychaete discussed in this chapter, however, the chemical identities of the metamorphic inducers are almost entirely unknown. The exceptions to this are tetrabromopyrrolle (TBP), involved in coral metamorphosis (Tebben et al. 2011; Sneed et al. 2014), and flagellin proteins involved in the settlement and metamorphosis of mussel larvae (Liang et al. 2020). Although more recently, Tebben et al. (2015) cast doubt on the ecological relevance of TBP as a metamorphic inducer. This lack of knowledge regarding the chemical identities of these cues reflects our lack of knowledge of the diversity of bacteria that stimulate larvae to settle, as well as, the mechanisms through which these bacteria act.

The potential for the bacteria—involved in induction of metamorphosis—to have an ongoing role in the life of the marine invertebrate is largely unexplored. There is evidence, however, that larvae and newly settled juveniles of the acorn barnacle *Amphibalanus (=Balanus) amphitrite* may incorporate elements of bacterial biofilms into their cements. Bacteria appear to be incorporated into the antennule cement plaques of cyprid larvae of *A. amphitrite* (Aldred et al. 2013). These plaques of bacteria remain associated with antennule cements and early stage spat, but bacteria in these biofilms are later killed by secretions from the juvenile barnacle as it continues to grow and secrete adhesives (Essock-Burns et al. 2017). To date, however, the link between metamorphosis-inducing bacteria and the bacteria that the adult invertebrate associate with remains to be elucidated.

1.5   Bacterial induction of metamorphosis in *Hydroides elegans*

The serpulid polychaete, *Hydroides elegans*, is a globally distributed member of the warm water fouling community (Bastida-Zavala and ten Hove 2002; Bastida-Zavala and ten Hove 2003; Sun et al. 2018). Larvae of *H. elegans* rapidly colonize newly submerged surfaces, where they can reach high densities, before being overgrown by other fouling organisms (Figure 1.2). Examinations of the bacterial communities that induce settlement of *H. elegans* have demonstrated that monospecific biofilms of taxonomically diverse bacteria, including both Gram-positive and Gram-negative species, isolated from wild biofilms, will induce larvae of *H. elegans* to settle and metamorphose (Unabia and Hadfield 1999; Lau and Qian 2001; Huang and Hadfield 2003; Hung et al. 2009; Freckelton et al. 2017). However, not all strains of bacteria isolated from the natural habitat of the worm have inductive properties (Unabia and Hadfield 1999; Lau and Qian 2001; Lau et al. 2005; Vijayan and Hadfield 2020). In one study, only eight of 18 isolated strains induced settlement and metamorphosis (Unabia and Hadfield 1999). These inductive strains had varied metabolic capabilities and were from diverse bacterial taxa. Due to these differences, the settlement-inducing cues produced by these bacteria are probably different. Due to its global distribution, it is not surprising that *H. elegans* evolved to respond to multiple bacterial cues as an adaption for colonizing submerged surfaces in harbors around the world.

Once a correct cue has been detected by a larva, metamorphosis begins within a few minutes and proceeds rapidly. As an early part of the metamorphic cascade in *H. elegans*, larvae secrete a proteinaceous primary tube that secures them to a surface (Figure 1.2; Carpizo-Ituarte and Hadfield 1998) throughout the initial stages of its metamorphosis. We have evidence that attachment to a biofilm increases the attachment strength of the primary tube significantly (Zardus et al. 2008). That is, the primary-tube cements that are secreted by a larva early in its metamorphosis adhere more firmly to biofilmed surfaces than clean ones (Figure 1.2e; Zardus et al. 2008). An interaction between cements in the primary tube and the biofilm may allow metamorphosing larvae to adhere more tightly to surfaces during this critical shift between a planktonic and a benthic lifestyle.

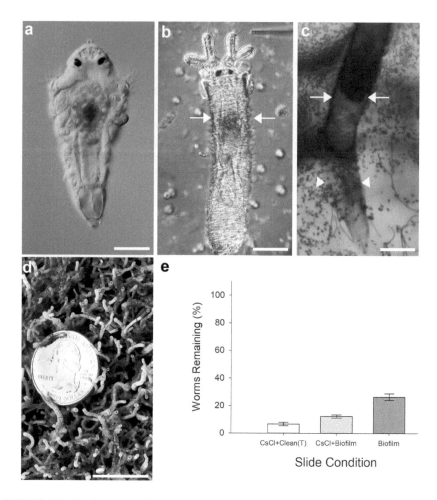

FIGURE 1.2   Settlement and metamorphosis of *Hydroides elegans*. (a) Competent larva (scale bar = 50 μm). (b) Juvenile worm having completed metamorphosis and secreted a calcified secondary tube (scale bar = 100 μm). (c) Primary and secondary tubes of *H. elegans* stained with crystal violet. Primary tube is proteinaceous and is secreted shortly after larva contacts inductive bacteria (scale bar = 100 μm). (d) Dense aggregation of *H. elegans* (scale bar = 2 cm). (e) Primary tubes of *H. elegans* settled on a natural biofilm have a higher resistance to detachment that those settled on clean glass when subjected to a turbulent-flow shear of 120 Pa for 4 min. (From Zardus, J. D., Nedved, B. T., Huang, Y., Tran, C., and Hadfield, M. G. 2008. *Biological Bulletin*, 214:91–98.)

Approximately eight hours after the induction of metamorphosis, larvae of *H. elegans* have completed metamorphosis. During this time, juvenile worms have lost several larva-specific organs (e.g., the prototroch) and begun to concentrate calcium carbonate from the surrounding seawater into a gland that secretes a calcified secondary tube (Figure 1.2b). The primary tube remains attached to the posterior end of the primary tube (Figure 1.2c), and thus, as future research, it will be intriguing to determine how bacteria and tube cements are interacting with each other during both the early and later stages of metamorphosis.

1.6   Identification of larval metamorphic cues from biofilm bacteria

A strain of the γ-proteobacterium, *Pseudoalteromonas luteoviolacea*, isolated from wild biofilms, was shown to induce metamorphosis of *H. elegans* in single-strain biofilms (Huang and Hadfield 2003). Random transposon-mediated mutagenesis produced two strains of *P. luteoviolacea* that were non-inductive and allowed determination of a set of genes that is essential for induction. In-frame deletion mutagenesis of these genes revealed their products to be structural elements derived from phage-tail gene sets termed tailocins. At least four of these genes are required for induction of metamorphosis in *H. elegans* (Figure 1.3; Huang et al. 2012; Shikuma et al. 2014). However, examination of other bacterial species that induce metamorphosis in larvae of *H. elegans* revealed that induction cannot be entirely explained by the presence of these tailocins (Freckelton et al. 2017; Vijayan and Hadfield 2020). Freckelton et al. (2017) examined other inductive bacteria (one Gram-negative: *Cellulophaga*

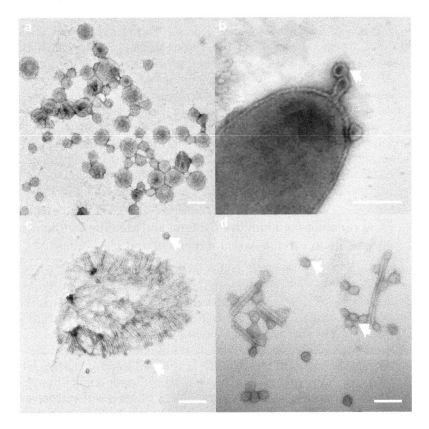

FIGURE 1.3   Negatively stained transmission electron micrographs of bacterial products involved in induction of metamorphosis of larvae of *H. elegans*. (a) Outer membrane vesicles (OMVs) from *Cellulophaga lytica* culture (scale bar = 100 nm); (b) OMVs budding from the outer membrane of *C. lytica* (scale bar = 200 nm); (c) Tailocin aggregate from *Pseudoalteromonas luteoviolacea* (scale bar = 200 nm); (d) OMVs and individual tailocins from *P. luteoviolacea* (scale bar=100 nm). White arrows indicate OMVs.

lytica, and two Gram-positive: *Bacillus aquimaris* and *Staphylcoccus warneri*) and determined that the cassette of genes encoding phage-tail bacteriocins is not encoded in the genomes of these bacteria. More recently, Vijayan and Hadfield (2020) isolated other inductive Pseudoalteromonads that also lacked tailocins.

Cellulophaga lytica was revealed to produce inductive cues by an entirely different mechanism: outer membrane vesicles (OMVs; Figure 1.3; Freckelton et al. 2017). OMVs are ubiquitously produced by Gram-negative bacteria and have frequently been associated with bacterial cell signaling (Kulp and Kuehn 2010; Deatherage and Cookson 2012; Lynch and Alegado 2017). OMVs are produced when small sections of the outer membrane first bulge outward and then pinch off from the bacterial cell. Consequently, OMVs can be highly diverse and include not only the outer membrane, with the bioactive molecules peptidoglycan and lipopolysaccharide (LPS), but also any compounds associated with the periplasmic space directly below the outer membrane. The OMVs thus included a list of possible components, including proteins (membrane bound or soluble), virulence factors, nucleic acids, or peptidoglycan (Deatherage and Cookson 2012).

To gain key insight into the molecular structure of the bacterial inducer produced by *C. lytica*, its OMV fractions were recently subjected to a battery of enzymatic and chemical treatments, and tested for loss of inductive activity using the settlement bioassay (Freckelton et al. 2020). These experiments asked whether OMV induction of metamorphosis was due to the presence of a protein, nucleic acid, or a lipid (Figure 1.4). OMV-induced metamorphosis was impacted only by treatment with lipases, indicating that either a lipid or an intact membrane is required for OMV induction (Freckelton et al. 2020). Assessment of the bacterial sacculus in combination with interrogation of OMVs with lysozyme, and finally isolation and purification of lipopolysaccharide (LPS), has led us to conclude that LPS is the molecule of interest for *C. lytica*-induced metamorphosis of *H. elegans* (Figure 1.4).

1.7 How variability of inductive bacteria and identified settlement cues relate to variable larval settlement and recruitment

The capacity to induce settlement in one invertebrate species by one strain of a bacterium species does not translate to the same inductive activity by all strains of that bacterium. This is not entirely unexpected, as strain-level differences in symbiotic ability have been observed in other classical symbioses as well (Bongrand and Ruby 2019). One possible explanation for strain specificity in bacteria-eukaryotic interactions is suggested by our recent work identifying the cell-surface lipopolysaccharide (LPS) as the relevant bacterial cueing molecule for larvae of *H. elegans*. The high level of structural variability of this macromolecule is closely tied to bacterial taxonomic lines and growth conditions (Aucken and Pitt 1993; Schletter et al. 1995; Zhang et al. 2006; Raetz et al. 2007; Leker et al. 2017), revealing that LPS may be responsible for induction of metamorphosis in a broad swathe of marine invertebrates. Most importantly, it would explain the selective settlement of different invertebrates into different habitats. If variations in the structure of LPS are responsible for variable inductive activity of different species and strains of bacteria, it would explain why many—but definitely not all—marine Gram-negative bacteria induce settlement and metamorphosis in the larvae of *H. elegans*. The taxonomic variability of LPS may

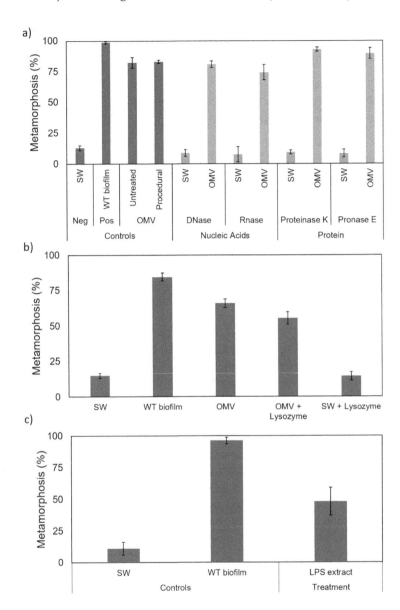

FIGURE 1.4   Lipopolysaccharide is the morphogen produced by *Cellulophaga lytica.*
(a) Outer membrane vesicles (OMVs) and seawater (SW) controls were exposed to either: DNase (20 U); RNase (20 U); trypsin (1 mg/mL) and Proteinase K (50 U) or trypsin and Pronase E, sonicated and then incubated for 2 hrs at 37°C. Prior to settlement testing, proteases were removed by ultracentrifugation (200,000 g, 4°C, 2 hrs). (b) OMVs and SW controls were exposed to lysozyme (20 U), sonicated, and then incubated for 2 hrs at 37°C. (c) Lipopolysaccharide isolated using the hot phenol method induced metamorphosis in the larvae of *H. elegans* after 24 hrs exposure. In all cases, metamorphosis was counted after 24 hrs exposure with SW as a negative control, a multispecies biofilm as a positive control, and an untreated OMV experimental control (Freckelton et al. 2020).

also explain why not all strains of a bacterium induce larvae of an invertebrate species to settle (Huang et al. 2012), as well as why larvae may settle selectively in some habitats and not others, even though the same bacterial species is present.

1.8   Lipopolysaccharide mediates both symbiotic and pre-symbiotic interactions

LPS was traditionally recognized as the prototypical class of Pathogen-Associated Molecular Patterns (PAMP). PAMPs are molecules that trigger the innate immune response of eukaryotes, and thereby notify host organisms of potential infection. With the recognition of increased benign and beneficial roles for host-associated bacteria, it has been argued that this terminology is restrictive (Koropatnick et al. 2004) and is more accurately portrayed by the term Microbe-Associated Molecular Pattern (MAMP). The ubiquity of a molecule such as LPS means that its detection by a host is entirely context dependent. Indeed, since MAMPs encompass a wide variety of molecular classes, many MAMPs are implicated in both pre-symbiotic associations, such as larval settlement (Tebben et al. 2015), and well-established symbiotic relationships, such as the Hawaiian Bobtail squid *Euprymna scolopes* and the bacterium *Vibrio fischeri* (Rader et al. 2012). In light of the growing perspective of benign and beneficial interactions between bacteria and eukaryotes, it is clear that examining the pre-symbiotic dealings between eukaryotes and bacteria will increase our understanding of the evolution of both facultative and obligate symbiotic relationships among animals and microbes.

1.9   Conclusion

In many instances, marine larvae are totally dependent upon the detection of bacterial cues for their settlement and development into adult forms. Just as in classical symbioses, these invertebrate–bacteria interactions require the host to adapt or develop mechanisms to distinguish "useful" bacteria from many others. Indeed, many of the same molecules, and undoubtedly their detection mechanisms, are involved in the processes of recognition and interaction in both symbiotic and more casual relationships between eukaryotes and bacteria, and thus research into one can only be enriched by research into the other. Few studies have yet examined if bacteria that induce settlement and metamorphosis of marine larvae persist in the core microbiomes of later adult stages. Understanding these relationships will certainly be an important future avenue of research to tie larval settlement to eukaryote:bacteria symbiosis.

References

Ainsworth, T. D., Fordyce, A. J., and Camp, E. F. 2017. The other microeukaryotes of the coral reef microbiome. *Trends in Microbiology*, 25:980–991.
Ainsworth, T. D., and Gates, R. D. 2016. Corals' microbial sentinels. *Science*, 352:1518–1519.
Aldred, N., Gohad, N. V., Petrone, L. et al. 2013. Confocal microscopy-based goniometry of barnacle cyprid permanent adhesive. *Journal of Experimental Biology*, 216:1969–1972.

Alegado, R. A., Brown, L. W., Cao, S. et al. 2012. A bacterial sulfonolipid triggers multicellular development in the closest living relatives of animals. *eLife*, 1:e00013.

Alfaro, A. C., Young, T., and Ganesan, A. M. 2011. Regulatory effects of mussel (*Aulacomya maoriana* Iredale 1915) larval settlement by neuroactive compounds, amino acids and bacterial biofilms. *Aquaculture*, 322:158–168.

Apprill, A. 2020. The role of symbioses in the adaptation and stress responses of marine organisms. *Annual Review of Marine Science*, 12:5.1–5.24.

Apprill, A., Marlow, H. Q., Martindale, M. Q., and Rappé, M. S. 2009. The onset of microbial association in the developing coral *Pocillopora meandrina*. *The ISME Journal*, 3:685–699.

Aucken, H. M., and Pitt, T. I. 1993. Lipopolysaccharide profile typing as a technique for comparative typing of Gram-negative bacteria. *Journal of Clinical Microbiology*, 31:1286–1289.

Bao, W. Y., Satuito, C. G., Yang, J. L., and Kitamura, H. 2007. Larval settlement and metamorphosis of the mussel *Mytilus galloprovincialis* in response to biofilms. *Marine Biology*, 150:565–574.

Bastida-Zavala, J. R., and ten Hove, H. A. 2002. Revisions of *Hydroides* Gunnerus, 1768 (Polychaeta: Serpulidae) from the Western Atlantic region. *Beaufortia*, 52:103–178.

Bastida-Zavala, J. R., and ten Hove, H. A. 2003. Revisions of *Hydroides* Gunnerus, 1768 (Polychaeta:Serpulidae) from the Eastern Pacific region and Hawaii. *Beaufortia*, 53:67–110.

Beemelmanns, C., Woznica, A., Alegado, R. A., Cantley, A. M., King, N., and Clardy, J. 2014. Synthesis of the Rosette-Inducing Factor RIF-1 and analogs. *Journal of the American Chemical Society*, 136:10210–10213.

Bongrand, C., and Ruby, E. G. 2019. Achieving a multi-strain symbiosis: Strain behavior and infection dynamics. *The ISME Journal*, 13:698.

Bordenstein, S. R., and Theis, K. R. 2015. Host biology in light of the microbiome: Ten principles of holobionts and hologenomes. *PLoS Biology*, 13:e1002226.

Bourne, D. G., Morrow, K. M., and Webster, N. S. 2016. Insights into the coral microbiome: Underpinning the health and resilience of reef ecosystems. *Annual Review Microbiology*, 70:317–640.

Cameron, R. A., and Hinegardner, R. T. 1974. Initiation of metamorphosis in laboratory cultured sea urchins. *Biological Bulletin*, 146:335–342.

Campbell, A. H., Meritt, D. W., Franklin, R. B., Boone, E. L., Nicely, C. T., and Brown, B. L. 2011. Effects of age and composition of field-produced biofilms on oyster larval setting. *Biofouling*, 27:255–265.

Carpizo-Ituarte, E., and Hadfield, M. G. 1998. Stimulation of metamorphosis in the polychaete *Hydroides elegans* Haswell (Serpulidae). *Biological Bulletin*, 194:14–24.

Chiu, J. M. Y., Thiyagarajan, V., Pechenik, J. A., Hung, O. S., and Qian, P. Y. 2007. Influence of bacteria and diatoms in biofilms on metamorphosis of the marine slipper limpet *Crepidula onyx*. *Marine Biology*, 151:1417–1431.

Deatherage, B. L., and Cookson, B. T. 2012. Membrane vesicle release in bacteria, eukaryotes, and archaea: A conserved yet underappreciated aspect of microbial life. *Infection and Immunity*, 80:1948–1957.

de Bary, A. 1879. *Die erscheinung der symbiose*. Strassburg, Germany: Verlag von Karl J. Trubner.

Essock-Burns, T., Gohad, N. V., Orihuela, B. et al. 2017. Barnacle biology before, during and after settlement and metamorphosis: A study of the interface. *The Journal of Experimental Biology*, 220:194–207.

Erwin, P. M., López-Legentil, S., González-Pech, R., and Turon, X. 2012. A specific mix of generalists: Bacterial symbionts in Mediterranean *Ircinia* spp. *FEMS Microbiology and Ecology*, 79:619–637.

Fan, L., Reynolds, D., Liu, M. et al. 2012. Functional equivalence and evolutionary convergence in complex communities of microbial sponge symbionts. *Proceedings of the National Academy of Science USA*, 109:e1878–e1887.

Fiore, C. L., Labrie, M., Jarett, J. K., and Lesser, M. P. 2015. Transcriptional activity of the giant barrel sponge, *Xestospongia muta* holobiont: Molecular evidence for metabolic interchange. *Frontiers in Microbiology*, 6:364.

Fisher, R. M., Henry, L. M., Cornwallis, C. K., Kiers, E. T., and West, S. A. 2017. The evolution of host-symbiont dependence. *Nature Communications*, 8:15973.

Freckelton, M. L., Nedved, B. T., and Hadfield, M. G. 2017. Induction of invertebrate larval settlement; different bacteria, different mechanisms? *Scientific Reports*, 7:42557.

Freckelton, M. L., Nedved B. T., Turano, H. et al. 2020. Bacterial lipopolysaccharide induces settlement and metamorphosis in a marine larva. Submitted to BioRXiv.

Hadfield, M. G. 2011. Biofilms and marine invertebrate larvae: What bacteria produce that larvae use to choose settlement sites. *Annual Review of Marine Science*, 3:453–470.

Hadfield, M. G., Asahina, A., Hennings, S., and Nedved, B. T. 2014. The bacterial basis of biofouling: A case study. *Indian Journal of Geo-Marine Sciences*, 43:2075–2084.

Hadfield, M. G., Unabia, C. C., Smith, C. M., and Michael, T. M. 1994. Settlement preferences of the ubiquitous fouler *Hydroides elegans*. In: Thompson M. F., Nagabhushanam R., Sarojini R., Fingerman M. (eds.) *Recent Developments in Biofouling Control*. New Delhi: Oxford and IBH Pub. Co., pp. 65–74.

Hamer, J. P., Walker, G., and Latchford, J. W. 2001. Settlement of *Pomatoceros lamarkii* (Serpulidae) larvae on biofilmed surfaces and the effect of aerial drying. *Journal of Experimental Marine Biology and Ecology*, 260:113–131.

Hentschel, U., Usher, K. M., and Taylor, M. W. 2006. Marine sponges as microbial fermenters. *FEMS Microbiology Ecology*, 55:167–177.

Hester, E. R., Barott, K. L., Nulton, J., Vermeij, M. J. A., and Rohwer, F. L. 2015. Stable and sporadic symbiotic communities of coral and algal holobionts. *The ISME Journal*, 10:1157–1169.

Huang, S., and Hadfield, M. G. 2003. Composition and density of bacterial biofilms determine larval settlement of the polychaete *Hydroides elegans*. *Marine Ecology Progress Series*, 260:161–172.

Huang, Y., Callahan, S., and Hadfield, M. G. 2012. Recruitment in the sea: Bacterial genes required for inducing larval settlement in a polychaete worm. *Scientific Reports*, 2:228.

Huggett, M., and Apprill, A. 2019. Coral Microbiome Database: Integration of sequences reveals high diversity and specificity of coral-associated microbes. *Environmental Microbiology Reports*, 11:372–285.

Huggett, M. J., Williamson, J. E., de Nys, R., Kjelleberg, S., and Steinberg, P. D. 2006. Larval settlement of the common Australian sea urchin *Heliocidaris erythrogramma* in response to bacteria from the surface of coralline algae. *Oecologia*, 149:604–619.

Hung, O. S., Lee, O. O., Thiyagarajan, V. et al. 2009. Characterization of cues from natural multi-species biofilms that induce larval attachment of the polychaete *Hydroides elegans*. *Aquatic Biology*, 4:253–262.

Kostic, A. D., Howitt, M. R., and Garrett, S. W. 2013. Exploring host-microbiota interactions in animal models and humans. *Genes & Development*, 27:701–718.

Klassen, J. L., Wolf, T., Rischer, M. et al. 2015. Draft genome sequences of six *Pseudoalteromonas* strains, P1–7a, P1-9, P1-13-1a, P1-16-1b, P1-25, and P1-26, which induce larval settlement and metamorphosis in *Hydractinia echinata*. *Genome Announcements*, 3:e01477–15.

Koropatnick, T. A., Engle, J. T., Apicella, M. A., Stabb, E. V., Goldman, W. E., and McFall-Ngai, M. J. 2004. Microbial factor-mediated development in a host-bacterial mutualism. *Science*, 306:1186–1188.

Kulp, A., and Kuehn, M. J. 2010. Biological functions and biogenesis of secreted bacterial outer membrane vesicles. *Annual Review of Microbiology*, 64:163–184.

Lackner, G., Peters, E. E., Helfrich, E. J. N., and Piel, J. 2017. Insights into the lifestyle of uncultured bacterial natural product factories associated with marine sponges. *Proceedings of the National Academy of Science USA*, 114:E347–E356.

La Marca, E.C., Catania, V., Quatrini, P., Milazzo, M., and Chemello, R. 2018. Settlement performance of the Mediterranean reef-builders *Dendropoma cristatum* (Biondi 1859) in response to natural bacterial films. *Marine Environmental Research*, 137:149–157.

Lau, S. C. K., and Qian, P. Y. 2001. Larval settlement in the serpulid polychaete *Hydroides elegans* in response to bacterial films: An investigation of the nature of putative larval settlement cue. *Marine Biology*, 138:321–328.

Lau, S. C. K., Thiyagarajan, V., Cheung, C. K., and Qian, P. Y. 2005. Roles of bacterial community composition in biofilms as a mediator for larval settlement of three marine invertebrates. *Aquatic Microbial Ecology*, 38:41–51.

Leker, K., Lozano-Pope, I., Bandyopadhyay, K., Choudhury, B. P., and Obonyo, M. 2017. Comparison of lipopolysaccharides composition of two different strains of *Helicobacter pylori*. *BMC Microbiology*, 17:226.

Lema, K. A., Constancias, F., Rice, S. A., and Hadfield, M. G. 2019. High bacterial diversity in near-shore and oceanic biofilms and their influence on larval settlement by *Hydroides elegans* (Polychaeta). *Environmental Microbiology*, 21:3472–3488.

Levin, T. C., and King, N. 2013. Evidence for sex and recombination in the choanoflagellate *Salpingoeca rosetta*. *Current Biology*, 23:2176–2180.

Liang, X., Zhang, X. K., Peng, L. H., Zhu, Y. T., Yoshida, A., Osatomi, K., and Yang, J. L. 2020. The flagellar gene regulates biofilm formation and mussel larval settlement and metamorphosis. *International Journal of Molecular Sciences*, 21:710.

Lynch, J. B., and Alegado, R. A. 2017. Spheres of hope, packets of doom: The good and bad of outer membrane vesicles in interspecies and ecological dynamics. *Journal of Bacteriology*, 199:e00012–17.

Margulis, L. 1971. Symbiosis and evolution. *Scientific American*, 225:48–61.

Moran, N. A., and Dunbar, H. E. 2006. Sexual acquisition of beneficial symbionts in aphids. *Proceedings of the National Academy of Science USA*, 103:12803–12806.

Moya, A., Peretó, J., Gil, R., and Latorre, A. 2008. Learning how to live together: Genomic insights into prokaryote–animal symbioses. *Nature Reviews Genetics*, 9:218.

Negri, A. P., Webster, N. S., Hill, R. T., and Heyward, A. J. 2001. Metamorphosis of broadcast spawning corals in response to bacteria isolated from crustose algae. *Marine Ecology Progress Series*, 223:121–131.

Nielsen, S. J., Harder, T., and Steinberg, P. D. 2015. Sea urchin larvae decipher the epiphytic bacterial community composition when selecting sites for attachment and metamorphosis. *FEMS Microbiology and Ecology*, 91:1–219.

Pachu, A. V., Rao, M. V., and Balaji, M. 2012. Recruitment response of larvae of *Teredo parksi* Bartsch (Teredinidae: Myoida: Bivalvia) to individual marine bacterial films. *Journal of the Indian Academy of Wood Science*, 9:160–164.

Pita, L., Fraune, S., and Hentschel, U. 2016. Emerging sponge models of animal-microbe symbioses. *Frontiers in Microbiology*, 7:2102.

Pita, L., Rix, L., Slaby, B. M., Franke, A., and Hentschel, U. 2018. The sponge holobiont in a changing ocean: From microbes to ecosystems. *Microbiome*, 6:46.

Pollock, F. J., McMinds, R., Smith, S. et al. 2018. Coral-associated bacteria demonstrate phylosymbiosis and cophylogeny. *Nature Communications*, 9:4921.

Rader, B. A., Kremer, N., Apicella, M. A., Goldman, W. E., and McFall-Ngai, M. J. 2012. Modulation of symbiont lipid A signaling by host alkaline phosphatases in the squid-vibrio symbiosis. *mBio*, 3:00093–12.

Raetz, C. R. H., Reynolds, C. M., Trent, M. S., and Bishop, R. E. 2007. Lipid A modification systems in Gram-negative bacteria. *Annual Review of Biochemistry*, 76:295–329.

Ritchie, K. B. 2006. Regulation of microbial populations by coral surface mucus and mucus-associated bacteria. *Marine Ecology Progress Series*, 322:1–14.

Roberts, B., Davidson, B., MacMaster, G. et al. 2007. A complement response may activate metamorphosis in the ascidian *Boltenia villosa*. *Development Genes and Evolution*, 217:449–458.

Rohwer, F., Seguritan, V., Azam, F., and Knowlton, N. 2002. Diversity and distribution of coral-associated bacteria. *Marine Ecology Progress Series*, 243:1–10.

Scarborough, C. L., Ferrari, J., and Godfray, H. C. J. 2005. Aphid protected from pathogen by endosymbiont. *Science*, 310:1781.

Schletter, J., Heine, H., Ulmer, A. J., and Rietschel, E. T. 1995. Molecular mechanisms of endotoxin activity. *Archives of Microbiology*, 164:383–389.

Sebesvari, Z., Neumann, R., Brinkhoff, T., and Harder, T. 2013. Single-species bacteria in sediments induce larval settlement of the infaunal polychaetes *Polydora cornuta* and *Streblospio benedicti*. *Marine Biology*, 160:1259–1270.

Seipp, S., Schmich, J., Kehrwald, T., and Leitz, T. 2007. Metamorphosis of *Hydractinia echinata*—natural versus artificial induction and developmental plasticity. *Development Genes and Evolution*, 217:385–394.

Sharp, K. H., Ritchie, K. B., Schupp, P. J., Ritson-Williams, R., and Paul, V. J. 2010. Bacterial acquisition in juveniles of several broadcast spawning coral species. *PLOS ONE*, 5:e10898.

Sharp, K. H., Sneed, J. M., Ritchie, K. B., McDaniel, L., and Paul, V. J. 2015. Induction of larval settlement in the reef coral *Porites astreoides* by a cultivated marine *Roseobacter* strain. *Biological Bulletin*, 228:98–107.

Shigenobu, S., and Wilson, A. C. 2011. Genomic revelations of a mutualism: The pea aphid and its obligate bacterial symbiont. *Cellular and Molecular Life Sciences*, 68:1297–1309.

Shikuma, N. J., Pilhofer, M., Weiss, G. L., Hadfield, M. G., Jensen, G. J., and Newman, D. K. 2014. Marine tubeworm metamorphosis induced by arrays of bacterial phage tail–like structures. *Science*, 343:529–533.

Shimeta, J., Cutajar, J., Watson, M.G., and Vlamis, T. 2012. Influences of biofilm-associated ciliates on the settlement of marine invertebrate larvae. *Marine Ecology Progress Series*, 449: 1–12.

Sneed, J. M., Sharp, K. H., Ritchie, K. B., and Paul, V. J. 2014. The chemical cue tetrabromopyrrole from a biofilm bacterium induces settlement of multiple Caribbean corals. *Proceedings of the Royal Society B: Biological Sciences*, 281:20133086.

Sun, Y., Wong, E., Ahyong, S. T., Williamson, J. E., Hutchings, P. A., and Kupriyanova, E. K. 2018. Barcoding and multi-locus phylogeography of the globally distributed calcareous tubeworm genus *Hydroides* Gunnerus, 1768 (Annelida, Polychaeta, Serpulidae). *Molecular Phylogenetics and Evolution*, 127:732–745.

Tebben, J., Motti, C. A., Siboni, N. et al. 2015. Chemical mediation of coral larval settlement by crustose coralline algae. *Scientific Reports*, 5:10803.

Tebben, J., Tapiolas, D. M., Motti, C. A., Abrego, D., Negri, A. P., Blackall, L. L., Steinberg, P. D., and Harder, T. 2011. Induction of larval metamorphosis of the coral *Acropora millepora* by tetrabromopyrrole isolated from a *Pseudoalteromonas* bacterium. *PLOS ONE*, 6:e19082.

Thomas, T., Moitinho-Silva, L., Lurgi, M. et al. 2016. Diversity, structure and convergent evolution of the global sponge microbiome. *Nature Communications*, 7:11870.

Thomas, T., Rusch, D., DeMaere, M. Z. et al. 2010. Functional genomic signatures of sponge bacteria reveal unique and shared features of symbiosis. *The ISME Journal*, 4:1557–67.

Tran, C., and Hadfield, M. G. 2011. Larvae of *Pocillopora damicornis* (Anthozoa) settle and metamorphose in response to surface-biofilm bacteria. *Marine Ecology Progress Series*, 433:85–96.

Unabia, C. R. C., and Hadfield, M. G. 1999. Role of bacteria in larval settlement and metamorphosis of the polychaete *Hydroides elegans*. *Marine Biology*, 133:55–64.

Vijayan, N., and Hadfield, M. G. 2020. Bacteria known to induce settlement of larvae of *Hydroides elegans* are rare in natural inductive biofilm. *Aquatic Microbial Ecology* 84:31–42.

Wahab, M. A. A., de Nys, R., Webster, N., and Whalan, S. 2014. Larval behaviours and their contribution to the distribution of the intertidal coral reef sponge *Carteriospongia foliascens*. *PLOS ONE*, 9:e98181.

Whalan, S., and Webster, N. S. 2015. Sponge larval settlement cues: The role of microbial biofilms in a warming ocean. *Scientific Reports*, 4:4072.

Wilson, M. C., Mori, T., Ruckert, C. et al. 2014. An environmental bacterial taxon with a large and distinct metabolic repertoire. *Nature*, 506:58–62.

Woese, C. R. 2004. A New Biology for a New Century. *Microbiology and Molecular Biology Reviews*, 68:173–186.

Woollacott, R. M., and Hadfield, M. G. 1996. Induction of metamorphosis in larvae of a sponge. *Invertebrate Biology*, 115:257–262.

Woznica, A., and King, N. 2018. Lessons from simple marine models on the bacterial regulation of eukaryotic development. *Current Opinion in Microbiology*, 43:108–116.

Yang, J. L., Shen, P. J., Liang, X., Li, Y. F., Bao, W. Y., and Li, J. L. 2013. Larval settlement and metamorphosis of the mussel *Mytilus coruscus* in response to monospecific bacterial biofilms. *Biofouling*, 29:247–259.

Zardus, J. D., Nedved, B. T., Huang, Y., Tran, C., and Hadfield, M. G. 2008. Microbial biofilms facilitate adhesion in biofouling invertebrates. *Biological Bulletin*, 214:91–98.

Zhang, Y., Arias, C., Shoemaker, C., and Klesius, P. 2006. Comparison of lipopolysaccharide and protein profiles between *Flavobacterium columnare* strains from different genomovars. *Journal of Fish Disease*, 29:657–663.

2 The language of symbiosis
Insights from protist biology

Morgan J. Colp and John M. Archibald

Contents

2.1   Introduction

The cell is the fundamental unit of life. In nature, however, even the simplest unicellular organisms do not exist in isolation. Within the eukaryotic domain, as animals and plants have evolved in close association with microbes (McFall-Ngai et al. 2013; Hassani et al. 2018), so too have the protists. The cellular and molecular biology, metabolisms, and even life histories of protists—eukaryotes other than land plants, animals, and true fungi—have been shaped by interactions with microbes living around, on, and within them. The most obvious manifestation of this fact is the existence of mitochondria and, in the case of algae and plants, plastids (chloroplasts). For much of 20th century biology the origins of these quintessentially eukaryotic organelles were mysterious. We now know that mitochondria and plastids are derived from α-proteobacterial and cyanobacterial endosymbionts, respectively (see Archibald 2015 for review). The results of decades of biochemical, molecular, and comparative genomic research have allowed us to reconstruct how endosymbionts become organelles. Extensive endosymbiont-to-host gene transfer occurs (Timmis et al. 2004); protein import machinery evolves to target the products of hundreds of nuclear genes back to the endosymbiont-turned-organelle (Gould et al. 2008; Wiedemann and Phanner 2017; Richardson and Schnell 2019); and membrane-localized metabolite transporters are coopted, facilitating the metabolic integration of host and nascent organelle (Becker and Wagner 2018; Marchand et al. 2018). The end result is a semi-autonomous, membrane-bound compartment that "communicates" with the cell in which it resides.

This is not to say that we fully understand how mitochondria and plastids evolved from once free-living bacteria—far from it. We are still woefully ignorant of the host–symbiont dialogues necessary and sufficient to set in motion the endosymbiont-to-organelle transition. The evolution of plastids from cyanobacteria happened a billion-plus years ago (Parfrey et al. 2011) and the mitochondrial endosymbiosis is even more ancient—perhaps coincident with the origin of the eukaryotic cell itself (see Roger et al. 2017, and references therein for discussion and debate). Consequently, extracting meaningful information about the earliest stages of mitochondrial and plastid evolution from the biology of modern-day organisms is exceedingly difficult. What kind of organism served as host for what was to become the mitochondrion? Was it a prokaryote, a eukaryote, or something in between? What metabolic interactions cemented the relationship between the two organisms, one inside the other? Was the environmental context for mitochondrial evolution aerobic or anaerobic? In the case of plastids, what were the selective pressures that led cyanobacteria to become permanent inhabitants in the cytoplasm of a heterotrophic protist? Was it beneficial for both partners or just one (and if the latter, which one)? These long-standing questions remain front and centre in the field of organelle evolution (Archibald 2015; López-García et al. 2017; Martin 2017; Roger et al. 2017; Gavelis and Gile 2018; Nowack and Weber 2018).

This is not a chapter focused on the origins and evolution of mitochondria and plastids. Rather, we explore the plethora of recently evolved protist–microbe symbioses that exist in nature. The literature on this topic is vast and we do not present a comprehensive overview; interested readers are encouraged to consult review articles by Gast et al. (2009), Nowack and Melkonian (2010), Dziallas et al. (2012), and Samba-Louaka et al. (2019) for deeper dives into the primary research. Our goal is to highlight the diverse ways in which protist hosts interact with symbionts, as well as less clearly defined members of the microbial communities in which they live (Figure 2.1). In some instances, these examples provide new ideas about the preconditions that might have existed when plastids and mitochondria evolved. In all cases they bolster our understanding of, and appreciation for, symbiosis as a driver of the evolution of life on our planet. As we shall see, there is no shortage of symbiotic novelty in the protist world, and more and more microbial symbioses are being developed into manipulatable model systems and studied in the laboratory. The results lead us to question long held assumptions about the extent to which symbioses are intrinsically mutualistic.

2.2   Cytoplasm as microcosm

In 1966, Kwang Jeon of the University of Tennessee discovered that one of his strains of *Amoeba proteus* had succumbed to a bacterial infection. Tens of thousands of rod-shaped bacteria had taken over the cytoplasm of this shape-shifting freshwater amoeba (Jeon and Lorch 1967); these so-called X-bacteria were capable of killing any new amoeba strain they encountered, and most of the amoebae in the original infected culture died. Most, but not all: *"Within a few years, the host xD amoebae became dependent on their newly acquired symbionts for survival"* (Jeon 1995). The X-bacteria turned out to be members of the Gram-negative gammaproteobacterial

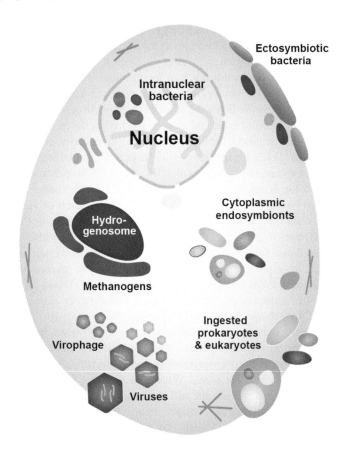

FIGURE 2.1 Overview of a generic single-celled eukaryote. Known symbiotic associations between organelles, endosymbionts and ectosymbionts are depicted. Note that hydrogenosomes are mitochondrion-derived organelles found in various microbial eukaryotes that have adapted to anaerobic environments.

genus *Legionella*, a lineage now famous for including *L. pneumophila*, the causative agent of Legionnaires' Disease (legionellosis). Jeon's decades-long investigation of the *Amoeba–Legionella* symbiosis raised more questions than it answered (see Jeon 2004 for review), but it contributed to the now widely appreciated notion that the innards of free-living amoebae are a playground for the development of symbiotic and pathogenic interactions.

The amoebae are not a monophyletic entity; they are a morphologically diverse grade of organisms occupying multiple branches on the eukaryotic tree of life (Simpson, Slamovits and Archibald 2017; Samba-Louaka et al. 2019). By virtue of their phagocytic lifestyle, amoebae and other heterotrophic protists are exposed to a wide array of microbes. For such organisms, the interior of a eukaryotic cell is an environment that can be inhabited and adapted to over relatively short timescales. In the symbiosis literature, protection is often described as one of the prime benefits afforded to an endosymbiont by its host. This is no doubt true in some circumstances—living inside another cell is certainly a good way to avoid being eaten—but the reality is probably more complicated. These complexities will be explored throughout the chapter.

In the wild, *Legionella* spends much of its time inside aquatic and soil amoebae such as *Acanthamoeba*, *Vermamoeba* (*Hartmanella*), and *Naegleria* (see Swart et al. 2018 and references therein). From the amoeba's perspective, the "goal" of ingestion is digestion; *L. pneumophila* avoids this fate by modifying the endomembrane system of its host. Once phagocytosed, the bacterium triggers the formation of a "*Legionella*-containing vacuole" (LCV) by injecting ~300 "effector" proteins into the cell using its type IV secretion system (Hoffmann et al. 2014). The LCV is resistant to acidification and fusion with the lysosome—it's the perfect place for the bacteria to hide and replicate before lysing the amoeba and heading off in search of another host. In essence, *Legionella* spp. have evolved to manipulate phagocytic cells; the same tricks *L. pneumophila* uses to persist inside amoebae are used to hide within macrophages in the lung and cause life-threatening pneumonia (Hoffmann et al. 2014). It is for this reason that free-living amoebae are often described as reservoirs or "training camps" for pathogenic bacteria (Samba-Louaka et al. 2019). Many advances in our understanding of Legionnaires' Disease and other respiratory illnesses have come from the use of *Acanthamoeba* as a model system (Swart et al. 2018).

Is the relationship between *Legionella* and its amoeba hosts symbiotic, pathogenic, or both? It depends on one's perspective. Our biomedical focus on the role of free-living amoebae in pathogen transmission is understandable, but it can prevent us from considering other vantage points. What is clear is that free-living amoebae are important predators in diverse aquatic and terrestrial ecosystems, and they harbor an impressive array of facultative and obligate endosymbionts from all three domains of life (see Samba-Louaka et al. 2019 for review). Alpha-, beta-, gamma-, and deltaproteobacteria, Bacteroidetes, Actinobacteria, and Chlamydiae—all have been found living inside diverse amoebae, and methanogenic archaea have been discovered in the anaerobic amoeba *Pelomyxa palustris* (Gutiérrez et al. 2017). Eukaryotic endosymbionts deserve special mention and are discussed in the following section.

Free-living amoebae are also breeding grounds for an astonishing diversity of viruses, some which are so large that they were initially mistaken for bacteria (see Colson et al. 2017 for review). The giant virus craze began with the isolation and characterization of Mimivirus (Mimi for "mimicking a microorganism") by co-culturing with *Acanthamoeba polyphaga* (La Scola et al. 2003; Raoult et al. 2004). Mimivirus is a so-called nucleocytoplasmic large DNA virus ~400 nm in diameter and with a genome of ~1.2 megabase pairs (Mbp), larger than that of many prokaryotes. Pandoravirus, which also infects *Acanthamoeba*, is ~1 μm long and packs ~2,500 genes into its 2.5 Mbp genome—this is larger and more gene rich than the smallest eukaryotic nuclear genomes (Philippe et al. 2013). Comparative genomic analyses of Mimivirus, Pandoravirus, and an ever-growing number of other giant viruses suggest that they have evolved on multiple occasions from smaller viruses by acquiring genes from their eukaryotic hosts as well as bacteria (e.g., Koonin and Yutin 2019). They also appear to play a role as mediators of gene exchange within and between prokaryotes and eukaryotes. Remarkably, giant viruses have viruses too: so-called virophages have been shown to infect various giant viruses inside their amoeba hosts and have since been found in other protists and algae (La Scola et al. 2008; Fischer and Suttle 2011; Colson et al. 2017). By negatively impacting the replication of the giant virus, the virophages benefit the host, bringing to mind the proverb "the enemy of my enemy is my friend."

Prokaryotic endosymbionts can exist free in the cytoplasm, inside vacuoles and vesicles, and even within the nucleus. The latter compartment is a particularly intriguing location for an endosymbiont, as it affords protection from degradative processes in the cytoplasm such as autophagy (Huang and Brumell 2014). A nuclear residence also provides endosymbionts with direct access to nuclear chromosomes, and thus the opportunity to secrete effectors that can directly influence host gene expression. As reviewed by Schulz and Horn (2015), intranuclear bacteria have been found in the amoebae *Naegleria clarki*, *Acanthamoeba polyphaga*, and *Hartmanella* sp., as well as ciliates (most notably *Paramecium*), dinoflagellate algae, and euglenoids. Again, our focus on human health and disease leads us to view eukaryote–prokaryote interactions primarily through the lens of pathogenesis. The reality is that in most instances we have little or no idea about the true nature of the "relationship" between partner cells. We simply know that it exists.

Occasionally, however, nature throws us a bone in the form of distinctive cellular or biochemical features that provide clues about the biology underlying a symbiotic partnership. One such example involves anaerobic ciliates. Pioneering work in the 1990s by Finlay, Fenchel, and Embley and colleagues showed that methanogenic archaea can be found inside ciliates that live in low oxygen environments (e.g., aquatic sediments and the rumen of cattle); when present, the archaea invariably form close associations with the host's "hydrogenosomes" (e.g., Fenchel and Finlay 1991a; Embley and Finlay 1994; Fenchel and Finlay 2010). The cell biology is often striking. In the marine species *Plagiopyla frontata*, for example, the ciliate's hydrogenosomes are flattened and arranged in an alternating fashion with the archaea like a stack of coins (Fenchel and Finlay 1991b) (Figure 2.1)! Over the years, the metabolic significance of this close physical contact has come to light. As their name suggests, the hydrogenosomes produce hydrogen by fermenting pyruvate, and the hydrogen is sucked up by the archaeon and used to fuel methanogenesis (Fenchel and Finlay 2010). (Note that hydrogenosomes were first discovered in another protist, the parabasalid flagellate *Tritrichomonas*, and have been shown to be mitochondrion-derived organelles; see Embley et al. 2002; Roger et al. 2017, and references therein.)

Elucidation of the biochemistry underlying the ciliate-methanogen syntrophy has been informative as well as inspiring. It contributed to the development of the hydrogen hypothesis by Müller and Martin (1998), which posits that the eukaryotic cell evolved as a symbiosis involving metabolic exchange between a carbon dioxide and hydrogen-producing α-proteobacterium (the symbiont) and a methane-producing archaeon (the host) (see Archibald 2015 and references therein). Metabolic syntrophies of various sorts between evolutionarily diverse ciliates and prokaryotes have evolved on numerous occasions, and as in other areas of symbiosis research, metagenomic and single-cell genomic tools are being used as hypothesis-generating tools (e.g., Boscaro et al. 2019; Rossi et al. 2019). The role of symbiosis in ciliate biology is worthy of a chapter all to itself; Dziallas et al. (2012) provide a useful launch-point into the primary literature.

Sometimes a symbiont is so distinctive that its raison d'être is immediately apparent. This is true in the case of an obscure freshwater thecate amoeba named *Paulinella chromatophora*. First discovered in the river Rhine more than a century ago (Lauterborn 1895; Melkonian and Mollenhauer 2005), the organism is noteworthy for the presence of one or two bright green "chromatophores" in its cytoplasm (Figure 2.2A). Plastids or endosymbiotic cyanobacteria? Surprisingly, the answer is somewhere in between. Gene sequence data have revealed that chromatophores are of α-cyanobacterial ancestry (specifically the *Synechococcus/Prochlorococcus* clade), in contrast to the β-cyanobacteria to which plastids are affiliated (Marin et al. 2005). Sequenced chromatophore genomes are ∼1 Mbp in size, an order of magnitude larger than most plastid genomes but a third the size of *Synechococcus* genomes

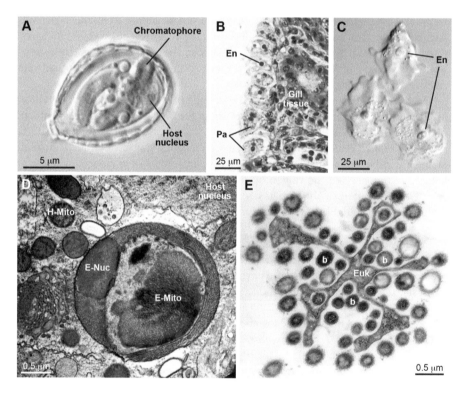

FIGURE 2.2   Protist symbioses. (A) The freshwater testate amoeba *Paulinella chromatophora* with cyanobacterium-derived photosynthetic chromatophores. (B) Histological section showing *Paramoeba* sp. (Pa) cells associated with *Salmo salar* gill tissue (material stained with haematoxylin and eosin). The *Perkinsela* sp. endosymbiont (En) can be seen in some of the amoeba cells. (C) Nomarski differential interference contrast microscopic image of cultured *Paramoeba pemaquidensis* cells with *Perkinsela* sp. endosymbionts. (D) Transmission electron microscopic image of *P. pemaquidensis* cell with *Perkinsela* sp. in its cytoplasm. The endosymbiont nucleus (E-nuc) and large mitochondrion (E-Mito) can be seen, as can host cell mitochondria (H-Mito). (E) Transmission electron microscopic image showing cross section of the anaerobic protist *Streblomastix strix* (Euk) and its numerous bacterial ectosymbionts (b). Images provided by Eva Nowack (A), Ivan Fiala (B–D) and Sebastian Treitli (E).

(Nowack et al. 2008; Reyes-Prieto et al. 2010). Molecular clock estimates suggest that the endosymbiosis that gave rise to the chromatophores of *P. chromatophora* occurred a mere 90–140 million years ago (e.g., Delaye et al. 2016)—a blink of an eye relative to the origin of plastids.

And yet research shows that the term "organelle" does apply. Integration of genomic (Nowack et al. 2008; Reyes-Prieto et al. 2010; Nowack et al. 2016) and chromatophore proteomic data (Singer et al. 2017) reveals that the metabolisms of *Paulinella* and the chromatophore are somewhat integrated, and that >400 chromatophore proteins are encoded by host nuclear genes. Intriguingly, only a small number (17 of 433) of these proteins are obviously related to α-cyanobacteria—most are of apparent host or unknown origin, and 26 appear to have been acquired by lateral (not endosymbiotic) gene transfer from bacteria other than the cyanobacterial progenitor of the chromatophore (Singer et al. 2017). On balance, the evidence suggests that the initial phase of chromatophore integration involved the "mixing and matching" of genes from various sources, not just the cyanobacterial endosymbiont itself.

The extent to which the *Paulinella*-chromatophore system truly recapitulates the evolution of canonical plastids may never be known. But it is interesting that a close relative of the phototrophic *P. chromatophora*, *P. ovalis*, does not have chromatophores but *does* make a living by eating and digesting *Synechococcus* (Johnson et al. 1988), the same type of cyanobacterium that gave rise to the chromatophore. This is consistent with phagotrophy-based models for early plastid evolution. And it is noteworthy that evolutionary mosaicism is a hallmark of the nuclear genomes and plastid proteomes of diverse algae and plants (see Ponce-Toledo et al. 2019 and references therein)—it is one of the reasons that the birth and spread of plastids across the eukaryotic tree has proven so difficult to reconstruct. Larkum, Howe and colleagues have put forth the "shopping bag" model of plastid evolution to account for mosaic genomes and organellar proteomes (Larkum et al. 2007; Howe et al. 2008). The model emphasizes genetic contributions from both endosymbionts and prey as being important during the establishment and maintenance of a nascent organelle. The idea has explanatory power, not just in the case of organelles, but for a wide range of obligate (endo)symbioses, including those involving animal hosts (e.g., Husnik and McCutcheon 2018; Husnik and Keeling 2019).

2.3   Eukaryotes inside eukaryotes (inside other eukaryotes)

Relative to prokaryotes, eukaryotic endosymbionts are few and far between. Or are they? We must preface this discussion by pointing out that one of the ways that photosynthesis has spread across the eukaryotic tree is by "secondary" endosymbiosis, that is, the engulfment of a "primary" plastid-bearing alga by a heterotrophic protist host. Whereas primary plastids appear to have evolved from cyanobacteria only once in the history of eukaryotic life (not counting the case of *Paulinella* discussed above), secondary plastids have arisen on multiple occasions by mergers between different hosts and both red and green algal endosymbionts (Gould et al. 2008; Archibald 2015; Ponce-Toledo et al. 2019). Numerous instances of tertiary endosymbiosis have also occurred, with mixotrophic dinoflagellates featuring prominently. These bloom-forming algae are sometimes referred to as the biological equivalent of Russian nesting dolls—they have captured, recaptured, and replaced their plastids on numerous occasions and show varying levels of genetic and cell biological integration (see Hehenberger et al. 2014 and references therein). A review of the complexities of plastid evolution is beyond the scope of this chapter. For our purposes, it is sufficient to say that while endosymbiotic events involving eukaryotic hosts and eukaryotic endosymbionts have been, and continue to be, important in nature, they are limited to algae and revolve around photosynthesis. We are aware of only one exception.

Members of the genus *Paramoeba* (*Neoparamoeba*) are opportunistic pathogens of marine animals (Nowak and Archibald 2018). Inside these economically significant microbes is a nucleus-associated "parasome" (or *Nebenkörper*) (Figure 2.2B-D). Observed for the first time in 1896, it took almost a century for the parasome to be recognized for what it is: an endosymbiotic eukaryote (see Sibbald et al. 2017 and references therein). The organism, now named *Perkinsela* sp., lies free in the cytoplasm and its relationship with the amoeba host is obligate—the organisms cannot be grown apart from one another. Molecular phylogenetic investigations show that *Perkinsela* sp. is a kinetoplastid protozoan, one that has lost its flagellum and has been coevolving for some time with its host amoeba (Dyková et al. 2008; Sibbald et al. 2017).

Recent genomic investigation has provided insight into the nature of this (apparently) unique eukaryote–eukaryote endosymbiosis. Tanifuji et al. (2017) sequenced the nuclear genomes of *Paramoeba pemaquidensis* (a fish gill-associated species) and the *Perkinsela* sp. living within it; consideration of their predicted biochemical capacities suggests metabolic and cellular interdependence between the two organisms. For example, amino acid and nucleotide biochemical pathways inferred from the endosymbiont genome are incomplete, suggesting that essential metabolites are imported from the host amoeba. Trypanothione biosynthesis, an essential kinetoplastid-specific pathway, also appears reliant on host-derived precursors (Tanifuji et al. 2017). Unexpectedly, electron microscopy suggests that endocytosis at the *Perkinsela* sp. plasma membrane is the mechanism by which it acquires metabolites from its host; this form of host–endosymbiont "communication" has not been observed before. Unlike the situation in primary and secondary plastids, endosymbiont-to-host gene transfer does not appear to have played a significant role in host–endosymbiont integration. In this sense, the symbiotic relationship between *Paramoeba* and *Perkinsela* is reminiscent of the nutritional symbioses that exist between sap-feeding insects and certain bacteria (Bennett and Moran 2015).

How *Perkinsela* sp. came to reside within *Paramoeba* species is unclear, but it is interesting that its closest known relatives are members of the genus *Ichthyobodo*, which are ectoparasites of fish (Todal et al. 2004). This observation, together with the fact that many kinetoplastid protists are notorious parasites (e.g., *Trypanosoma* and *Leishmania*), it would not be surprising if the initial *Paramoeba–Perkinsela* association was parasitic in nature. Regardless, what we know is that today the organisms depend on one another for life.

Another interesting (and puzzling) facet of the *Paramoeba–Perkinsela* endosymbiosis is the role of bacteria. Taken from nature, bacteria are invariably cocultured with *Paramoeba* spp., and attempts to produce bacteria-free amoeba cultures have been unsuccessful (Jellet and Scheibling 1988) (amoebae are typically fed *E. coli* in the lab whilst growing on a rich marine agar). In the case of the sea urchin pathogen *Paramoeba invadens*, reduced growth rates and loss of virulence occurred much more quickly when the organism was grown in monoxenic culture than when the amoeba was grown polyxenically (Jellet and Scheibling 1988). The reason(s) for this are not known. What is known is that bacteria can persist inside amoeba cells for extended periods of time; their role in amoeba biology and pathogenicity is an important question for the future.

Other documented instances of eukaryotes living inside other microbial eukaryotes are considered parasitic in nature. For example, the moss-associated amoebozoan *Thecamoeba quadrilineata* harbors a "fungal-like" intranuclear parasite called *Nucleophaga amoebae* (Michel et al. 2009; Corsaro et al. 2014a). Similar cases have been described in related amoebae isolated from natural and artificial freshwater environments (including tap water), including *Vanella* and *Saccamoeba* (Corsaro et al. 2014b). The cell biology and life histories of *Nucleophaga*-like endoparasites is similar to microsporidians, which is not surprising given that these well-studied animal parasites are also related to fungi (James et al. 2013). The taxonomy applied to these parasites is very much a work in progress (see Corsaro et al. 2016 and Bass et al. 2018 for discussion), and presently very little is known about the host ranges of *Nucleophaga*-like endobionts across protist and animal diversity.

2.4   Ectosymbiosis: It's a jungle out there

Thus far we have focused on endosymbiosis: coevolving hosts and symbionts, the latter inside the former. However, ectosymbioses are extremely common in the protist world and we should not be surprised—the surface of a cell is a substrate, one particularly ripe for colonization given that it has the potential to provide protection from grazers and direct and/or indirect access to nutrition. One of the most (in) famous examples of ectosymbiosis in biology involves the protist *Mixotricha*, a giant (up to 0.5 mm long) parabasalid that lives in the gut of the wood-eating Australian termite *Mastotermes*. The movement of *Mixotricha* is facilitated by the coordinated beating of tens of thousands of spirochaete bacteria attached to its surface, each with its own specialized attachment site (Cleveland and Grimstone 1964). It was this "motility symbiosis" that inspired Lynn Margulis to postulate that eukaryotic flagella (which she called undulipodia) were of exogenous origin, that is, they evolved early in eukaryotic evolution from ectosymbiotic spirochaete bacteria (Margulis 1970, 1981).

A pioneering microscopy-based study of low oxygen, high sulphide marine sediments at the bottom of the Santa Barbara Basin by Bernhard et al. (2000) provided a fascinating glimpse of the prevalence of ectosymbioses in nature. Diverse and novel protist lineages belonging to the ciliates, euglenozoans, and foraminiferans were identified, most of which were found to harbor bacterial endo- and/or ectosymbionts. Over the past 20 years, the application of new molecular and genomic tools to the study of ectosymbiosis has brought the fascinating world glimpsed at by Bernhard et al. into much sharper focus. As summarized by Gast et al. (2009), the nature of, and apparent reasons for, ectosymbiotic relationships between protists and bacteria are diverse. Here, we discuss but a few case studies in the realm of metabolism and nutrition.

The first example is the symbiontids, which are euglenozoans that live in anaerobic marine sediments (including those of the Santa Barbara Basin). A mixture of rod-shaped and spherical proteobacteria typically festoon the surface of these long, flagellated cells; microscopic evidence suggests that the epibiotic bacteria are metabolically connected to the protists hydrogenosome-like mitochondria, which lie immediately beneath the plasma membrane, presumably to maximize opportunities for metabolite exchange (Edgcomb et al. 2011; Yubuki and Leander 2018). Evidence supporting this hypothesis has recently come from Monteil et al. (2019) who used metagenomics to study the symbiontid-associated bacterial communities. A near-complete genome sequence from an ectosymbiotic deltaproteobacterium suggests that the organism is a chemolithoautotroph capable of reducing sulfate with the hydrogen coming from the protists subsurface hydrogenosome-like organelles. It was also suggested that organic matter may make its way into the protist host by diffusion and/or ectosymbiont ingestion. Remarkably, this symbiotic consortium—the protist and its ectosymbionts—is capable of magnetotactic motility. The nonmotile bacteria produce "magnetosomes" which render the flagellated, motile host capable of magnetoreception (Monteil et al. 2019), a stunning example of the innovative potential of symbiosis.

Treitli et al. (2019) recently used genomics and fluorescence *in situ* hybridization to characterize rod-shaped bacteria associated with the surface of a bizarre protist called *Streblomastix strix* (Figure 2.2E). This organism, an oxymonad flagellate belonging to the Excavata, has long been known to live anaerobically in the hindgut of certain termites (Cleveland 1925) and appears reliant on its bacterial epibionts for life. When antibiotics are used to reduce the number of bacteria on the cell (Dexter-Dyer and Khalsa 1993; Leander and Keeling 2004), the protist changes from being a long spindle-shaped cell to one that is teardrop-shaped (Leander and Keeling 2004). Using single-cell genomic techniques, Treitli et al. (2019) demonstrated the existence of multiple distinct Bacteroidetes bacteria on each *S. strix* cell, most of which lack the gene for the essential glycolytic enzyme enolase. The authors suggest that this deficiency is overcome by metabolite exchange between the protist host and the bacteria. For their part, the bacteria secrete numerous glycosyl hydrolases, which degrade the cellulose in the hindgut environment into monomers usable by the metabolism of the protist (and its animal host). Some of the bacteria also appear capable of nitrogen fixation and provision of essential amino acids and cofactors that *S. strix* cannot itself synthesize. Other protists in the hindgut of "lower" termites, namely parabasalid flagellates such as *Pseudotrichonympha grassii*, have both ecto- and endosymbionts that contribute to the breakdown of cellulose and lignocellulose, thereby providing the animal with carbon and energy, and allowing it to survive on wood as its sole source of food (Hongoh et al. 2008; Yuki et al. 2015). Not surprisingly, there is growing interest in harnessing the collective metabolic potential of the termite "microbial bioreactor" for the purposes of biofuels production from lignocellulosic feedstock (Brune 2014).

A recurring theme in microbial symbiosis is the role of anaerobiosis as a driver of protist–microbe symbioses. We have already discussed several examples, including anaerobic ciliates and the methanogenic archaea that snuggle up to their hydrogen-producing hydrogenosomes. Moving beyond the termite hindgut, other examples include a newly discovered facultative symbiosis between bacteria and the flagellated unicell *Lenisia limosa*. This protist is a so-called "breviate" (Brown et al. 2013);

it branches on the eukaryotic tree of life near animals, fungi, and amoebozoans and lives in marine sediments. The surface of *Lenisia* can be colonized by the epsilonproteobacterium *Arcobacter* (curiously, with only one to three epibionts per cell); comparative genomic and proteomic investigation reveals that this relationship is driven by protist-to-bacterium hydrogen transfer (Hamann et al. 2016).

What exactly the protist gains from having *Arcobacter* on its surface is not entirely clear, but growth experiments show that in the absence of the bacterium, *Lenisia* suffers a significant reduction in fitness. And the presence of the epibiont alters the expression of protist genes associated with anaerobic hydrogen production—clearly the two organisms are communicating with one another. Interestingly, *Arcobacter* species and their kin are best known as pathogens of humans and other animals; they are capable of producing harmful toxins and persisting for extended periods of time in diverse oxic and anoxic environments, including those induced by food processing and storage (Ferreira et al. 2016). The apparently mutualistic symbiosis between *Lenisia* and *Arcobacter* is thus mysterious. Clearly, there is still a lot to learn about the biology of epsilonproteobacteria like *Arcobacter* and the nature of their interactions with single-celled and multicellular eukaryotes.

Should we consider ecto- and endosymbiosis to be mutually exclusive? Not necessarily. In the case of the above-mentioned termite gut inhabitant *Streblomastix strix*, electron microscopy reveals that its surface-associated bacteria can be ingested and digested, thus conceivably providing the protist with an alternate way to access fixed carbon and metabolic intermediates (Treitli et al. 2019). A conceptually similar example can be seen in a marine dinoflagellate protist named *Ornithocercus magnificus* (Taylor 1971), where cyanobacteria grow ectosymbiotically within a cellulosic chamber (the "crown") located at one end of the cell. The two organisms have clearly coevolved for some time: the ectobionts are inherited vertically from mother to daughter chambers during cell division (Decelle et al. 2015), and the recently sequenced genome of the cyanobacterium shows evidence of genome reduction (Nakayama et al. 2019), as is typical of symbionts that are adapting to a specialized environment (McCutcheon and Moran 2012). Interestingly, partially digested cyanobacteria can be seen in the digestive vacuoles of *Ornithocercus* cells (Lucas 1991). This has led to the intriguing hypothesis that the ectosymbionts are being "farmed" for their photosynthate, which becomes available to the host upon ingestion (Lucas 1991; Tarangkoon, Hansen and Hansen 2010). Where better for the dinoflagellate to store its food than close at hand?

2.5   Microbial symbioses: Power struggles in time and space

We suggest that endosymbiotic interactions are best thought of not as mutualistic "happily ever-after" stories, but instead as "use it up and cast it off" situations that are stable for variable lengths of time.

Keeling and McCutcheon 2017

The science of symbiosis has a rich and complex history (Lewis 1985; Sapp 1994; Archibald 2014). Different fields of biology view the issue of "living together" from different angles and through different lenses, constrained by the nature of the specific organisms they study and the history of their subdisciplines. Keeling and McCutcheon (2017) recently noted that symbioses—and endosymbioses in particular—are often described as (or assumed to be) mutualisms, despite the fact that the benefits for both partners are often far from obvious. They also argue that endosymbioses are intrinsically antagonistic, and more often than not are seeded by a host–pathogen relationship. Many of the symbioses discussed in this chapter are consistent with this hypothesis, but there is often little or no evidence for or against the idea. Fortunately, an increasing number of microbial symbioses are being developed into model systems with which to ask the question "who benefits, host and/or symbiont?" The consistent answer emerging from laboratory experimentation is "it depends."

One of the most experimentally tractable systems for the study of endosymbiosis involves the ciliate *Paramecium bursaria*. This aquatic organism harbors facultative green algal endosymbionts belonging to the genus *Chlorella*, which are housed within a specialized, host-derived compartment called the perialgal vacuole (Dziallas et al. 2012; Fujishima and Kodama 2012). The alga provides the ciliate with maltose and oxygen, and in return the ciliate gives the alga organic nitrogen. It sounds fair but it is not. In 2016, Lowe et al. performed a cost-benefit analysis of the system by growing the organisms separately and together (i.e., endosymbiotically) under varying degrees of light and food availability, which impact the growth of the photosynthetic endosymbiont and the heterotrophic host, respectively. As expected, the host benefited the most in high light, low food conditions. In contrast, algal abundance *decreased* under high light—not what one would expect from a card-carrying phototroph. The experiments suggest that under high light conditions the host reduces its endosymbiont load, possibly by limiting the flow of nitrogen to the alga. Lowe et al. (2016) describe the symbiosis between the two organisms as a case of "controlled exploitation."

Given the spectacular ecological and evolutionary successes of plastid-bearing eukaryotes, it is tempting to assume that phototrophic endosymbionts are inherently beneficial to hosts. But the reality is not so simple. Uzuka et al. (2019) recently used coculturing experiments to show that protist predators experience oxidative stress upon ingestion of photosynthetic prey. When exposed to light, significant changes in host stress-related gene expression were observed, notably upregulation of a laterally acquired gene involved in chlorophyll degradation/detoxification. Illumination also triggered a reduction in the rate of phagocytosis as well as increased digestion of photosynthetic prey cells (Uzuka et al. 2019). In the case of the "green" *Paramecium* discussed above, high light conditions can actually result in the expulsion of *Chlorella* symbionts from the host, presumably as a stress reduction response to algal-derived reactive oxygen species (see Kawano et al. 2010 and references therein).

Another informative example is the farming symbiosis that exists in the social soil amoeba *Dictyostelium*. Many amoeba strains in nature engage in what has been described as "husbandry of bacteria:" upon starvation, they incorporate food bacteria into their fruiting bodies and, after spore dispersal, seed a fresh crop of bacteria in a new environment (Brock et al. 2011). Who benefits, how, and under what conditions? Recent work has shown that farmed *Burkholderia* can have both pathogenic and mutualistic effects on *Dictyostelium* depending on environmental conditions, and that the bacteria are themselves capable of initiating the farming symbiosis (DiSalvo et al. 2015). At the same time, soil microcosm experiments show that in the presence of *Dictyostelium*, different species of *Burkholderia* symbionts exhibit variable fitness costs and benefits (Garcia et al. 2019). The picture emerging from the study of this complex system is that both host and symbiont genotypes can contribute to the "success" or "failure" of a mutualism—it depends on the context.

2.6   Conclusion

Symbiosis is a fact of life. The past half century of exploration of the microbial biosphere has opened our eyes to the remarkable extent to which organisms interact and coevolve. The specific examples of microbe–microbe interactions discussed herein cover the full spectrum from parasitism to mutualism; they reflect discipline-specific biases in the way that organisms are examined, experiments are designed, and results are interpreted. Modern technologies allow us to study the biology of symbiosis using reductionist methodology. Microbial symbiotic relationships that on the surface appear to be mutualistic are increasingly being portrayed as context-dependent power struggles. But we must be alert to the possibility that in our efforts to "turn the dials" on symbiosis in the lab we oversimplify it to the point that biological realism is lost. The more complex the symbiosis in nature, the more likely it is that important nuances will be lost. Cross-fertilization and integration of experimental results from across the full spectrum of symbiosis research, including the animal– and plant–microbe systems discussed elsewhere in this book, will be important if meaningful progress in our understanding of the role of symbiosis in biology is to be made.

Acknowledgments

We thank Ivan Fiala, Eva Nowack, and Sebastian Treitli for kindly providing microscopic images. Research in the Archibald lab on symbiosis and genome evolution is supported by the Natural Sciences and Engineering Research Council of Canada (NSERC; RGPIN 05871-2014) and the Gordon and Betty Moore Foundation (GBMF5782). MC is supported by a graduate student scholarship from NSERC.

References

Archibald, J.M. 2014. *One Plus One Equals One: Symbiosis and the Evolution of Complex Life*. Oxford University Press.

Archibald, J.M. 2015. Endosymbiosis and eukaryotic cell evolution. *Curr. Biol.* 25, R911–921.

Bass, D. et al. 2018. Clarifying the relationships between Microsporidia and Cryptomycota. *J. Eukaryot. Microbiol.* 65, 773–782.

Becker, T., and Wagner, R. 2018. Mitochondrial outer membrane channels: Emerging diversity in transport processes. *BioEssays*. 40, 1800013.

Bennett, G.M., and Moran, N.A. 2015. Heritable symbiosis: The advantages and perils of an evolutionary rabbit hole. *Proc. Natl. Acad. Sci. USA*. 112, 10169–10176.

Bernhard, J.M., Buck, K.R., Farmer, M.A., and Bowser, S.S. 2000. The Santa Barbara Basin is a symbiosis oasis. *Nature*. 403, 77–80.

Boscaro, V., Husnik, F., Vannini, C., and Keeling, P.J. 2019. Symbionts of the ciliate *Euplotes*: Diversity, patterns and potential as models for bacteria-eukaryote endosymbioses. *Proc. R. Soc. B.* 286, 2019063.

Brock, D.A., Douglas, T.E., Queller, D.C., and Strassmann, J.E. 2011. Primitive agriculture in a social amoeba. *Nature.* 469, 393–396.

Brown, M.W. et al. 2013. Phylogenomics demonstrates that breviate flagellates are related to opisthokonts and apusomonads. *Proc. R. Soc. B.* 280, 20131755.

Brune, A. 2014. Symbiotic digestion of lignocellulose in termite guts. *Nature Microbiol.* 12, 168–180.

Cleveland, L.R. 1925. The effects of oxygenation and starvation on the symbiosis between the termite *Termopsis* and its intestinal flagellates. *Biol. Bull.* 48, 309–325.

Cleveland, L.R., and Grimstone, A.V. 1964. The fine structure of the flagellate *Mixotricha paradoxa* and its associated micro-organisms. *Proc. R. Soc. B.* 159, 668–686.

Colson, P., La Scola, B., and Raoult, D. 2017. Giant viruses of amoebae: A journey through innovative research and paradigm changes. *Annu. Rev. Virol.* 4, 61–85.

Corsaro, D. et al. 2014a. Rediscovery of *Nucleophaga amoebae*, a novel member of the Rozellomycota. *Parasitol. Res.* 113, 4491–4498.

Corsaro, D. et al. 2014b. Microsporidia-like parasites of amoebae belong to the early fungal lineage Rozellomycota. *Parasitol. Res.* 113, 1909–1918.

Corsaro, D. et al. 2016. Molecular identification of *Nucleophaga terricolae* sp. nov. (Rozellomycota), and new insights on the origin of the Microsporidia. *Parasitol. Res.* 115, 3003–3011.

Decelle, J., Colin, S., and Foster, R.A. 2015. Photosymbiosis in marine planktonic protists. In *Marine Protists: Diversity and Dynamics*, S. Ohtsuka, T. Suzaki, T. Horiguchi, N. Suzuki, and F. Not (eds.), pp. 465–500. Springer, Tokyo, Japan.

Delaye, L., Valadez-Cano, C., and Pérez-Zamorano, B. 2016. How really ancient is *Paulinella Chromatophora*? *PloS. Curr: Tree Life.* 8, doi: 10.1371/currents.tol. e68a099364bb1a1e129a17b4e06b0c6b.

Dexter-Dyer, B., and Khalsa, O. 1993. Surface bacteria of *Streblomastix strix* are sensory symbionts. *Biosystems.* 31, 169–180.

DiSalvo, S., Haselkorn, T.S., Bashir, U., Jimenez, D., Brock, D.A., Queller, D.C., and Strassmann, J.E. 2015. *Burkholderia* bacteria infectiously induce the proto-farming symbiosis of *Dictyostelium* amoebae and food bacteria. *Proc. Natl. Acad. Sci. USA.* 112, E5029–E5037.

Dyková, I., Fiala, I., and Pecková, H. 2008. *Neoparamoeba* spp. and their eukaryotic endosymbionts similar to Perkinsela amoebae (Hollande, 1980): Coevolution demonstrated by SSU rRNA gene phylogenies. *Eur. J. Protistol.* 44, 269–277.

Dziallas, C., Allgaier, M., Monaghan, M.T., and Grossart, H.-P. 2012. Act together— implications of symbioses in aquatic ciliates. *Frontiers Microbiol.* Doi: 10.3389/ fmicb.2012.3, 288.

Edgcomb, V.P., Breglia, S.A., Yubuki, N., Beaudoin, D., Patterson, D.J., Leander, B.S., and Bernhard, J.M. 2011. Identity of epibiotic bacteria on symbiontid euglenozoans in O_2-depleted marine sediments: Evidence for symbiont and host co-evolution. *ISME J.* 5, 231–243.

Embley, T.M., and Finlay, B.J. 1994. The use of small subunit rRNA sequences unravel the relationships between anaerobic ciliates and their methanogen endosymbionts. *Microbiology.* 140, 225–235.

Embley, T.M., van der Giezen, M., Horner, D.S., Dyal, P.L., and Foster, P.G. 2002. Mitochondria and hydrogenosomes are two forms of the same fundamental organelle. *Phil. Trans. R. Soc. Lond. B.* 358, 191–203.

Fenchel, T., and Finlay, B.J. 1991a. Endosymbiotic methanogenic bacteria in anaerobic ciliates: Significance for the growth efficiency of the host. *J. Protozool.* 38:18–22.

Fenchel, T., and Finlay, B.J. 1991b. Synchronous division of an endosymbiotic methanogenic bacterium in the anaerobic ciliate *Plagiopyla frontata* Kahl. *J. Protozool.* 38, 22–28.

Fenchel T., and Finlay B.J. 2010. Free-living protozoa with endosymbiotic methanogens. In *(Endo)symbiotic Methanogenic Archaea*, J.H.P. Hackstein (ed.), pp. 35–53. Microbiology Monographs 19, Springer-Verlag, Heidelberg, Berlin.

Ferreira, S., Queiroz, J.A., Oleastro, M., and Domingues, F.C. 2016. Insights in the pathogenesis and resistance of *Arcobacter*: A review. *Crit. Rev. Microbiol.* 42, 364–383.

Fischer, M.G., and Suttle, C.A. 2011. A virophage at the origin of large DNA transposons. *Science.* 332, 231–234.

Fujishima, M., and Kodama, Y. 2012. Endosymbionts in *Paramecium. Eur. J. Protistol.* 48, 124–137.

Garcia, J.R., Larsen, T.J., Queller, D.C., and Strassmann, J.E. 2019. Fitness costs and benefits vary for two facultative *Burkholderia* symbionts of the social amoeba, *Dictyostelium discoideum. Ecol. Evol.* 9, 9879–9890.

Gast, R.J., Sanders, R.W., and Caron, D.A. 2009. Ecological strategies of protists and their symbiotic relationships with prokaryotic microbes. *Trends Microbiol.* 17, 563–569.

Gavelis, G.S., and Gile, G.H. 2018. How did cyanobacteria first embark on the path to becoming plastids?: Lessons from protist symbioses. *FEMS Microbiol. Lett.* 365, doi: 10.1093/femsle/fny209.

Gould, S.B., Waller, R.F., and McFadden, G.I. 2008. Plastid evolution. *Annu. Rev. Plant Biol.* 59, 491–617.

Gutiérrez, G., Chistyakova, L.V., Villalobo, E., Kostygov, A.Y., and Frolov, A.O. 2017. Identification of *Pelomyxa palustris* endosymbionts. *Protist.* 168, 408–424.

Hamann, E. et al. 2016. Environmental *Breviata* harbour mutualistic *Arcobacter* epibionts. *Nature.* 534, 254–258.

Hassani, M.A., Durán, P., and Hacquard S. 2018. Microbial interactions within the plant holobiont. *Microbiome.* 6, 58.

Hehenberger, E., Imanian, B., Burki, F., and Keeling, P.J. 2014. Evidence for the retention of two evolutionary distinct plastids in dinoflagellates with diatom endosymbionts. *Genome Biol. Evol.* 6, 2321–2334.

Hoffmann, C., Harrison, C.F., and Hilbi, H. 2014. The natural alternative: Protozoa as cellular models for *Legionella* infection. *Cell. Microbiol.* 16, 15–26.

Hongoh, Y. et al. 2008. Genome of an endosymbiont coupling N2 fixation to cellulolysis within protist cells in termite gut. *Science.* 322, 1108–1109.

Howe, C., Barbrook, A., Nisbet, R., Lockhart, P., and Larkum, A. 2008. The origin of plastids. *Philos. Trans. R. Soc. B.* 363, 2675–2685.

Huang, J., and Brumell, J.H. 2014. Bacteria-autophagy interplay: A battle for survival. *Nat. Rev. Microbiol.* 12, 101–114.

Husnik, F., and Keeling, P.J. 2019. The fate of obligate endosymbionts: Reduction, integration, or extinction. *Curr. Op. Genet. Dev.* 58–59, 1–8.

Husnik, F., and McCutcheon, J.P. 2018. Functional horizontal gene transfer from bacteria to eukaryotes. *Nat. Rev. Microbiol.* 16, 67–79.

James, T.Y. et al. 2013. Shared signatures of parasitism and phylogenomics unite Cryptomycota and Microsporidia. *Curr. Biol.* 23, 1548–1553.

Jellet, J.F., and Scheibling, R.E. 1988. Virulence of *Paramoeba invadens* Jones (Amoebida, Paramoebidae) from monoxenic and polyxenic culture. *J. Protozool.* 35, 422–424.

Jeon, K.W. 1995. The large, free-living amoebae: Wonderful cells for biological studies. *J. Eukaryot. Microbiol.* 42, 1–7.

Jeon, K.W. 2004. Genetic and physiological interactions in the amoeba-bacteria symbiosis. *J. Eukaryot. Microbiol.* 51, 502–508.

Jeon, K.W., and Lorch, I.J. 1967. Unusual intra-cellular bacterial infection in large, free-living amoebae. *Exp. Cell. Res.* 48, 236–240.

Johnson, P.W., Hargraves, P.E., and Sieburth, J.M. 1988. Ultrastructure and ecology of *Calycomonas ovalis* Wulff, 1919, (Chrysophyceae) and its redescription as a testate rhizopod, *Paulinella ovalis* N. Comb. (Filosea: Euglyphina). *J. Protozool.* 35, 618–626.

Kawano, T., Irie, K., and Kadono, T. 2010. Oxidative stress-mediated development of symbiosis in green paramecia. In *Symbiosis and Stress: Joint Ventures in Biology*, J. Seckbach, and M. Grube (eds.), pp. 177–195. Springer, Netherlands, Dordrecht.

Keeling, P.J., and McCutcheon, J.P. 2017. Endosymbiosis: The feeling is not mutual. *J. Theoret. Biol.* 434, 75–79.

Koonin, E.V., and Yutin, N. 2019. Chapter five – evolution of the large nucleocytoplasmic DNA viruses of eukaryotes and convergent origins of viral gigantism. *Adv. Virus Res.* 103, 167–202.

La Scola, B. et al. 2003. A giant virus in amoebae. *Science.* 299, 2033.

La Scola, B. et al. 2008. The virophage as a unique parasite of the giant mimivirus. *Nature.* 455, 100–104.

Larkum, A., Lockhart, P.J., Howe, C.J. 2007. Shopping for plastids. *Trends Plant Sci.* 12, 189–195.

Lauterborn, R. 1895. Protozoenstudien II. *Paulinella chromatophora* nov gen., nov., spec., ein beschalter Rhizopode des Süßwassers mit blaugrünen chromatophorenartigen Einschlüssen. *Protozoan. Z. Wiss. Zool.* 59, 537–544.

Leander, B.S., and Keeling, P.J. 2004. Symbiotic innovation in the oxymonad *Streblomastix strix. J. Eukaryot. Microbiol.* 51, 291–300.

Lewis, D.H. 1985. Symbiosis and mutualism: Crisp concepts and soggy semantics. In *The Biology of Mutualism: Ecology and Evolution*, D.H. Boucher (ed.), London.

López-García, P., Eme, L., and Moreira, D. 2017. Symbiosis in eukaryotic evolution. *J. Theoret. Biol.* 434, 20–33.

Lowe, C.D., Minter, E.J., Cameron, D.D., and Brockhurst, M.A. 2016. Shining a light on exploitative host control in a photosynthetic endosymbiosis. *Curr. Biol.* 26, 207–211.

Lucas, I.A.N. 1991. Symbionts of the tropical dinophysiales (Dinophyceae). *Ophelia.* 33, 213–224.

Marchand, J., Heydarizadeh, P., Schoefs, B., and Spetea, C. 2018. Ion and metabolite transport in the chloroplast of algae: Lessons from land plants. *Cell. Mol. Life Sci.* 75, 2153–2176.

Margulis, L. 1970. *Origin of Eukaryotic Cells.* Yale University Press, New Haven.

Margulis, L. 1981. *Symbiosis in Cell Evolution.* W.H. Freeman and Company, San Francisco.

Marin, B., Nowack, E.C., and Melkonian, M. 2005. A plastid in the making: Evidence for a second primary endosymbiosis. *Protist.* 156, 425–432.

Martin, W.F. 2017. Physiology, anaerobes, and the origin of mitosing cells 50 years on. *J. Theoret. Biol.* 434, 2–10.

Martin, W., and Müller, M. 1998. The hydrogen hypothesis for the first eukaryote. *Nature.* 392, 37–41.

McCutcheon, J.P., and Moran N.A. 2012. Extreme genome reduction in symbiotic bacteria. *Nat. Rev. Microbiol.* 10, 13–26.

McFall-Ngai, M. et al. 2013. Animals in a bacterial world, a new imperative for the life sciences. *Proc. Natl. Acad. Sci. USA.* 110, 3229–3236.

Melkonian, M., and Mollenhauer, D. 2005. Robert Lauterborn (1869–1952) and his *Paulinella chromatophore. Protist.* 156, 253–262.

Michel, R., Hauröder, B., and Zöller, L. 2009. Isolation of the amoeba *Thecamoeba quadrilineata* harbouring intranuclear spore forming endoparasites considered as fungus-like organisms. *Acta Potozool.* 48, 41–49.

Monteil, C.L. et al. 2019. Ectosymbiotic bacteria at the origin of magnetoreception in a marine protist. *Nature Microbiol.* 4, 1088–1095.

Nakayama, T. et al. 2019. Single-cell genomics unveiled a cryptic cyanobacterial lineage with a worldwide distribution hidden by a dinoflagellate host. *Proc. Natl. Acad. Sci. USA.* 116, 15973–15978.

Nowack, E.C., Price, D.C., Bhattacharya, D., Singer, A., Melkonian, M., and Grossman, A.R. 2016. Gene transfers from diverse bacteria compensate for reductive genome evolution in the chromatophore of *Paulinella chromatophora. Proc. Natl. Acad. Sci. USA.* 113, 12214–12219.

Nowack, E.C.M., and Melkonian, M. 2010. Endosymbiotic associations within protists. *Phil. Trans. R. Soc. B.* 365, 699–712.

Nowack, E.C.M., Melkonian, M., and Glöckner, G. 2008. Chromatophore genome sequence of *Paulinella* sheds light on acquisition of photosynthesis by eukaryotes. *Curr. Biol.* 18, 410–418.

Nowack, E.C.M., and Weber, E.P.M. 2018. Genomics-informed insights into endosymbiotic organelle evolution in photosynthetic eukaryotes. *Ann. Rev. Plant Biol.* 69, 51–84.

Nowak, B.F., and Archibald, J.M. 2018. Opportunistic but lethal: The mystery of paramoebae. *Trends Cell Biol.* 34, 404–419.

Parfrey, L.W., Lahr, D.J., Knoll, A.H., and Katz, L.A. 2011. Estimating the timing of early eukaryotic diversification with multigene molecular clocks. *Proc. Natl. Acad. Sci. USA.* 108, 13624–13629.

Philippe, N. et al. 2013. Pandoraviruses: Amoeba viruses with genomes up to 2.5 Mb reaching that of parasitic eukaryotes. *Science.* 341, 281–286.

Ponce-Toledo, R.I., López-García, P., and Moreira, D. 2019. Horizontal and endosymbiotic gene transfer in early plastid evolution. *New Phytol.* 224, 618–624.

Raoult, D. et al. 2004. The 1.2-megabase genome sequence of Mimivirus. *Science.* 306, 1344–1350.

Reyes-Prieto, A. et al. 2010. Differential gene retention in plastids of common recent origin. *Mol. Biol. Evol.* 27, 1530–1537.

Richardson, L.G.L., and Schnell, D.J. 2019. Origins, function, and regulation of the TOC-TIC general protein import machinery of plastids. *J. Exp. Bot.* 71, 1226–1238.

Roger, A.J., Muñoz-Gómez, S.A., and Kamikawa, R. 2017. The origin and diversification of mitochondria. *Curr. Biol.* 27, R1177–R1192.

Rossi, A., Bellone, A., Fokin, S.I., Voscaro, V., and Vannini, C. 2019. Detecting associations between ciliated protists and prokaryotes with culture-independent single-cell microbiomics: A proof-of-concept study. *Microbial Ecol.* 78, 232–242.

Samba-Louaka, A., Delafont, V., Rodier, M.-H., Cateau, E., and Hechard, Y. 2019. Free-living amoebae and squatters in the wild: Ecological and molecular features. *FEMS Microbiol. Rev.* 43, 415–434.

Sapp, J. 1994. *Evolution by Association (A History of Symbiosis).* Oxford University Press.

Schulz, F., and Horn, M. 2015. Intranuclear bacteria: Inside the cellular control center of eukaryotes. *Trends Cell Biol.* 25, 339–346.

Sibbald, S.J. et al. 2017. Diversity and evolution of *Paramoeba* spp. and their kinetoplastid endosymbionts. *J. Eukaryot. Microbiol.* 64, 598–607.

Simpson, A.G.B., Slamovits, C.H., and Archibald, J.M. 2017. Protist diversity and eukaryote phylogeny. In *Handbook of the Protists (Second Edition of the Handbook of Protoctista by Margulis et al.),* J.M. Archibald, A.G.B. Simpson, and C.H. Slamovits (eds.), pp. 1–21. Springer International Publishing.

Singer, A. et al. 2017. Massive protein import into the early-evolutionary-stage photosynthetic organelle of the amoeba *Paulinella chromatophora.* *Curr. Biol.* 27, 2763–2773.

Swart, A.L., Harrison, C.F., Eichinger, L., Steinert, M., and Hilbi, H. 2018. *Acanthamoeba* and *Dictyostelium* as cellular models for *Legionella* infection. *Front. Cell. Infect. Microbiol.* 8, 61.

Tanifuji, G. et al. 2017. Genome sequencing reveals metabolic and cellular interdependence in an amoeba-kinetoplastid symbiosis. *Scientific Rep.* 7, 11688.

Tarangkoon, W., Hansen, G., and Hansen, P. 2010. Spatial distribution of symbiont-bearing dinoflagellates in the Indian Ocean in relation to oceanographic regimes. *Aquat. Microb. Ecol.* 58, 197–213.

Taylor, F.J.R. 1971. Scanning electron microscopy of thecae of the dinoflagellate genus *Ornithocercus.* *J. Phycol.* 7, 249–258.

Timmis, J.N., Ayliffe, M.A., Huang, C.Y., and Martin, W. 2004. Endosymbiotic gene transfer: Organelle genomes forge eukaryotic chromosomes. *Nat. Rev. Genet.* 5, 123–135.

Todal, J.A. et al. 2004. *Ichthyobodo necator* (Kinetoplastida)–a complex of sibling species. *Dis. Aquat. Org.* 58, 9–16.

Treitli, S.C., Kolisko, M., Husnik, F., Keeling, P.J., and Hampl, V. 2019. Revealing the metabolic capacity of *Streblomastix strix* and its bacterial symbionts using single-cell genomics. *Proc. Natl. Acad. Sci. USA.* 116, 19675–19684.

Uzuka, A. et al. 2019. Responses of unicellular predators to cope with the phototoxicity of photosynthetic prey. *Nature Comm.* 10, 5606.

Wiedemann, N., and Phanner, N. 2017. Mitochondrial machineries for protein import and assembly. *Annu. Rev. Biochem.* 86, 685–714.

Yubuki, N. and Leander, B.S. 2018. Diversity and evolutionary history of the Symbiontida (Euglenozoa). *Front. Ecol. Evol.* 6, 100.

Yuki, M. et al. 2015. Dominant ectosymbiotic bacteria of cellulolytic protists in the termite gut also have the potential to digest lignocellulose. *Environ. Microbiol.* 17, 4942–4943.

3 Trichoplax and its bacteria

How many are there? Are they speaking?

Michael G. Hadfield and Margaret J. McFall-Ngai

Contents

3.1 Introduction

The phylum Placozoa, a group of small animals, ∼0.3–3 mm in diameter with 10s to 100s of thousands of cells, is nested among the basal groups of the animal kingdom (Figure 3.1). Although they have been studied with increasing intensity for more than four decades, many aspects of their biology remain controversial. Their precise phylogenetic position is in dispute, but many biologists studying this clade place them as diverging basal to the Cnidaria and bilaterians (Figure 3.1C) (Eitel et al. 2018; DuBuc et al. 2019; Osigus et al. 2019). Further, its longtime standing as a phylum of only a single species, *Trichoplax adhaerens*, has been reconfigured to include many different haplotypes of this species (Voigt et al. 2004; Eitel and Schierwater 2010), and now two additional species, each in a new genus, have been recognized (Eitel et al. 2018; Osigus et al. 2019). The simple body plan of three cell layers has remained dogma for the phylum. However, while six basic cell types had been recognized since the 1970s (Figure 3.1B), a recent microscopic and "-omics" study suggests that more cell types may be present (Sebé-Pedrós et al. 2018).

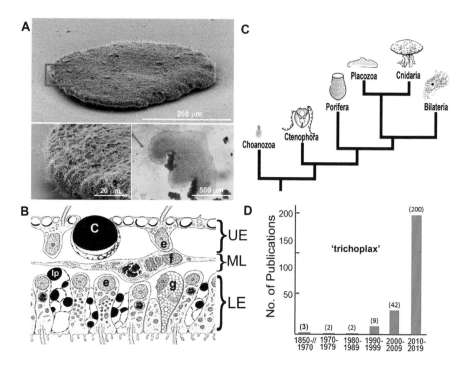

FIGURE 3.1 *Trichoplax adhaerens.* (A) Upper, scanning electron micrograph of entire animal; lower left, magnified segment of the margin; lower right, a living animal. (B) Diagrammatic cross section showing body layers and cell types: UE, upper epithelium, ML, middle layer, LE, lower epithelium; c, "crystal cell"; e, ciliated epithelial cell; f, fiber cell; lp, lipophilic cell; g, gland cell. (Modified from Grell, K. G. 1972. *Zoomorphology* 73 (4):297–314.) (C) Hypothetical phylogeny of the metazoa. (D) Publications on Placozoa by decades.

Although known since the late 1800s, *T. adhaerens* was "put on the map" by Karl Grell and colleagues in the 1970s. While clearly establishing that *T. adhaerens* was indeed a whole animal with its own suite of unique characters, Grell and Benwitz (1971) published very clear transmission electron micrographs (TEM) that showed bacteria within the fiber cells lying between the upper and lower epithelia. The presence of these deep intracellular bacterial symbionts was mostly ignored until the first genomes of *T. adhaerens* were sequenced. Among the genes were those found uniquely in the bacterial phylum Rickettsiales (Driscoll et al. 2013). Although a few subsequent papers present TEM evidence for bacteria in the fiber cells, no other studies were performed to understand the nature of the symbiotic relationship between *T. adhaerens* and its intracellular symbionts prior to that of Gruber-Vodicka et al. (2019). This is somewhat amazing, given the intense proliferation of papers in the last decade dealing with the biology of Placozoa (e.g., Pearse and Voigt 2007; Smith et al. 2014; DuBuc et al. 2019), their geographical distribution (Eitel and Schierwater 2010), and the phylogenetic position of this clade (e.g., Laumer et al. 2018) (Figure 3.1D).

It is the goal of this brief review to examine the evidence for true symbiotic relationships between placozoans and the microbial world. We first ask how many different bacteria are resident in *T. adhaerens,* where they reside, and if they are present in all placozoan haplotypes and species. Secondly, we look at structural data that may help clarify the nature of the relationships and then explore the molecular-genetic data that may define the nature of the symbioses. We next consider two important questions: what is the significance of the giant mitochondrial complexes in the same cells that harbor numerous rickettsial bacteria; and, how are symbiotic bacteria obtained by new generations of placozoans? We conclude with a set of precise questions and potential research directions that may significantly enhance our understanding of important microbe–animal symbioses at the very base of metazoan evolution.

This review includes new microscopical observations. The sources of the placozoans and the methods employed are detailed in the Box 3.1.

Box 3.1   Organisms and approaches

Collecting and culturing Trichoplax

All specimens were collected from the biofouling communities on piers in Pearl Harbor and Rainbow Marina located in Keʻehi Lagoon, Oʻahu, Hāwaiʻi or from aquaria supplied with a constant flow of unfiltered seawater at the Kewalo Marine Laboratory, Honolulu, HI. We gather the assemblage of sponges, ascidians, polychaetes, bivalves, and algae from the piers, bring it to the marine laboratory and place it in plastic colanders, which are positioned over a vertical drainpipe in the center of a deep tank. Constantly flowing ocean water is run through hoses into the tank so that the water must flow up through the colanders to spill over the rim of the drainpipe. In this way, the organisms of the fouling community are kept aerated and healthy. We placed cleaned microscope slides among the organisms for periods of at least three days and then examined them with the aid of a dissecting microscope. Placozoans were removed and established on slides that had been suspended in flowing seawater long enough, ten days or longer, to accumulate a natural marine biofilm. These Trichoplax cultures were hung in 2 L beakers of seawater supplemented with the microalga *Isochrysis galbana* Tahitian Strain. The beakers were maintained in front of fluorescent lights with constant bubbling. We did not determine the haplotypes of the specimens of *Trichoplax adhaerens* that we examined microscopically. Haplotypes H1, H2, H4, H6, and H8 have been reported in Hawaiʻi (Pearse and Voigt 2007; Ward 2008; Eitel and Schierwater 2010), and we might have sampled several of them.

Electron microscopy

Before fixing Trichoplax for electron microscopy, they were transferred to gelatin-coated Petri dishes and allowed to settle and assume their normal configurations. They can be much more easily removed from these dishes. Next, the placozoans were briefly chilled at $-22°C$, nearly all seawater was removed, and the dishes were flooded with the fixative solution. The fixation solution included 2% paraformaldehyde and 2% glutaraldehyde in 0.1 M sodium cacodylate buffer (pH 7.4) for 1–2 hr at room temperature. The specimens were washed in 0.1 M sodium cacodylate buffer, two changes, and then transferred 1% OsO_4 in 0.1 M cacadolyte buffer. The fixed placozoans were dehydrated through a graded ethanol series (10%, 20%, 30%, 50%, 70%, 85%, and 95%), with three changes of 3–4 min at each dilution. They were then transferred to 100% ethanol for 3 changes of 10 min each.

Scanning electron microscopy

The fixed specimens were dried in a Tousimis Samdri-795 critical point dryer. Specimens were mounted on aluminum stubs and sputter coated with gold/palladium in a Hummer 6.2 sputter coater. Specimens were viewed with a Hitachi S-4800 Field Emission Scanning Electron Microscope at an accelerating voltage of 5.0 kV.

Transmission electron microscopy

After embedding the specimens in LX-112 epoxy resin from LADD, ultrathin (60–80 nm) sections were obtained on an RMC Powertome ultramicrotome, double stained with uranyl acetate and lead citrate, viewed on a Hitachi HT7700 TEM at 100 kV, and photographed with an AMT XR-41B 2k × 2k CCD camera.

3.2   How many symbionts are known to be present and where do they occur?

As noted above, Grell and Benwitz (1971) clearly demonstrated intracellular bacteria in the fiber cells of *T. adhaerens*. Further, Driscoll et al. (2013), using the genomic data of Srivastava et al. (2008), identified a prevalent bacterium in *T. adhaerens* as a member of the Rickettsiales. In the study by Gruber-Vodicka et al. (2019), sequencing detected two consistent bacterial phylotypes, named Cand. *Grellia incantans* (Rickettsiales) and Cand. *Ruthmannia eludens* (Margulisbacteria), in all five individuals of *Trichoplax adhaerens* haplotype H2 that were analyzed. In this same study, however, sequencing detected eight additional phylotypes that were present in at least 1000 out of a million reads in some, but not all, individuals. The authors noted high levels of variation in occurrence across the ten bacterial phylotypes. Three of the five individuals of *T. adhaerens* that were sampled had reads for all ten phylotypes, with ratios among the phylotypes varying widely between individuals. For example, while *G. incantans* and *R. eludens*, together, had the highest average percentages of the bacterial reads (62%) among the five individuals, these two phylotypes, while always present, represented a range of between 16% and 91% of the total bacterial reads across the five samples. Further, a given bacterial phylotype was frequently present with only a few reads in one individual placozoan, but was among the most abundant reads in another.

Caution must be taken with metagenomic sequence data, as they do not account for the possibility that common bacterial phylotypes occur as abundant members of the biofilms on which placozoans feed; thus, sequencing might include a common bacterial food source across the five samples. Or, if the bacteria are common to the animals' environment, some of these phylotypes could be fouling the dorsal surface of the placozoans; however, high magnification images we made never revealed large numbers of bacteria on an animal's surface. In addition, sequencing does not account for dead bacteria in a sample.

Taken together, the data presented by Gruber-Vodicka et al. (2019) provide evidence that in *T. adhaerens* haplotype H2: (i) *G. incantans* and *R. eludens* represent the core, coevolved microbiome, that is, shared by all individuals; and (ii) many other bacterial phylotypes are shared by most, but not all, individuals. Sharing of additional bacterial phylotypes by a subset of the placozoan population could reflect important genetic differences between individuals, or the availability of a specific subset of bacterial phylotypes during acquisition. However, all five individuals of *T. adhaerens* sampled by Gruber-Vodicka et al. (2019) came from the same culture of an H2 clone. As such, in this instance, we cannot explain the variation among the samples as a result of being acquired from different environments.

Until recently, all mention of symbionts in the placozoans restricted their presence to the fiber cells. However, in the recent paper of Gruber-Vodicka et al. (2019) on *T. adhaerens* H2, TEM and fluorescent *in situ* hybridization (FISH) identified *G. incantans* only in the fiber cells and *R. eludens* only inside cells in the ventral epithelium.

3.3 Do all placozoans harbor both *G. incantans* and *R. eludens*?

The full extent to which the rickettsial species *G. incantans* and the margulisbacterium *R. eludens* occur in other placozoan haplotypes remains to be determined. While sequences for a rickettsial species are found in both H1 and H2 haplotypes, comparisons of nucleic acid and derived peptide sequences between the two bacteria in the two Trichoplax haplotypes indicate that they are divergent enough to belong to different species, and maybe even genera (Gruber-Vodicka et al. 2019). In contrast, no sequences identifiable as a margulisbacterium were found in haplotype H1 (Driscoll et al. 2013; Gruber-Vodicka et al. 2019). Furthermore, several papers on *T. adhaerens* haplotype H1 have high quality TEM images of the animal cells, and endosymbionts were not detected in ventral epithelial cells (see e.g., Grell and Benwitz 1971, 1981; Smith et al. 2014; Smith and Reese, 2016), nor did we observe them in TEMs of placozoans of unidentified haplotypes collected in Hawai'i)..

In a recent study, another placozoan genome, haplotype H13, was sequenced and found to be so divergent that it was designated a new species and a member of a new genus, *Hoilungia* (Eitel et al. 2018). TEMs of the new species, *H. hongkongensis*, revealed bacteria in the host fiber cells, similar in appearance to the rickettsial endosymbionts of *T. adhaerens* haplotypes H1 and H2. Unfortunately, the genomic sequences of the new species, were enriched by "contamination screening," which eliminated bacterial sequences from the analysis; as such, a symbiont was not identified. However, high resolution TEMs in this paper reveal no evidence for bacterial symbionts in any ventral epithelial cells. Finally, another genus and species of placozoan, *Polyplacotoma mediterranea,* was recently reported (Osigus et al. 2019). Although its morphology is distinctly different from that of all other species, its phylogenetic designations were derived from mitochondrial gene sequences, which eliminated not only host genomic DNA, but also that of any endosymbiont. The species has not been examined with the transmission electron microscope, so morphological information that might reveal symbionts is not available.

3.4   Intracellular locations of the placozoan symbionts

G. incantans, which belongs to the alphaproteobacterial phylum Rickettsiales, appeared to be restricted to the rough endoplasmic reticulum (rER) in the *T. adhaerens* haplotype H2 (Gruber-Vodicka et al. 2019). In studies of other placozoan species, bacteria were also reported to occur in the rER (Eitel et al. 2018) and, in our earlier unpublished TEM studies of unidentified Hawaiian haplotypes of *T. adhaerens*, we likewise observed bacteria in the rER (Figure 3.2). While the bacteria were not identified in these other studies, the data suggest that the presence of rickettsial symbionts in the rER may be a shared derived character of the Placozoa. Our TEMs also revealed instances of bacteria appearing to reside either inside the nuclear membrane or within the nucleus (Figures 3.2 and 3.3). Schulz and Horn (2015) discuss the occurrences of intranuclear bacteria in many protists and animals, noting advantages to such a position in protection from cytoplasmic defense mechanisms and access to nutrients.

TEM imaging captured ultrastructural features on the *G. incantans* surface that resemble outer membrane vesicles (OMVs) (Figure 3.2). OMVs are blebs on the cell surface of bacteria that contain a wide variety of biomolecules, including proteins and nucleic acids. Recent extensive research on OMVs has demonstrated that they are common features in Gram-negative bacteria used for communication between bacteria and between symbiotic bacteria and their host (Lynch and Alegado 2017). As such, they may provide a mechanism by which *G. incantans* interacts across the rER membrane with the host placozoan.

We also observed bacteria in other regions of the fiber cells in our TEMs; they occurred free in the cytoplasm associating with or distant from the mitochondria (Figure 3.2). Rickettsial pathogens are known to occur in other regions of the cell, but the identity of the bacteria that we observed remains to be determined; thus, the bacteria in our images may be Rickettsiales of the same species, of a different species, or non-rickettsial bacteria entirely. Further, we visualized extracellular bacteria among the fiber cells by both SEM and TEM (Figure 3.3). Because members of the phylum Rickettsiales are obligate intracellular pathogens, it is likely that these extracellular bacteria do not belong to this group.

Single or multiple cells *of Ruthmannia eludens* were visualized in vacuoles of ventral epithelial cells (Gruber-Vodicka et al. 2019). TEM images also suggested that this species produces OMVs, and surface structures resembling fimbriae (or pili) were also observed. Bacterial fimbriae are structures by which bacterial cells attach to substrates (Proft and Baker 2009; Hospenthal et al. 2017). Although the fimbriae of *R. eludens* in the placozoan vacuoles were larger in diameter and more structured than is typical of bacterial fimbriae, they did appear to connect the bacterial cell to the host membrane. The functional significance of these structures remains to be elucidated.

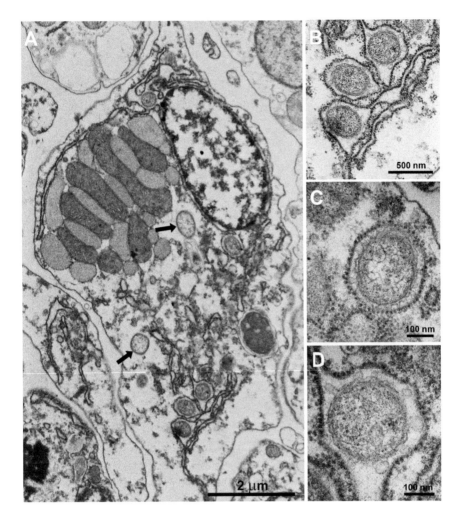

FIGURE 3.2   Fiber cells and intracellular bacteria. (A) Section through a fiber cell illustrating a large mitochondrial complex, nucleus and bacteria both in the cytoplasm and within rough endoplasmic reticulum (rER). (B) Higher magnification showing sections through three bacteria in rER. (C) Single bacterium in rER. (D) Single bacterium producing outer membrane vesicles around its entire body.

3.5   Unusual mitochondria in placozoan fiber cells and their possible relationship to symbiosis

The mitochondria of placozoans have a genome structure and size (Signorovitch et al. 2005; Eitel et al. 2018) and a proteome (Muthye and Lavrov 2018) that are divergent from these features of other metazoans. In addition, a shared character of the placozoan fiber cell is the carriage of large "mitochondrial complexes" as well as a population of rickettsial symbionts (Driscoll et al. 2013; Gruber-Vodicka et al. 2019) (Figure 3.4). Whether the bacteria influence the production of these complexes, or

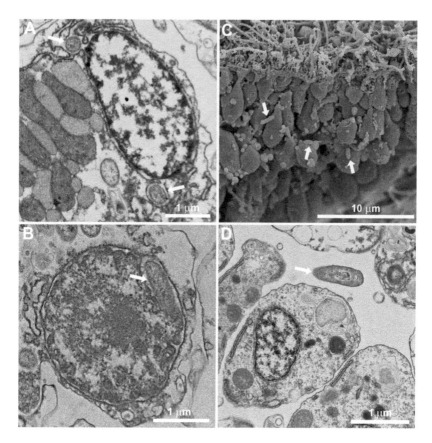

FIGURE 3.3   (A, B) Transmission electron micrographs reveal bacteria inside the nuclear membrane of fiber cells (white arrows). (C, D) Extracellular bacteria among fiber cells inside *T. adhaerens*: (C) scanning electron micrograph of a fractured specimen of *T. adhaerens* revealing extracellular bacteria (white arrows); (D) transmission electron micrograph showing bacteria between cells (white arrow).

the mitochondrial complexes enable carriage of the bacteria, or no relationship exists between the two remain to be determined. A rich literature, however, describes the profound influence of intracellular pathogens on mitochondrial structure and function (Spier et al. 2019), so it would not be unprecedented to find that the host mitochondria and *G. incantans* interact.

TEM images of the mitochondrial complexes in placozoan fiber cells reveal strong variation in their appearance, ranging from relatively disorganized to highly organized (Figure 3.4). Highly organized complexes have alternating layers of more and less electron dense areas. This type of structure is reminiscent of some hydrogenosomes, which are double-membrane-bound organelles thought to be derived from mitochondria (Biagini et al. 1997; Benchimol 2009). Inconsistent with these complexes having hydrogenosome components is that hydrogenosomes are typically associated with protists or animals in anaerobic environments. For

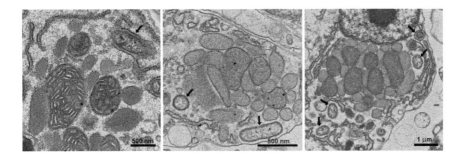

FIGURE 3.4 Variation in the mitochondrial complexes of the placozoan fiber cells. This array of images illustrates the varied appearance of the mitochondrial complexes in the host cells harboring *Grellia incantans* (arrows). In some instances (left) the cristae of the mitochondria are well defined, whereas in other instances they are not (middle, right). In addition, the features that can be identified as mitochondria are also in association with membrane-bound structures of similar size that have low internal complexity, reminiscent of hydrogenosomes in their ultrastructural characteristics. See also Figure 3.2A.

example, certain species of the animal phylum Loricifera that live in deep-sea anoxic sediments have been described to carry hydrogenosomes (Danovaro et al. 2010), as has a protist in the hindgut of a certain cockroaches (Boxma et al. 2005). Interestingly, ultrastructural studies of the loriciferans and protists reveal that they also have intracellular symbionts (Gijzen et al. 1991; Danovaro et al. 2010). Whether these complexes in placozoans are, in fact, hydrogenosomes and—if they are—what their function might be, remain to be determined. It seems unlikely that, in cells harboring twenty or more bacteria and strange and large mitochondrial complexes, the phenomena are unrelated.

3.6 Molecular inferences on the nature of the Trichoplax-bacteria symbioses

Because, thus far, the bacterial partners of placozoans are not culturable outside of the symbioses, biologists are left to rely on inferences from "-omics" data about the mechanism of host recognition of the symbiont and the associated signaling pathways, as well as about the basis of the symbiotic association (e.g., nutritional, defense, etc.). Studies of the host genome provide evidence that the placozoans have a robust innate immune system (Kamm et al. 2019). Bioinformatic analyses have indicated that several gene families associated with innate immunity are expanded in the Placozoa. For example, they have numerous genes with molecular signatures of pattern-recognition receptors (PRRs), with motifs of Toll-like receptors and NOD-like receptors for the recognition of bacterial "microbe associated molecular patterns" (MAMPs) (Takeuchi and Akira 2010). The isoforms of such receptors recognize specific biomolecules, for example, derivatives of bacterial lipopolysaccharide and peptidoglycan, bacterial flagella, and CpG DNA (Takeuchi and Akira 2010). Whether any given receptor recognizes any of these ligands will be difficult to determine until robust experimental methods for analyses of these systems are developed. Symbioses of the "-omics" data available for the Placozoa have revealed that they are missing many of the typical downstream pathway elements associated with host responses to interactions with microbes. However, the data do indicate that members of this phylum have some subset of the innate-immune genes conserved among animals.

Studies of the genome and transcriptome of *T. adhaerens* haplotype H2 have provided insights into the nature of the symbiotic associations this host has with *G. incantans* and *R. eludens* (Gruber-Vodicka et al. 2019). With *G. incantans*, the rickettsian symbiont of the fiber cells, it was not surprising to find the genes associated with energy parasitism, which is the basis of the symbiosis in parasitic rickettsians. However, analysis of the transcriptomes of *G. incantans* revealed that these genes are not expressed; rather the bacterium appears to make its own ATP, exhibiting a strongly expressed ATP synthase. In addition, bacterial genes encoding proteins of the TCA cycle and electron-transport chain are also expressed, suggesting that ATP synthesis occurs by oxidative phosphorylation. For its contribution to the symbiosis, the host provides the substrate metabolites for the TCA cycle, notably alpha keto acids and C4-dicarboxylates. And for its part, the symbiont supplies the host with riboflavin (vitamin B_2); the genes required for synthesis of this vitamin are missing in the host. These inferences provide the best evidence for a mutualistic symbiosis between the placozoan and its rickettsian partner.

The genomic and transcriptomic data provide evidence that *R. eludens*, the symbiont of the ventral epithelia in *T. adhaerens* H2, derives nutritional benefit from the host by the uptake of host-degraded lipids (Gruber-Vodicka et al. 2019). The lipids are thought to come from the algae upon which the host feeds. The host genome encodes lipases, which, it is theorized, convert these lipids to glycerol and fatty acids, which are taken up and used by the symbiont. The symbionts have high levels of expression of downstream genes, notably fatty acid coenzyme A ligase that would covert the fatty acids to acetyl CoA, which could be used in the TCA cycle. Notably, the symbiont genome has genes encoding lipases, but the transcripts for these proteins were not found in the transcriptomes, suggesting that they are not translated. The genome of *R. eludens* includes genes for the production of nine essential amino acids of animals, although expression of no specific transporters for these amino acids was detected. One possible mechanism of transfer might be the OMVs that were noted in TEMs (Gruber-Vodicka et al. 2019). The data provided by "-omics" analysis strongly suggest a tight relationship between host and symbiont, where both partners rely on goods and services from the other. These analyses also suggest a long coevolutionary history of these partners.

3.7 How are the bacterial symbionts of placozoans transmitted between generations?

Given that most placozoans are asexually produced by simple fission, it is very likely that cells containing bacteria remain with each portion of each new animal. A microscopical study of the budding of free-swimming swarmers in *T. adhaerans* revealed that the swarmers contain all cellular elements of these benthic placozoans (Thiemann and Ruthmann 1991). Thus, whether new placozoan individuals arise by fission or budding, their microbial symbionts are vertically transmitted. However, eggs and embryonic cleavage stages were observed in the spaces between the upper and lower epithelia of *T. adhaerens* by Grell (1972) when he mixed together individuals from two different clones he had established. Although Grell never observed sperm, he conjectured that the presence of an apparent fertilization membrane around the eggs was evidence that they were fertilized. Subsequent papers by Grell (1984) and Eitel et al. (2011) provided evidence that egg production can be induced both by warming and by starving cultures of *T. adhaerens* H2. Additionally, both Grell and Eitel et al. noted that all individuals producing eggs were in a degenerative state. Signorovitch et al. (2005), based on sequence comparisons across ten field-collected individuals, deduced that sexual reproduction must have been the basis for mixes of homozygotes and heterozygotes with the same sequences. Sexual reproduction was further confirmed by the genetic studies of Eitel et al. (2011) who recorded microscopical evidence for the presence of ova and early cleavage stages within the middle layer in *T. adhaerens* haplotypes H1, H2, and H16. Most importantly, Eitel et al. (2011) record the presence of endosymbiotic bacteria in oocytes and embryos of H2. While not identifying the source of the oocytes, the authors note the transfer of bacteria from fiber cells to developing oocytes. Additionally, phylotype H2 was observed by Gruber-Vodicka et al. (2019) to have intracellular bacteria within cells of the lower epithelium, providing another potential source for the intra-oocyte bacteria observed by Eitel et al. (2011).

3.8 Some big questions remaining and suggestions for their resolution

1. Do rickettsian symbionts occur in the fiber cells of all strains and species of placozoans? Given the recognition of Gruber-Vodicka et al. (2019) that the rickettsian symbionts of *T. adhaerens* haplotypes H1 and H2 differ at the genus level, how consistent are any of these rickettsians across the placozoan strains and species? To resolve these questions, full genomes must be developed for more placozoan haplotypes and genera and analyzed for the presence of bacterial genes.

2. Is *Ruthmannia eludens* present in the ciliated ventral cells of different strains and species and genera of *T. adhaerens,* or only in haplotype H2? *R. eludens* appears not to be present in H1. Resolution of this question will require additional genomic analyses and careful TEM examination of other haplotypes, species, and genera, plus application of the FISH probe developed for *R. eludens* by Gruber-Vodicka et al. (2019).

3. As noted above, "-omics" analyses of the relationships between *R. eludens* and *T. adhaerans* and their host strongly suggest mutualistic relationships where both partners rely on goods and services from the other. Additional investigation of the association with tools such as CRISPR-Cas 9 gene editing and development of methods to culture the symbionts, provide an exciting horizon for continuing studies of the fascinating symbiotic systems of the placozoans.

4. Exactly which ventral epithelial cells are occupied by *R. eludens*? Gruber-Vodicka et al. (2019) report only that they are in "epithelial cells," leaving unclear which of the three long-recognized cell types that make up that

epithelium harbors the bacteria. Recently, Sebé-Pedrós et al. (2018) provided data from single-cell transcriptomics that suggest more cell types are present in the ventral epithelium of *T. adhaerens* H1, which expands the question. This question might be addressed by developing useful *in situ* probes for specific cell types based on the transcriptomic data of Sebé-Pedrós et al. (2018), and treating sections of *T. adhaerens* simultaneously with these probes and the FISH probes developed by Gruber-Vodicka et al. (2019) for *R. eludens*.

5. How stable are the relationships between placozoans and their bacterial endosymbionts? Assuming the limited benefits to the host shown by Gruber-Vodicka et al. (2019) and a requisite host habitat for the bacteria, what maintains the balance? I.e., what keeps the host from killing its symbionts and the symbionts from overwhelming the cells they reside in? That is, what are they saying to each other to keep the relationships stable?

Acknowledgments

The authors are grateful to many members of their laboratories who support their research. For this chapter, Dr. Brian Nedved prepared our figures, and Dr. Marnie Freckelton gave very useful advice on an earlier draft of the manuscript. Dr. Vicki Pearse provided an expert review and useful suggestions for improving this chapter. Research on the Placozoa was supported in the Hadfield lab by grants from the Gordon and Betty Moore Foundation and the U.S. Office of Naval Research.

References

Benchimol, M. 2009. Hydrogenosomes under microscopy. *Tissue and Cell* 41:151–168. doi: 10.1016/j.tice.2009.01.001.

Biagini, G. A., B. J. Finlay, and D. Lloyd. 1997. Evolution of the hydrogenosome. *FEMS Microbiology Letters* 155 (2):133–140. doi: 10.1111/j.1574-6968.1997.tb13869.x.

Boxma, B., R. M. de Graaf, G. W. van der Staay, T. A. van Alen, G. Ricard, T. Gabaldón, A. H. Van Hoek, S. Y. Moon-Van Der Staay, W. J. Koopman, and J. J. van Hellemond. 2005. An anaerobic mitochondrion that produces hydrogen. *Nature* 434 (7029):74. doi: 10.1038/nature03343. PMID:15744302.

Danovaro, R., A. Dell'Anno, A. Pusceddu, C. Gambi, I. Heiner, and R. M. Kristensen. 2010. The first metazoa living in permanently anoxic conditions. *BMC Biology* 8 (1):30. PMID:20370908.

Driscoll, T., J. J. Gillespie, E. K. Nordberg, A. F. Azad, and B. W. Sobral. 2013. Bacterial DNA sifted from the *Trichoplax adhaerens* (Animalia: Placozoa) genome project reveals a putative rickettsial endosymbiont. *Genome Biology and Evolution* 5 (4):621–645. doi: 10.1093/gbe/evt036. PMID:23475938.

DuBuc, T. Q., J. F. Ryan, and M. Q. Martindale. 2019. "Dorsal–ventral" genes are part of an ancient axial patterning system: Evidence from *Trichoplax adhaerens* (Placozoa). *Molecular Biology and Evolution* 36 (5):966–973. doi: 10.1093/molbev/msz025. PMID:30726986.

Eitel, M., W. R. Francis, F. Varoqueaux, J. Daraspe, H.-J. Osigus, S. Krebs, S. Vargas, H. Blum, G. A. Williams, and B. Schierwater. 2018. Comparative genomics and the nature of placozoan species. *PLoS Biology* 16 (7):e2005359. doi: 10.1371/journal.pbio.2005359. PMID:30063702.

Eitel, M., L. Guidi, H. Hadrys, M. Balsamo, and B. Schierwater. 2011. New insights into placozoan sexual reproduction and development. *PLOS ONE* 6 (5):e19639. doi: 10.1371/journal.pone.0019639. PMID:21625556.

Eitel, M., and B. Schierwater. 2010. The phylogeography of the Placozoa suggests a taxon-rich phylum in tropical and subtropical waters. *Molecular Ecology* 19 (11):2315–2327. doi: 10.1111/j.1365-294X.2010.04617.x. PMID:20604867.

Gijzen, H. J., C. Broers, M. Barughare, and C. K. Stumm. 1991. Methanogenic bacteria as endosymbionts of the ciliate *Nyctotherus ovalis* in the cockroach hindgut. *Applied and Environmental Microbiology* 57 (6):1630–1634. PMID:1908205.

Grell, K. G. 1972. Eibildung und furchung von *Trichoplax adhaerens* F. E. Schulze (Placozoa). *Zoomorphology* 73 (4):297–314.

Grell, K. G. 1984. Reproduction of Placozoa. In: Engels W, ed. *Advances in Invertebrate Reproduction*. Elsevier, pp. 541–546.

Grell, K. G., and G. Benwitz. 1971. Die Ultrastruktur von *Trichoplax adhaerens* F. E. Schulze. *Cytobiologie* 4 (2):216–240.

Grell, K. G. and G. Benwitz. 1981. Ergänzende Untersuchungen zur Ultrastruktur von *Trichoplax adhaerens* F.E. Schulze (Placozoa). *Zoomorphology* (1981) 98:47–67.

Gruber-Vodicka, H. R., N. Leisch, M. Kleiner, T. Hinzke, M. Liebeke, M. McFall-Ngai, M. G. Hadfield, and N. Dubilier. 2019. Two intracellular and cell type-specific bacterial symbionts in the placozoan *Trichoplax* H2. *Nature Microbiology* 1. doi: 10.1038/s41564-019-0475-9. PMID:31182796.

Hospenthal, M.K., T.R.D. Costa, and G. Waksman. 2017. A comprehensive guide to pilus biogenesis in Gram-negative bacteria. *Nature Reviews Microbiology* 2017;15(6):365–379. doi: 10.1038/nrmicro.2017.4. PMID:28496159.

Kamm, K., B. Schierwater, and R. DeSalle. 2019. Innate immunity in the simplest animals–placozoans. *BMC Genomics* 20 (1):5. doi: 10.1186/s12864-018-5377-3. PMID:30611207.

Laumer, C. E., H. Gruber-Vodicka, M. G. Hadfield, V. B. Pearse, A. Riesgo, J. C. Marioni, and G. Giribet. 2018. Support for a clade of Placozoa and Cnidaria in genes with minimal compositional bias. *Elife* 7:e36278. doi: 10.7554/eLife.36278. PMID:30373720.

Lynch, J. B., and R. A. Alegado. 2017. Spheres of hope, packets of doom: The good and bad of outer membrane vesicles in interspecies and ecological dynamics. *Journal of Bacteriology* 199 (15):e00012–e00017. doi: 10.1128/JB.00012-17. PMID:28416709.

Muthye, V., and D. V. Lavrov. 2018. Characterization of mitochondrial proteomes of nonbilaterian animals. *IUBMB Life* 70 (12):1289–1301. doi: 10.1002/iub.1961. PMID:30419142.

Osigus, H.-J., S. Rolfes, R. Herzog, K. Kamm, and B. Schierwater. 2019. *Polyplacotoma mediterranea* is a new ramified placozoan species. *Current Biology* 29 (5):R148–R149. doi: 10.1016/j.cub.2019.01.068. PMID:30836080.

Pearse, V. B., and O. Voigt. 2007. Field biology of placozoans (Trichoplax): Distribution, diversity, biotic interactions. *Integrative and Comparative Biology* 47 (5):677–692. doi: 10.1093/icb/icm015. PMID:21669749.

Proft, T., and E. N. Baker. 2009. Pili in Gram-negative and Gram-positive bacteria—Structure, assembly and their role in disease. *Cellular and Molecular Life Sciences* 66(4):613–635. doi: 10.1007/s00018-008-8477-4. PMID:18953686.

Schulz, F., and M. Horn. 2015. Intranuclear bacteria: Inside the cellular control center of eukaryotes. *Trends in Cell Biology* 15(6):339–346. doi: 10.1016/j.tcb.2015.01.002.

Sebé-Pedrós, A., E. Chomsky, K. Pang, D. Lara-Astiaso, F. Gaiti, Z. Mukamel, I. Amit, A. Hejnol, B. M. Degnan, and A. Tanay. 2018. Early metazoan cell type diversity and the evolution of multicellular gene regulation. *Nature Ecology & Evolution* 2 (7):1176. doi: 10.1038/s41559-018-0575-6. PMID:29942020.

Signorovitch, A. Y., S. L. Dellaporta, and L. W. Buss. 2005. Molecular signatures for sex in the Placozoa. *Proceedings of the National Academy of Sciences* 102 (43):15518–15522. doi: 10.1073/pnas.0504031102. PMID:16230622.

Smith, C. L., and T. S. Reese. 2016. Adherens junctions modulate diffusion between epithelial cells in *Trichoplax adhaerens*. *The Biological Bulletin* 231 (3):216–224. doi: 10.1086/691069. PMID:28048952.

Smith, C. L., F. Varoqueaux, M. Kittelmann, R. N. Azzam, B. Cooper, C. A. Winters, M. Eitel, D. Fasshauer, and T. S. Reese. 2014. Novel cell types, neurosecretory cells, and body plan of the early-diverging metazoan *Trichoplax adhaerens*. *Current Biology* 24 (14):1565–1572. doi: 10.1016/j.cub.2014.05.046. PMID:24954051.

Spier, A., F. Stavru, and P. Cossart. 2019. Interaction between intracellular bacterial pathogens and host cell mitochondria. *Microbiology Spectrum* 7 (2). doi: 10.1128/microbiolspec. BAI-0016-2019. PMID:30848238.

Srivastava, M., E. Begovic, J. Chapman, N. H. Putnam, U. Hellsten, T. Kawashima, A. Kuo, T. Mitros, A. Salamov, and M. L. Carpenter. 2008. The Trichoplax genome and the nature of placozoans. *Nature* 454 (7207):955. doi: 10.1038/nature07191. PMID:18719581.

Takeuchi, O., and S. Akira. 2010. Pattern recognition receptors and inflammation. *Cell* 140:805–820. doi: 10.1016/j.cell.2010.01.022. PMID:20303872.

Thiemann, M., and A. Ruthmann. 1991. Alternative modes of axsexual reproduction in *Trichoplax adhaerens* (Placozoa). *Zoomorphology* 110:165–174.

Voigt, O., A. G. Collins, V. B. Pearse, J. S. Pearse, A. Ender, H. Hadrys, and B. Schierwater. 2004. Placozoa–no longer a phylum of one. *Current Biology* 14 (22):R944–R945. doi: 10.1016/j.cub.2004.10.036. PMID:15556848.

Ward, J. 2008. Diversity and Biogeography of the Unique, Tropical Phylum Placozoa. *Master of Science*, University of Hawaii at Manoa (4338).

4 Decoding cellular dialogues between sponges, bacteria, and phages

*Lara Schmittmann, Martin T. Jahn,
Lucía Pita, and Ute Hentschel*

Contents

4.1 Introduction

The evolution of multicellularity has not only enabled the specialization of eukaryote cell types, but also provided stable confined habitats for microbes to engage in symbiotic associations with metazoans. Animal–microbe interactions presented new challenges, such as self/nonself recognition, but also new opportunities that have shaped the evolution and diversification of holobionts. Sponges (Porifera), as one of the most basal animals, provide a fundamental resource to decipher key mechanisms of animal–microbe interactions with implications for our understanding of the interactions between more complex invertebrates/vertebrates and microbes. Significant progress has been made in our comprehension of the metabolic interactions and (meta-)genomic repertoires of sponge symbioses and the reader is kindly referred to several excellent reviews in the field (Taylor et al. 2007; Webster et al. 2016; Pita et al. 2018b). In this chapter, we summarize the current knowledge on cellular dialogues within sponge holobionts by taking a close look at the different players and interactions that make sponges one of the most diverse and successful marine animal groups.

Sponges have a fossil record dating back to ∼600 Mya (Yin et al. 2015) and around 9,000 extant species have been described (Van Soest et al. 2012). Despite their high taxonomic diversity, all sponges possess a sessile, filter-feeding lifestyle (exceptions: carnivorous sponges and the pelagic larval phase). Sponges continuously pump water and consume large amounts of microbial cells as well as dissolved organic matter (DOM) (De Goeij et al. 2013). Specialized flagellated cells (choanocytes) capture particles from the surrounding water and transfer them into the sponge interior, the sponge extracellular matrix (mesohyl). Once inside the mesohyl, the particles are digested by phagocytotically active, amoeboid cells (archaeocytes). While seawater bacteria constitute one of the main sponge food sources, the mesohyl also harbors dense bacterial symbiotic communities within the sponge host (Thomas et al. 2016; Moitinho-Silva et al. 2017b).

Comprehensive knowledge on sponge microbiome diversity and functions has been gained from 16S rRNA gene sequencing and high-throughput sequencing technologies such as metagenomics, metatranscriptomics, and single-cell genomics (Horn et al. 2016; Thomas et al. 2016; Moitinho-Silva et al. 2017a; Podell et al. 2019). An extremely high diversity of sponge-associated bacteria has been discovered with members of >60 bacterial phyla so far (Thomas et al. 2016; Moitinho-Silva et al. 2017b). Although a fraction of sponge-associated microbes also occurs in the surrounding environment, each sponge species maintains a specific and stable microbial community (Thomas et al. 2016). Notably, the sponge-specific symbionts are adapted to live within the sponge habitat (Siegl et al. 2011; Jahn et al. 2016), making sponges a refuge for novel biodiversity. The symbionts are either maintained by vertical transmission from adults to offspring (Sharp et al. 2007; Schmitt et al. 2008; Webster et al. 2010; Sipkema et al. 2015; Björk et al. 2018; Russell 2019) or are acquired horizontally from the seawater (Björk et al. 2018). Based on the composition and abundance of microbes in their tissues, sponges can be defined in high and low microbial abundance (HMA/LMA) sponges (Gloeckner et al. 2014). Those two lifestyles can be differentiated based on microbial microscopy and taxonomy (Gloeckner et al. 2014; Moitinho-Silva et al. 2017c), as well as on sponge physiology and pumping rates (Weisz et al. 2008).

In contrast to most other animals, microbes in sponges mainly occur extracellularly, in close vicinity to sponge cells (but note exceptions where bacteria are enclosed in bacteriocytes e.g. Burgsdorf et al. 2019; Tianero et al. 2019). Bacterial cell densities can reach up to 10^9 cells per cm^3 of sponge tissue and outnumber sponge cell abundance by orders of magnitude (Taylor et al. 2007). Thus, these morphological basal animals constitute one of the most complex holobionts with several types of sponge cells and a large diversity of microbial symbiont lineages coexisting in the same matrix. In this chapter, we focus on three main types of interactions: (i) the dialogue between sponge cells and bacteria, (ii) the dialogue between bacterial cells, and (iii) the tripartite interaction between sponge cells, bacteria, and bacteriophages (Figure 4.1). In the first section, we will discuss the current knowledge on host mechanisms for microbial recognition as well as microbial features to promote tolerance. In the second section, we will present recent literature on bacteria–bacteria interactions in the context of

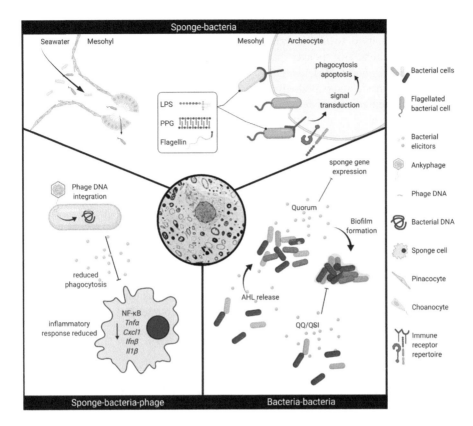

FIGURE 4.1 Schematic presentation of the cellular interactions on the platform that the sponge mesohyl provides (central circle): sponge and bacterial cells (upper panel); bacterial cells (right panel); sponge and bacterial cells with bacteriophages. (left panel; adapted from Jahn, M. T. et al. 2019. *Cell Host Microbe* 26: 542–550.e5.) LPS, lipopolysaccharides; PPG, peptidoglycan; AHL, N-acyl homoserine lactones; QQ/QSI, quorum sensing/quorum sensing inhibition. (The figure was created with the visualization tool BioRender.com)

quorum sensing/quenching. In the third section, we will present a recent discovery on how bacteriophages can foster sponge–bacteria symbiosis. Finally, we will highlight emerging topics in sponge–microbe research.

4.2   Host–bacteria dialogue

Already in the early 1980s, Wilkinson et al. (1979) reported that sponges can distinguish between seawater bacteria and their symbionts by feeding tritium-labelled bacteria to sponges followed by high-resolution radioautography of sponge tissue. While most cells of the bacterial symbionts passed through the sponge unharmed and were expelled via the exhalant water, the seawater bacteria (*Vibrio alginolyticus*) were retained by the sponge and were digested. Later, Wehrl et al. (2007) confirmed that feeding rates of sponges on seawater bacteria are higher than on sponge symbionts. How does the sponge differentiate between food bacteria and symbionts?

To address this question in an experimental way, differential gene expression analyses was used to characterize the molecular response of sponges towards microbial elicitors. Two sponges that are representatives of the HMA/LMA dichotomy (*Aplysina aerophoba,* HMA, *Dysidea avara,* LMA) were exposed to a cocktail of lipopolysaccharide and peptidoglycan as signals (Pita et al. 2018a). We hypothesized that the different microbial densities in these sponges would affect the host's responses towards bacterial elicitors. Both species responded to microbial stimuli by increasing the expression of a subset of immune receptors (such as NLRs in *D. avara*, SRCR and GPCRs in *A. aerophoba*) and activating kinase cascades likely yielding apoptotic and phagocytotic processes (Figure 4.2). Moreover, the magnitude of the transcriptionally-regulated response (in terms of number of differentially expressed genes) was more complex in *A. aerophoba* (HMA) than in *D. avara* (LMA). We propose that the HMA species requires a more fine-tuned regulated response to deal with conflicting signals coming from the microbial stimuli versus those from the symbionts.

Other studies support the role of the sponge immune system in the crosstalk with microbes. The sponge *Petrosia ficiformis* displayed an increased expression of a gene containing the conserved SRCR domain when living in symbiosis with a cyanobacterium, in comparison to the aposymbiotic status (Steindler et al. 2007). In juvenile *Amphimedon queenslandica*, bacterial encounter involved regulation of

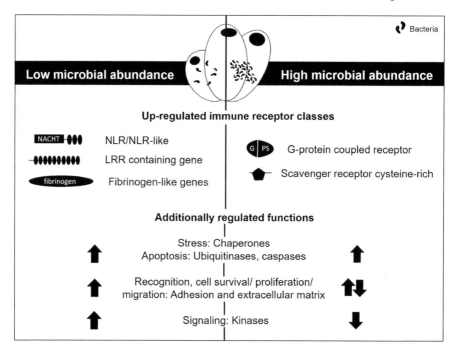

FIGURE 4.2 Differentially expressed genes in one LMA (*Dysidea avara*, left side) and one HMA (*Aplysina aerophoba*, right side) sponge after exposure to bacterial elicitors (LPS and peptidoglycan). (Adapted from Pita et al. 2018a. *Sci. Rep.* 8: 1–15.) The upregulated immune receptors with characteristic, conserved domains are depicted, as well as additional regulated functions (up- and/or downregulation represented by arrows).

SRCR-containing genes, but the downstream signaling response differed depending on the origin of the bacteria (Yuen 2016). In particular, the transcription factors *FoxO* and *NFκβ* were upregulated upon exposure to own symbionts, but not to a bacterial fraction from another sponge species (Yuen 2016). Finally, the components of the TLR pathway such as *MyD88* were activated in response to microbial signals in different sponge species (Wiens et al. 2005; Yuen 2016).

4.2.1   Sponge immune receptors

The sponge cellular immune response was studied already in the 19th century by Nobel laureate Elias Metchnikoff and colleagues (Metchnikoff 1893). While these first studies were not focused on the sponge's response to microbes, the observed cell behaviors suggested that two cell types are the key players for mediating the interactions with microbes: choanocytes (representing the first barrier for external microbes) and archeocytes (representing the patrol of the sponge matrix). Despite this promising takeoff, our understanding of sponge cellular immunity is still at the beginning. The publication of the first sponge genome, that of the Great Barrier Reef species *Amphimedon queenslandica*, brought a complex and expanded repertoire of immune receptors into light (Srivastava et al. 2010). This repertoire included several extracellular (i.e., scavenger receptor cystein-rich, SRCR, domain), membrane-bound (immunoglobulin-like domains), and intracellular (NOD-like receptor, NLR, domains) receptors (Srivastava et al. 2010; reviewed in Hentschel et al. 2012). These gene families were identified at sequence level because the domains (sequence patterns) were arranged in a particular architecture that is conserved from early metazoans to vertebrates. The conserved gene structure suggests a conserved function, which is not always clear. For example, Toll-like receptors (TLRs) are transmembrane receptors characterized by several extracellular leucin-rich repeat (LRR) motifs and an intracellular Toll/interleukin-1 receptor (TIR) domain. In vertebrates, the extracellular LRR motifs recognize the ligand (e.g. bacteria) and transduce the signal via the TIR domain. All components of the signaling cascade induced by TLRs are present in *A. queenslandica*, but not the conventional TLR. The genome of *A. queenslandica* contained a TIR domain-containing gene, homolog of the TIR-domain in vertebrates TLRs, which was combined with extracellular Ig domains rather than LRR motifs. Therefore, its role in bacteria recognition remains to be validated.

A striking feature of the *A. queenslandica* genome was the high diversification of two other immune receptor families (Hentschel et al. 2012). The NLR family are defined according to the presence of a nucleotide-binding domain combined with a leucine-rich repeat domain (Ting et al. 2008). The genetic animal models *Drosophila melanogaster* and *C. elegans* lost this receptor family and, therefore, it was long thought that these receptors had their origin in the teleost. An interesting feature of *A. queenslandica* NLRs is their enormous diversity: *A. queenslandica* genome comprises 135 genes, which is in stark contrast to 20 NLR genes in humans (Yuen et al. 2014). Similarly, the family of scavenger receptors cysteine rich (SRCRs) in *A. queenslandica* is also highly expanded (ca. 300 genes) when compared to vertebrates (e.g. 16 genes in humans) and other invertebrates (Buckley and Rast 2015). All cell types in *A. queenslandica* adults express genes with SRCR domains, but they are significantly enriched in choanocytes (Sebé-Pedrós et al. 2018). The evolutionary forces driving a high diversity of pattern recognition receptors (PRRs) are suggested to be related to the specific recognition of a wide variety of microbial compounds and has been proposed as a mechanism for specificity in sponges and other invertebrates (Schulenburg et al. 2007; Messier-Solek et al. 2010; Buckley and Rast 2015; Degnan 2015).

Since the publication of *A. queenslandica* genome in 2010, further sponges have been sequenced at genome and transcriptome level. Most of these reference genomes and transcriptomes are incomplete, yet they are adding new knowledge to our understanding on sponge molecular repertoire of immunity. Poriferan TLR/IL-1R-like receptors as well as their downstream signaling cascades were detected in other sponge genomes and transcriptomes (Riesgo et al. 2014; Germer et al. 2017; Pita et al. 2018a). These new data confirmed the complex and expanded repertoire of Poriferan immune receptors, notably NLRs and SRCRs (Germer et al. 2017; Pita et al. 2018a). However, there are also differences among sponge species that may be related to their symbiotic status (HMA or LMA). Ryu et al. (Ryu et al. 2016) detected different enrichment in immune domains depending on symbiont densities within the mesohyl when comparing the genomes of LMA sponges *A. queenslandica* and *Stylissa carteri* versus the HMA sponge *Xestospongia testudinaria*. Along similar lines, we detected 80 *bona fide* NLRs in the reference transcriptome of the LMA sponge *D. avara*; whereas, using the same experimental setup, we found only one *bona fide* NLR gene in *A. aerophoba* (HMA) reference transcriptome (Pita et al. 2018a). The reference transcriptome of the HMA sponge *Vaceletia* sp. contained no NLR (Germer et al. 2017). These distinct signatures in the HMA and LMA genomic repertoires of immune receptors support the different evolutionary trajectories imposed by the symbiosis with microbes.

Apart from the above-mentioned receptors, lectins are also diversified in sponges. This class of soluble or membrane-bound proteins recognizes carbohydrates and mediates cell adherence, self/nonself recognition, and symbiotic relationships (Brown et al. 2018; Dinh et al. 2018). A few dozen lectins from sponges are known so far of which some may aid in bacterial recognition by responding to carbohydrates from gram-positive (peptidoglycan) as well as from gram-negative (lipopolysaccharides) bacteria (reviewed in Gardères et al. 2012). In a growth assay, a lectin from *Halichondria panicea* stimulated bacterial proliferation of sponge derived bacterial strains (Müller et al. 1981).

4.2.2   Microbe associated molecular patterns (MAMPs)

Immune receptors detect microorganisms via molecules that are present in prokaryotes but absent in eukaryotes, the so-called "pathogen associated molecular patterns" (PAMPs). PAMPs include components of bacterial cell walls and membranes such as peptidoglycans (PPGs) of gram-positive bacteria and lipopolysaccharides (LPS) from gram-negative bacteria. If such PAMPs are recognized by an immune receptor, they will trigger a signaling cascade yielding the elimination of the microbial invader (e.g. via phagocytosis). It was soon recognized that PAMPs are not exclusive for pathogens, but are also present in bacterial symbionts (Koropatnick et al. 2004), leading to the alternative use of the term "microbe associated molecular patterns", or MAMPs. Therefore, host recognition of microorganisms must be specific enough to yield an appropriate response, which may be either to eliminate or tolerate microorganisms.

We have limited experimental evidence for which symbiont MAMPs may be recognized by poriferan receptors. However, genetic features enriched or depleted in sponge-associated versus free-living bacteria revealed interesting patterns. In this context, it is remarkable that sponge-associated microbes lack flagella (Siegl et al. 2011). Flagellin is known as a powerful immune stimulator that initiated, for example, signal transduction mediated by TLR5 receptors (Hayashi et al. 2001). The absence of flagella (thus, flagellin) could allow microbes to evade host immunity and persist within the sponge holobiont. On a different note, a common sponge symbiont, "*Candidatus* Synechococcus spongiarum" (Cyanobacteria) presents a modified O-antigen in its LPS, as compared to free-living *Synechococcus* relatives (Burgsdorf et al. 2015). This modification could represent another mechanism for recognition. Thus, modifications in MAMP structure could help microbes to escape recognition as "nonself" by the host. Altogether, these studies provide evidence of adaption to symbiosis in both the host and the microbial side.

4.3   Bacteria–bacteria dialogue

Within the sponge holobiont, bacterial cells do not only interact with the sponge cells, but also with bacteria of the same or other species. Figure 4.3 is an illustration of the density of microbes within the sponge mesohyl matrix. Competition for space and resources, or initiation of biofilm formation, as well as secondary metabolite production are among the well-studied topics of bacterial communication (Abisado et al. 2018). Thereby, competition and cooperation are facilitated within the proximity

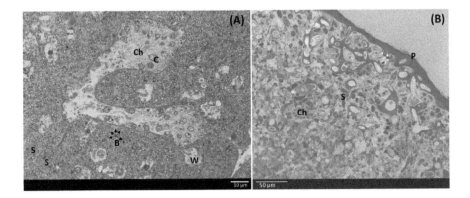

FIGURE 4.3   The sponge extracellular matrix is densely populated by sponge cells, as well as by diverse and abundant bacterial symbionts. Light microscopy images of semi-thin sponge tissue sections (A) of *Plakortis simplex* stained with Richardson solution and (B) of *Acantheurypon spinispinosum*. An epithelium-like outer cell layer, (pinacoderm, visible in (B)) surrounds the extracellular matrix (mesohyl) and the inorganic skeleton made up of spicules. Only pinacocytes are connected by tight junctions (Draper et al. 2019). C, choanocytes in Ch, choanocyte chamber; P, pinacoderm; B, bacterial cells in mesohyle; S, sponge cell; W, water canal. (Images kindly provided by Kathrin Busch [GEOMAR Helmholtz Centre for Ocean Research Kiel].)

of micrometers and are often a matter of balance (Nadell et al. 2016; Rakoff-Nahoum et al. 2016). Modeling can aid in predicting the metabolic interactions between bacteria, either based on co-occurrence models from relative abundance data (Thomas et al. 2016) or from metabolic models as inferred from metagenomic data (Slaby et al. 2017).

4.3.1 Quorum sensing

Bacteria–bacteria communication within the sponge extracellular matrix is mediated by quorum-sensing (QS). QS is a universal principle that aids interbacterial communication and is known from free-living as well as host-associated marine bacteria (reviewed in Hmelo 2017). Quorum sensing relies on the use of diffusible chemical signals in a population density-dependent manner. With increasing bacteria population size, the concentration of released QS molecules increases accordingly and eventually reaches a level (quorum) that initiates coordinated responses at the population level. QS mediates cellular mechanisms such as cell division, secondary metabolite production, plasmid transfer, and biofilm formation (Fuqua et al. 1994; Venturi and Subramoni 2009).

One of the most studied QS active molecule classes are N-acyl homoserine lactones (AHLs). AHLs are produced via the synthase LuxI family and interact with the LuxR cognate receptor proteins to initiate transcriptional activators and gene expression (Fuqua et al. 1994). The AHLs were first discovered in the marine bacterium *Vibrio fisherii* (Fuqua et al. 1994), and were detected in bacteria isolated from sponges for the first time in 2004 (Taylor et al. 2004). A great variety of AHLs have been recovered from sponge-derived bacterial isolates of different phylogenetic affiliations, including Gammaproteobacteria, Alphaproteobacteria, Firmicutes, and Flavobacteria (Mohamed et al. 2008; Bin Saidin et al. 2017; Mangano et al. 2018).

While bacterial isolates allow a thorough characterization of AHLs and their producer, only a small fraction of the sponge symbionts is cultivable, making it difficult to interpret the relevance of AHLs within the holobiont. Metagenomic data from *Theonella swinhoei* revealed an AHL synthase of an uncultured member of the Rhodobacterales family (Britstein et al. 2016). When heterologously expressed in *E. coli*, the synthase produced three different AHLs, demonstrating its function *in vitro* (Britstein et al. 2016). The first *in vivo* evidence of AHL production within the sponge comes from a study on *Suberites domuncula* (Gardères et al. 2012). AHLs were found in extracts of the whole sponge but not in extracts from the sponge cells, suggesting that the sponge itself does not produce AHLs (Gardères et al. 2012).

AHL production seems to be dependent on host species and varies over time (Britstein et al. 2018). Out of four investigated sponge species, one showed AHL production year-round, one showed no production at all, and two species displayed periodic production of AHLs (Britstein et al. 2018). For the sponge with constant AHL production, 14 different AHL molecules were identified, while only nine were present in the three replicate individuals. However, it is still unclear what drives this diversity. AHL patterns were neither related to LMA/HMA dichotomy nor correlated to microbiome composition (Britstein et al. 2018). One possibility is that constantly expressed AHLs derive from the core microbiome while varying AHL molecules result from transient seawater bacteria. Britstein et al. (2018) propose that microbial activity (i.e. gene expression) rather than microbial composition could account for AHL variability.

On the contrary, a single sponge-associated bacterium can produce a high diversity of AHLs, as in the case of a member of the family Rhodobacteraceae (Alphaproteobacteria), *Paracoccus sp.* Ss63 isolated from *Sarcotragus sp.* (Saurav et al. 2016). *Paracoccus sp.* Ss63 is present in low abundances in seawater, sediment, and other sponges. A diverse array of AHL molecules may provide the possibility to sense various environmental cues aiding the free-living versus host-associated lifestyle (Girard et al. 2019). For example, the pH gradient between seawater and sponges might benefit the accumulation of AHLs within the host and thus aid symbiosis establishment (Saurav et al. 2016). The clear role of AHLs within the sponge holobiont remains unknown, however, they are likely relevant for the bacteria–bacteria dialogue within the sponge holobiont.

4.3.2   Quorum quenching

Quorum quenching (QQ) and quorum sensing inhibition (QSI) refer to mechanisms by which QS molecules are degraded or inactivated and communication is interrupted (reviewed in Borges and Simoes 2019). Sponges have recently been mined for both, QSI and QQ molecules (reviewed in Saurav et al. 2017), while several sponge extracts showed QSI or QQ activity which inhibited biofilm formation and/or population growth (Annapoorani et al. 2012; Mai et al. 2015; Britstein et al. 2016; Gutiérrez-Barranquero et al. 2017). In some cases, these molecules were able to disrupt established biofilms (Gutiérrez-Barranquero et al. 2017). Plakofuranolactone (γ-lactone) is one of the few well-described QQ active molecules and was isolated from the sponge *Plakortis cf. lita* (Costantino et al. 2017). The bacterial origin of this molecule has been proven, but the microbial producer remains unidentified. A dual QS/QSI activity was described for bacteria isolated from sponges (Gutiérrez-Barranquero et al. 2017), highlighting the complexity of bacterial interactions within the sponge holobiont.

The exchange of molecules between bacteria might ultimately interfere with the communication between sponge cells, or between sponge and bacteria. QS was shown to not only work in bacteria–bacteria interactions, but also to be involved in interkingdom communication in both animals and plants (González and Venturi 2013; Pietschke et al. 2017; Weiland-Bräuer et al. 2019). In primmorph-cultures and adult *Suberitus domuncula* sponges, short-term stimulation with bacterial N-3-oxododecanoyl-L-homoserine lactone affected gene expression of the sponge host, while cell viability and morphology remained unaffected (Gardères et al. 2014). More specifically, genes related to immunity and apoptosis were downregulated, as assessed by qRT-PCR, potentially aiding the sponge to monitor and regulate bacterial populations (Gardères et al. 2014). This is a fascinating example of the interlinked dialogue between sponges and bacteria, as molecules that have originally evolved for bacteria–bacteria interactions may eventually be adopted by the sponge as a means to detect and respond to microorganisms.

4.4   Phage-bacteria–host dialogue

Phages are the most abundant and diverse entities in the oceans (Rohwer 2003) and, along with their role as major bacterial killers, significantly impact global biochemical cycles (Suttle 2007), bacterial fitness, and diversity (King et al. 2018). In terms of numbers, each milliliter of seawater contains on average about 10 million virus particles. As filter-feeding animals, sponges pump up to 24,000 liters of seawater through their system per day (Weisz et al. 2008), exposing them to up to an estimated $\sim 2.4 \times 10^{13}$ viruses daily. The very high exposure to viruses prompts the question whether viruses interact in any way with either the sponge host or with its associated microbial symbionts. Interestingly, defense mechanisms against invading phages were identified previously as enriched features of microbial sponge symbionts by metagenomics (Fan et al. 2012; Slaby et al. 2017). These defense mechanisms are based on self–nonself-discrimination (i.e., restriction-modification system) or prokaryotic adaptive immunity (i.e., CRISPR-Cas system), representing major strategies against viral infection. While sponges are clearly exposed to massive amounts of viruses, little is known about their potential dialogue with the sponges and its associated microbial symbionts.

4.4.1   Phage diversity and host-specificity

The presence of virus-like particles within sponge tissues was already described in 1978 (Vacelet and Gallissian 1978) and was confirmed recently by electron-microscopy (Pascelli et al. 2018). In order to capture the molecular diversity of the viral associates, sponge virome sequencing was performed on several Great Barrier Reef sponges (Laffy et al. 2018). Interestingly, the identified patterns indicated species-specific viral signatures. Taxonomically, many of the recovered sponge associated viruses were dominated by clades of bacteriophages such as by tailed bacteriophages of the order *Caudovirales* (dsDNA) and *Microviridae* (ssDNA), as well as viruses including members of *Megavirales* and *Parvoviridae* (Laffy et al. 2018). High viral diversity and novelty was also found in a recent study by Jahn et al. (2019) who used metagenomics to characterize the viral diversity of three Mediterranean sponge species along with seawater controls. The extent of novelty in the sponge viromes was astonishing: only 3% were known on the taxonomic family level. The identified virome signatures ("fingerprints") were highly specific to their host sponges in that each individual displayed its own unique virome signature (Jahn et al. 2019). The observation of viruses being individual specific is consistent with similar findings in humans (Moreno-Gallego et al. 2019).

4.4.2   Ankyphages aid symbionts in immune evasion

Jahn et al. (2019) further described a group of phages (hereafter termed "Ankyphages") that suppresses immune cell function and phagocytosis in eukaryotic cells (reviewed in Leigh 2019). These Ankyphages encode a novel symbiont phage-encoded protein, ANKp, that modulates eukaryote–bacterium interaction by altering the eukaryotes' physiology in response to bacteria. Specifically, it appears that the phage-encoded Ankyrin protein is secreted from the bacterial cell and downregulates eukaryotic proinflammatory cytokines and phagocytosis in response to ANKp. These experiments were performed in murine cell lines as an experimentally tractable model for sponge–microbe interactions is still lacking. Murine macrophages display many features of a major class of sponge cells (archaeocytes) which, much like macrophages, are single, amboeboid cells that patrol the sponge matrix in search for bacteria to be phagocytosed. Moreover, the major elements involved in mammalian immune signaling were found to be present in sponges. The resulting data show, to our knowledge for the first time, that phage ANKp modulates the eukaryote response to bacteria by downregulating proinflammatory signaling along with reduced phagocytosis rates.

Surprisingly, homology searches revealed that Ankyphages are widely distributed in host-associated environments, including the human oral cavity, gut, and stomach. It is thus tempting to speculate that the role of Ankyphages in mediating the dialogue between bacteria and animal hosts is much more widespread than in the context of marine sponges. In summary, ANKp represents the first secreted phage effector protein that downregulates eukaryote immunity upon exposure to bacteria. This is of relevance to host–microbe symbiosis research in that it provides the functional underpinnings for tripartite phage–bacteria–eukaryote dialogue. Moreover, this finding is of interest in the context of phage therapy as mechanisms to temper host immune responses are urgently sought-after in clinical and medical settings.

4.5   Conclusions and future perspectives

Sponge holobionts represent astonishingly complex ecosystems that consist of different types of host cells, and a high diversity of microbes existing in close proximity to each other. We suggest phagocytizing cells, like choanocytes and archaeocytes, to be of special interest to unravel the sponge–microbe dialogue on the cellular level. Current efforts to develop experimental models for sponge symbioses promise a more functional understanding of the cellular interactions in the sponge holobiont. Further, they will inform how sponge–microbe interactions shape and maintain the performance of the holobiont and allow evolutionary insights.

Acknowledgments

We thank the members of the CRC1182 for stimulating dialogues on metaorganism research. Funding was provided by the DFG CRC1182 and our national and international colleagues "Origin and Function of Metaorganisms" to U. H. (TPB1) and to M. T. J. (Young Investigator Award). L.S. was funded by the International Max Planck Research School for Evolutionary Biology.

References

Abisado, R. G., Benomar, S., Klaus, J. R., Dandekar, A. A., and Chandler, J. R. 2018. Bacterial quorum sensing and microbial community interactions. *MBio* 9: 1–13.

Annapoorani, A., Jabbar, A. K. K. A., Musthafa, S. K. S., Pandian, S. K., and Ravi, A. V. 2012. Inhibition of quorum wensing mediated virulence factors production in urinary pathogen *Serratia marcescens PSI* by marine sponges. *Indian J. Microbiol.* 52: 160–166.

Bin Saidin, J., Abd Wahid, M. E., and Le Pennec, G. 2017. Characterization of the *in vitro* production of *N*-acyl homoserine lactones by cultivable bacteria inhabiting the sponge *Suberites domuncula. J. Mar. Biol. Assoc. United Kingdom* 97: 119–127.

Björk, J. R., Díez-Vives, C., Astudillo-Garcia, C., Archie, E., and Montoya, J. M. 2018. Vertical transmission of sponge microbiota is inconsistent and unfaithful. *BioRxiv* 3: 1172–1183.

Borges, A., and Simoes, M. 2019. Quorum sensing inhibition by marine bacteria. *Mar. Drugs* 17: 1–25.

Britstein, M., Devescovi, G., Handley, K. M. et al. 2016. A new *N*-acyl homoserine lactone synthase in an uncultured symbiont of the red sea sponge *Theonella swinhoei. Appl. Environ. Microbiol.* 82: 1274–1285.

Britstein, M., Saurav, K., Teta, R. et al. 2018. Identification and chemical characterization of *N*-acyl-homoserine lactone quorum sensing signals across sponge species and time. *FEMS Microbiol. Ecol.* 94: 1–7.

Brown, G. D., Willment, J. A., and Whitehead, L. 2018. C-type lectins in immunity and homeostasis. *Nat. Rev. Immunol.* 18: 374–389.

Buckley, K. M., and Rast, J. P. 2015. Diversity of animal immune receptors and the origins of recognition complexity in the *Deuterostomes. Dev. Comp. Immunol.* 49: 179–189.

Burgsdorf, I., Handley, K. M., Bar-Shalom, R., Erwin, P. M., and Steindler, L. 2019. Life at home and on the roam: Genomic adaptions reflect the dual lifestyle of an intracellular, facultative symbiont. *MSystems* 4.

Burgsdorf, I., Slaby, B. M., Handley, K. M. et al. 2015. Lifestyle evolution in cyanobacterial symbionts of sponges. *MBio* 6: e00391–15.

Costantino, V., Sala, G. D., Saurav, K. et al. 2017. Plakofuranolactone as a quorum quenching agent from the Indonesian sponge *Plakortis* cf. *lita. Mar. Drugs* 15: 1–12.

De Goeij, J. M., Van Oevelen, D., Vermeij, M. J. A. et al. 2013. Surviving in a marine desert: The sponge loop retains resources within coral reefs. *Science* 342(6154): 108–110.

Degnan, S. M. 2015. The surprisingly complex immune gene repertoire of a simple sponge, exemplified by the NLR genes: A capacity for specificity? *Dev. Comp. Immunol.* 48: 269–274.

Dinh, C., Farinholt, T., Hirose, S., Zhuchenko, O., and Kuspa, A. 2018. Lectins modulate the microbiota of social amoebae. *Science* 361(6400): 402–406.

Draper, G. W., Shoemark, D. K., and Adams, J. C. 2019. Modelling the early evolution of extracellular matrix from modern Ctenophores and Sponges. *Essays Biochem.* 63(3): 389–405.

Fan, L., Reynolds, D., Liu, M. et al. 2012. Functional equivalence and evolutionary convergence in complex communities of microbial sponge symbionts. *Proc. Natl. Acad. Sci.* 109: E1878–E1887.

Fuqua, W. C., Winans, S. C., and Greenberg, E. P. 1994. Quorum sensing in bacteria: The LuxR-LuxI family of cell density-responsive transcriptional regulators. *J. Bacteriol.* 176: 269–275.

Gardères, J., Henry, J., Bernay, B. et al. 2014. Cellular effects of bacterial *N*-3–oxo-dodecanoyl-*L*-homoserine lactone on the sponge Suberites domuncula (Olivi, 1792): Insights into an intimate inter-kingdom dialogue. *PLOS ONE* 9: 1–10.

Gardères, J., Taupin, L., Bin-Saïdin, J., Dufour, A., and Le Pennec, G. 2012. *N*-acyl homoserine lactone production by bacteria within the sponge *Suberites domuncula* (Olivi, 1792) (Porifera, Demospongiae). *Mar. Biol.* 159: 1685–1692.

Germer, J., Cerveau, N., and Jackson, D. J. 2017. The holo-transcriptome of a calcified early branching metazoan. *Front. Mar. Sci.* 4: 1–19.

Girard, L., Lantoine, F., Lami, R., Vouvé, F., Suzuki, M. T., and Baudart, J. 2019. Genetic diversity and phenotypic plasticity of AHL-mediated Quorum sensing in environmental strains of *Vibrio mediterranei. ISME J.* 13: 159–169.

Gloeckner, V., Wehrl, M., Moitinho-Silva, L. et al. 2014. The HMA-LMA dichotomy revisited: An electron microscopical survey of 56 Sponge Species. *Biol. Bull.* 227: 78–88.

González, J. F., and Venturi, V. 2013. A novel widespread interkingdom signaling circuit. *Trends Plant Sci.* 18: 167–174.

Gutiérrez-Barranquero, J. A., Reen, F. J., Parages, M. L., McCarthy, R., Dobson, A. D. W., O'Gara, F. 2017. Disruption of *N*-acyl-homoserine lactone-specific signalling and virulence in clinical pathogens by marine sponge bacteria. *Microb. Biotechnol.* 12: 1049–1063.

Hayashi, F., Smith, K. D., Ozinsky, A. et al. 2001. The innate immune response to bacterial flagellin is mediated by Toll-like receptor 5. *Nat. Lett.* 410: 1–6.

Hentschel, U., Piel, J., Degnan, S. M., and Taylor, M. W. 2012. Genomic insights into the marine sponge microbiome. *Nat. Rev. Microbiol.* 10: 641–654.

Hmelo, L. R. 2017. Quorum sensing in marine microbial environments. *Ann. Rev. Mar. Sci.* 9: 257–281.

Horn, H., Slaby, B. M., Jahn, M. T. et al. 2016. An Enrichment of CRISPR and other defense-related features in marine sponge-associated microbial metagenomes. *Front. Microbiol.* 7.

Jahn, M. T., Arkhipova, K., Markert, S. M. et al. 2019. A phage protein aids bacterial symbionts in eukaryote immune evasion. *Cell Host Microbe* 26: 542–550.e5.

Jahn, M. T., Markert, S. M., Ryu, T. et al. 2016. Shedding light on cell compartmentation in the candidate phylum Poribacteria by high resolution visualisation and transcriptional profiling. *Sci. Rep.* 6: 1–9.

King, K. C., Zelek, M., Gray, C., Betts, A., and MacLean, R. C. 2018. High parasite diversity accelerates host adaptation and diversification. *Science* 360: 907–911.

Koropatnick, T. A., Engle, J. T., Apicella, M. A., Stabb, E. V., Goldman, W. E., and McFall-Ngai, M. J. 2004. Microbial factor-mediated development in a host-bacterial mutualism. *Science* 306: 1186–1188.

Laffy, P. W., Wood-Charlson, E. M., Turaev, D. et al. 2018. Reef invertebrate viromics: Diversity, host specificity and functional capacity. *Environ. Microbiol.* 20: 2125–2141.

Leigh, B. A. 2019. Cooperation among conflict: Prophages protect bacteria from phagocytosis. *Cell Host Microbe* 26: 450–452.

Mai, T., Tintillier, F., Lucasson, A. et al. 2015. Quorum sensing inhibitors from *Leucetta chagosensis* Dendy, 1863. *Lett. Appl. Microbiol.* 61: 311–317.

Mangano, S., Caruso, C., Michaud, L., and Lo Guidice, A. 2018. First evidence of quorum sensing activity in bacteria associated with *Antarctic sponges*. *Polar Biol.* 41: 1435–1445.

Messier-Solek, C., Buckley, K. M., and Rast, J. P. 2010. Highly diversified innate receptor systems and new forms of animal immunity. *Semin. Immunol.* 22: 39–47.

Metchnikoff, E. 1893. *Lectures on the Comparative Pathology of Inflammation*. London, Kegan Paul, Trench, Trubner & Co.

Mohamed, N. M., Cicirelli, E. M., Kan, J., Chen, F., Fuqua, C., and Hill, R. T. 2008. Diversity and quorum-sensing signal production of Proteobacteria associated with marine sponges. *Environ. Microbiol.* 10: 75–86.

Moitinho-Silva, L., Díez-Vives, C., Batani, G., Esteves, A. I. S., Jahn, M. T., and Thomas, T. 2017a. Integrated metabolism in sponge-microbe symbiosis revealed by genome-centered metatranscriptomics. *ISME J.* 11: 1651–1666.

Moitinho-Silva, L., Nielsen, S., Amir, A. et al. 2017b. The sponge microbiome project. *Gigascience* 6: 1–7.

Moitinho-Silva, L., Steinert, G., Nielsen, S. et al. 2017c. Predicting the HMA-LMA status in marine sponges by machine learning. *Front. Microbiol.* 8: 1–14.

Moreno-Gallego, J. L., Chou, S.-P., Di Rienzi, S. C. et al. 2019. Virome diversity correlates with intestinal microbiome diversity in adult monozygotic twins. *Cell Host Microbe* 25: 261–272.e5.

Müller, W. E., Zahn, R. K., Kurelec, B., Lucu, C., Müller, I., and Uhlenbruck, G. 1981. Lectin, a possible basis for symbiosis between bacteria and sponges. *J. Bacteriol.* 145: 548–558.

Nadell, C. D., Drescher, K., and Foster, K. R. 2016. Spatial structure, cooperation and competition in biofilms. *Nat. Rev. Microbiol.* 14: 589–600.

Pascelli, C., Laffy, P. W., Kupresanin, M., Ravasi, T., and Webster, N. S. 2018. Morphological characterization of virus-like particles in coral reef sponges. *PeerJ* 6: e5625.

Pietschke, C., Treitz, C., Forêt, S. et al. 2017. Host modification of a bacterial quorum-sensing signal induces a phenotypic switch in bacterial symbionts. *Proc. Natl. Acad. Sci.* 114(40): E8488–E8497.

Pita, L., Hoeppner, M. P., Ribes, M., and Hentschel, U. 2018a. Differential expression of immune receptors in two marine sponges upon exposure to microbial-associated molecular patterns. *Sci. Rep.* 8: 1–15.

Pita, L., Rix, L., Slaby, B. M., Franke, A., and Hentschel, U. 2018b. The sponge holobiont in a changing ocean: From microbes to ecosystems. *Microbiome* 6(46): 1–18.

Podell, S., Blanton, J. M., Neu, A. et al. 2019. Pangenomic comparison of globally distributed Poribacteria associated with sponge hosts and marine particles. *ISME J.* 13: 468–481.

Rakoff-Nahoum, S., Foster, K. R., and Comstock, L. E. 2016. The evolution of cooperation within the gut microbiota. *Nature* 533: 255–259.

Riesgo, A., Farrar, N., Windsor, P. J., Giribet, G., and Leys, S. P. 2014. The analysis of eight transcriptomes from all poriferan classes reveals surprising genetic complexity in sponges. *Mol. Biol. Evol.* 31: 1102–1120.

Rohwer, F. 2003. Global phage diversity. *Cell.* 113(2): 141.

Russell, S. L. 2019. Transmission mode is associated with environment type and taxa across bacteria-eukaryote symbioses: A systematic review and meta-analysis. *FEMS Microbiol. Lett.* 366: 430–439.

Ryu, T., Seridi, L., Moitinho-Silva, L. et al. 2016. Hologenome analysis of two marine sponges with different microbiomes. *BMC Genomics* 17: 1–11.

Saurav, K., Burgsdorf, I., Teta, R. et al. 2016. Isolation of Marine *Paracoccus* sp. Ss63 from the Sponge *Sarcotragus* sp. and characterization of its quorum-sensing chemical-signaling molecules by LC-MS/MS analysis. *Isr. J. Chem.* 56: 330–340.

Saurav, K., Costantino, V., Venturi, V., and Steindler, L. 2017. Quorum sensing inhibitors from the sea discovered using bacterial *N*-acyl-homoserine lactone-based biosensors. *Mar. Drugs* 15: 53.

Schmitt, S., Angermeier, H., Schiller, R., Lindquist, N., and Hentschel, U. 2008. Molecular microbial diversity survey of sponge reproductive stages and mechanistic insights into vertical transmission of microbial symbionts. *Appl. Environ. Microbiol.* 74: 7694–7708.

Schulenburg, H., Boehnisch, C., and Michiels, N. K. 2007. How do invertebrates generate a highly specific innate immune response? *Mol. Immunol.* 44: 3338–3344.

Sebé-Pedrós, A., Chomsky, E., Pang, K. et al. 2018. Early metazoan cell type diversity and the evolution of multicellular gene regulation. Nat. *Ecol. Evol.* 2: 1176–1188.

Sharp, K. H., Eam, B., Faulkner, D. J., and Haygood, M. G. 2007. Vertical transmission of diverse microbes in the tropical sponge *Corticium* sp. *Appl. Environ. Microbiol.* 73: 622–629.

Siegl, A., Kamke, J., Hochmuth, T. et al. 2011. Single-cell genomics reveals the lifestyle of Poribacteria, a candidate phylum symbiotically associated with marine sponges. *ISME J.* 5: 61–70.

Sipkema, D., de Caralt, S., Morillo, J. A. et al. 2015. Similar sponge-associated bacteria can be acquired via both vertical and horizontal transmission. *Environ. Microbiol.* 17: 3807–3821.

Slaby, B. M., Hackl, T., Horn, H., Bayer, K., and Hentschel, U. 2017. Metagenomic binning of a marine sponge microbiome reveals unity in defense but metabolic specialization. *ISME J.* 11: 2465–2478.

Srivastava, M., Simakov, O., Chapman, J. et al. 2010. The Amphimedon queenslandica genome and the evolution of animal complexity. *Nature* 466: 720–726.

Steindler, L., Schuster, S., Ilan, M., Avni, A., Cerrano, C., Beer, S. 2007. Differential gene expression in a marine sponge in relation to its symbiotic state. *Mar. Biotechnol.* 9: 543–549.

Suttle, C. A. 2007. Marine viruses—major players in the global ecosystem. *Nat. Rev. Microbiol.* 5: 801.

Taylor, M. W., Schupp, P. J., Baillie, H. J. et al. 2004. Evidence for acyl homoserine lactone signal production in bacteria associated with marine sponges. *Appl. Environ. Microbiol.* 70: 4387–4389.

Taylor, M. W., Thacker, R. W., and Hentschel, U. 2007. Evolutionary insights from sponges. *Source Sci. New Ser.* 316: 1854–1855.

Thomas, T., Moitinho-Silva, L., Lurgi, M. et al. 2016. Diversity, structure and convergent evolution of the global sponge microbiome. *Nat. Commun.* 7: 11870.

Tianero, M. D., Balaich, J. N., and Donia, M. S. 2019. Localized production of defence chemicals by intracellular symbionts of *Haliclona* sponges. *Nat. Microbiol.* 4: 1149–1159.

Ting, J. P. Y., Lovering, R. C., Alnemri, E. S. et al. 2008. The NLR Gene Family: A Standard Nomenclature. *Immunity* 28: 285–287.

Vacelet, J., and Gallissian, M. F. 1978. Virus-like particles in cells of the sponge *Verongia cavernicola* (demospongiae, dictyoceratida) and accompanying tissues changes. *J. Invertebr. Pathol.* 31: 246–254.

Van Soest, R. W. M., Boury-Esnault, N., Vacelet, J. et al. 2012. Global diversity of sponges (porifera). *PLOS ONE* 7: e35105.

Venturi, V., and Subramoni, S. 2009. Future research trends in the major chemical language of bacteria. *HFSP J.* 3: 105–116.

Webster, N. S., Taylor, M. W., Behnam, F. et al. 2010. Deep sequencing reveals exceptional diversity and modes of transmission for bacterial sponge symbionts. *Environ. Microbiol.* 12: 2070–2082.

Webster, N. S., and Thomas, T. 2016. Defining the sponge hologenome. *Am. Soc. Microbiol.* 7(2): 1–14.

Wehrl, M., Steinert, M., and Hentschel, U. 2007. Bacterial uptake by the marine sponge *Aplysina aerophoba*. *Microb. Ecol.* 53: 355–365.

Weiland-Bräuer, N., Fischer, M. A., Pinnow, N., and Schmitz, R. A. 2019. Potential role of host-derived quorum quenching in modulating bacterial colonization in the moon jellyfish *Aurelia aurita*. *Sci. Rep.* 9: 1–12.

Weisz, J. B., Lindquist, N., and Martens, C. S. 2008. Do associated microbial abundances impact marine demosponge pumping rates and tissue densities? *Oecologia* 155: 367–376.

Wiens, M., Korzhev, M., Krasko, A. et al. 2005. Innate immune defense of the sponge *Suberites domuncula* against bacteria involves a MyD88-dependent signaling pathway: Induction of a perforin-like molecule. *J. Biol. Chem.* 280: 27949–27959.

Wilkinson, C. R., Garrone, R., and Vacelet, J. 1979. Marine sponges discriminate between food bacteria and bacterial symbionts: Electron microscope radioautography and *in situ* evidence. *Proc R Soc L. B* 205: 519–528.

Yin, Z., Zhu, M., Davidson, E. H., Bottjer, D. J., Zhao, F., and Tafforeau, P. 2015. Sponge grade body fossil with cellular resolution dating 60 Myr before the Cambrian. *Proc. Natl. Acad. Sci.* 112: 1453–1460.

Yuen, B. 2016. Deciphering the genomic toolkit underlying animal–bacteria interactions-insights through the demosponge *Amphimedon queenslandica*. PhD diss., University of Queensland.

Yuen, B., Bayes, J. M., and Degnan, S. M. 2014. The characterization of sponge NLRs provides insight into the origin and evolution of this innate immune gene family in animals. *Mol. Biol. Evol.* 31(1): 106–120.

5 Symbiotic interactions in the holobiont *Hydra*

Jay Bathia and Thomas C. G. Bosch

Contents

5.1 Introduction

Life has evolved in a microbial world. The event of successful symbiosis was a pivotal step that resulted in the origin of a eukaryotic cell. Organisms, therefore, are considered metaorganisms comprised of the macroscopic host and synergistic interdependence with bacteria, archaea, fungi and numerous other microbial and eukaryotic species including algal symbionts (Bosch and McFall-Ngai 2011). The interactions between all the partners are affected by the prevailing environmental conditions. Understanding the mechanisms and the causality of the interactions in such a metaorganism requires model organisms where the contribution of each partner can be disentangled and functionally analyzed.

Aptly named, *Hydra* is known for its remarkable capacity of regeneration and non-senescence. For decades, *Hydra* has been used as a model organism to study the cellular pathways governing body patterning and ageing by evolutionary conserved mechanisms (Galliot 2012). However, like all other metaorganisms, *Hydra* hosts numerous bacterial, viral and algal symbionts (Habetha et al. 2003; Fraune and Bosch

2007; Grasis et al. 2014). For various reasons, *Hydra* turns out to be an excellent model system to study the interactions between these partners (Bosch 2013). *Hydra* being diploblastic, possess a simple body plan composed to two stem cell epithelial layers – ectoderm and endoderm, separated by a thin layer of connective tissue, mesoglea. The third stem cell lineage, the interstitial cells, is localized in the ectoderm and gives rise to the nervous system, nematocytes, gland cells and germ line cells. Towards the outside on the ectoderm, a multilayered glycocalyx harbours the bacterial symbionts.

The extensive genomic and transcriptomic data of various *Hydra* species can be utilized to map known pathways and look for evolutionary conserved genes. Moreover, it is also possible to genetically manipulate the host in loss/gain of function studies (Wittlieb et al. 2006; Klimovich et al. 2019). The transparent tissue facilitates visualization of cellular components without rigorous fixing and bleaching methods. Large populations of *Hydra* can be easily cultivated and maintained under lab conditions. Clonal propagation ensures the genetic homogeneity of the entire population. In addition, since *Hydra* belongs to the phylum Cnidaria which is the sister group of Bilaterians, it is an informative model to study evolutionary conserved mechanisms in metazoans. Being able to separate all the partners of the metaorganism makes *Hydra* suitable to study the underlying principles of the complex multiplayer interactions (Li et al. 2015, 2017; Wein et al. 2018). In this chapter, we summarize the state of the art understanding of the *Hydra* metaorganism, the interactions between the host and algal symbiont in *Hydra viridissima*, and also the interactions of the *Hydra* host with the bacterial symbionts.

5.2   Interactions between *Hydra viridissima* and the *Chlorella* photobiont

5.2.1   Location and transmission of the photobiont

Symbiotic *Chlorella* algae are present intracellularly in the endodermal epithelial cells. This location points to a high level of co-dependence and a rather complex cellular machinery involved in the maintenance of the symbiosis. Inside the *Hydra* endothelial cells, the algae are held towards the base of the cells (Jolley and Smith 1978) (Figure 5.1A). This compartmentalizes an endodermal cell into two zones: digestive zone, which is the apical part of the cell responsible for phagocytosis and digestion of the food vesicles and the symbiotic zone, which is at the base of the cell where the symbiotic algae are present (McNeil 1981; McNeil et al. 1981). Inside the host cell, each algal cell is surrounded by a host-derived vesicle membrane that acts as an interface for the biotrophic exchange of nutrients. This 'peri-symbiotic' membrane is the result of the phagocytosis by which the symbiont is taken up (McNeil 1981; Davy et al. 2012).

The peri-symbiotic membrane is congruent to the membrane of a phagosome. The resulting structure is often referred to as a 'symbiosome' and the space between the peri-symbiotic membrane and the algae is called 'peri-symbiotic space' (Davy et al. 2012). Being present intracellularly demands a need for successful transfer of the symbiont to the next generation. *Chlorella* algae can be transmitted horizontally as well as vertically. The host has a capacity to expel excess of living and preferentially dividing algae actively (Baghdasarian and Muscatine 2000). Such algae can be

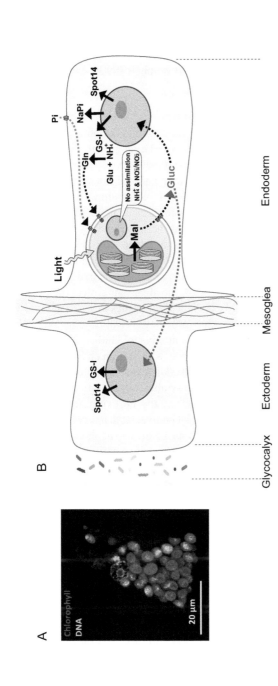

FIGURE 5.1 *Hydra viridissima/Chlorella* **symbiosis.** (A) Each endothelial epithelial cell harbours about 30 *Chlorella* algae (red autofluorescence) in the basal part below the host nucleus (stained green). (Image courtesy: Jay Bathia and Alexander Klimovich, Zoological Institute, Kiel University.) (B) Summary of symbiotic interactions between *Hydra* and *Chlorella* A99. During light conditions, *Chlorella* A99 performs photosynthesis and produces maltose (Mal), which is secreted into the *Hydra* symbiosome where it is possibly digested to glucose (Gluc), shown in red. The sugar induces expression of *Hydra* genes encoding glutamine synthetase (GS), Na/Pi transporter (NaPi) and Spot14. GS catalyses the condensation of glutamate (Glu) and ammonium (NH4+) to form glutamine (Gln), which is used by *Chlorella* as a nitrogen source. Since the sugar also upregulates the NaPi gene, which controls intracellular phosphate levels, it might be involved in the supply of phosphorus to *Chlorella* as well (blue broken line). The sugar is transported also to the ectoderm (red broken line) and there induces the expression of GS and Spot14. (Modified from Hamada, M., et al. 2018. *ELife* 7. doi: 10.7554/eLife.35122.)

later be taken up from the environment by an aposymbiotic host to reestablish symbiosis. Vertical transmission takes place during the asexual reproduction phase because the endodermal cell layer of the mother polyp is in continuation with that of the developing bud, which ensures the transfer of algae in the resulting bud (Jolley and Smith 1980). Interestingly, symbiotic algae can also be transmitted vertically to the offspring through germ-line maternal cells (Campbell 1990).

5.2.2 Mutual benefits

The mutual exchange of metabolites between the host and the photobiont results in a phenotypic benefit for both the partners. The population growth rate of symbiotic animals is faster than the aposymbiotic animals (Habetha et al. 2003; Ishikawa et al. 2016; Hamada et al. 2018). Moreover, the symbiotic animals can survive for a longer period under starvation as compared to the aposymbiotic animals (Ishikawa et al. 2016). It has also been observed that the presence of algae increases the frequency of host oogenesis (Habetha et al. 2003). Since oogenesis is typically an energetically expensive process, the transport of sugar from the algae may aid in overcoming the cost of oogenesis. Interestingly, any attempt to culture the endosymbiotic algae *in vitro* consistently failed. This underlines the strong dependence of the algae on the host. Apart from being nutritionally beneficial, the algae can protect the host under environmental stress. A recent study (Ye et al. 2019) indicates that if symbiotic algae were acclimated to higher temperatures, they provide survival benefits to the *Hydra* holobiont at higher temperatures. There are a number of studies that utilized the ability of *Hydra viridissima* to establish symbiosis with foreign endosymbiotic algae such as *Chlorella* NC64A (Dorling et al. 1997; Kodama and Fujishima 2015; Hamada et al. 2018). *Chlorella* NC64A is a native endosymbiont of the protist *Paramecium bursaria*. *Chlorella* NC64A can release sugar to the host only in low amounts, resulting in lower population growth rate, almost equivalent to aposymbiotic animals (Hamada et al. 2018), compared to polyps harbouring the native photobiont *Chlorella* A99. This further indicates the importance of harbouring a native symbiont.

5.2.3 Establishment and maintenance of the *Chlorella-Hydra* symbiosis

To establish a new symbiosis, the aposymbiotic host engulfs the algae through phagocytosis. This results in the formation of a phagosome enclosing the algae. It is acidified by the fusion with acidosomes (Allen and Fok 1983). Only when the engulfed algae is a true symbiont is the fusion of lysosome inhibited, as indicated by the absence of acid phosphatase activity in the late phagosome stage (Hohman et al. 1982). The recognition of symbiotic algae occurs probably by the presence of microbe-associated molecular patterns (MAMPs). Recent comparative transcriptomic studies indicate the putative involvement of the Toll/Interleukin-1 receptor (TIR), a known pattern recognition receptor (PRR) and C-type lectin, to be involved in recognizing the MAMPs of a symbiotic algae (Hamada et al. 2018).

The physiological conditions, and in particular the pH, experienced by an endosymbiotic alga are very different from those experienced by a free-living alga. The pH of the peri-symbiotic space is maintained between 4 and 5. This is crucial because only at this low pH is the algae shown to be able to release maltose (Cernichiari et al. 1969; Dorling et al. 1997). The maltose releasing capacity is unique to endosymbiotic algae. In plants and free-living algae, during nighttime, the stored starch in chloroplasts is mobilized to the cytoplasm in the form of maltose, but never secreted to the extracellular space (Lu and Sharkey 2006). This may point to a maltose transport machinery on the membrane of the symbiotic algae similar to the membrane of the chloroplast. Another important characteristic of a successful mutualistic relationship is to control the proliferation of the endosymbiont. There is evidence that *Hydra* control the number of algae per cell by regulating the pH of the peri-symbiotic space (Muscatine and Neckelmann 1981; McAuley 1986; Dunn 1987). Moreover, the host is capable to actively expel any excess number of algae (Baghdasarian and Muscatine 2000).

5.2.4 Molecular mechanisms involved in maintaining the symbiosis

A elaborated comparative transcriptomics and genomic study of *Hydra viridissima* revealed the key components and pathways involved in maintaining the host–algal symbiosis (Hamada et al. 2018). To identify the host genes involved in maintaining symbiosis with the native algae, this unbiased high-throughput study (Hamada et al. 2018) included a comparison of the symbiotic host associated with the native symbiotic *Chlorella* A99, the aposymbiotic host and the host bearing foreign symbiotic algae *Chlorella* NC64A.

Comparing the native symbiotic animals with aposymbiotic animals revealed the upregulation of a number of conserved cellular pathways in symbiotic animals, which maintain the symbiosis. The transcriptomics data revealed the up-regulation of a host V-type ATPase subunit and carbonic anhydrase, involved in maintaining the low pH of the perisymbiotic space and facilitating CO_2 transport across the symbiosome respectively. Moreover, there is an upregulation of GLUT8-like transporters, putatively involved in transportation of the photosynthate across the peri-symbiotic membrane. The study also uncovered the up-regulation of cell-adhesion genes encoding rhamnospondin and fibrillin. These molecules might be involved in retaining the algae to the base of the endothelial cell. Moreover, a high level of peroxidase, methionine-r-sulfoxide reductase/selenoprotein and glutaredoxin might help to combat the surge in oxidative stress generated by the constant oxygen influx from the algae to the host.

A comparison of *Hydra* polyps, harbouring *Chlorella* A99, with *Hydra* polyps colonized by the non-native *Chlorella* NC64A algae showed that the interactions between the host and the algae are mainly metabolic in nature (summarized in Figure 5.1B). The native photobiont provides the photosynthetically fixed carbon to the host in the form of maltose. This maltose is likely to be digested to its monomer glucose, which is transported to the host by transporter proteins in the symbiosome membrane. This excess sugar shifts the host metabolism to ammonium assimilation. In response to this, there is an up-regulation of host glutamine synthetase (GS-I) which is involved in the production of glutamine from ammonium and glutamate. GS-I is specific to the symbiotic state, as other isoforms are unaffected by the presence or absence of the algae. As another important stabilizer of the *Hydra-Chlorella* symbiosis, the sodium-dependent phosphate transport protein (NaPi) can increase the available phosphate for the host and the algae. Apart from these, a Spot_14 like domain-containing protein with a potential role in lipid metabolism, and two taxonomically restricted genes (TRGs) are also up-regulated. Exposing polyps to the photosynthesis inhibitor DCMU or culturing them at constant darkness, results in a drastically reduced expression level of these genes, indicating that the expression of these symbiosis-specific genes is dependent on the photosynthetic activity of the symbiont.

Genome analysis of the symbiotic *Chlorella* A99 algae and the comparison to the genome of the free-living counterparts revealed interesting features, which helped to explain the strong dependence of the symbiont on the host (Hamada et al. 2018). Most strikingly, the symbiotic algae have lost their key nitrite/nitrate and ammonium assimilation genes. Symbiotic algae, therefore, depend on the host for fixed nitrogen in the form of glutamine. Further genomic analyses of the algae also indicated an increase in the number of amino acid transporters, supporting amino acid uptake. Finally, although all attempts to culture these algae *in vitro* failed, the growth of the algae can be temporarily maintained by adding CAS amino acids or glutamine as a source of nitrogen to the culture medium. This reflects the strong metabolic co-dependence of both partners and a unique adaptation of the symbiotic algae (Hamada et al. 2018).

5.3 Interactions between *Hydra* and symbiotic bacteria

In addition to host–photobiont interactions in *Hydra viridissima*, all *Hydra* species are stably associated with a species-specific microbiome. Even after over 30 years of culturing in the lab under artificial conditions, *Hydra* species maintain their distinct bacterial composition (Fraune and Bosch 2007; Franzenburg et al. 2013b) which resembled the microbiota of polyps freshly collected from the wild. This, together with the cellular and molecular accessibility of the *Hydra* polyp for functional approaches, allows investigating the interaction between microbes and *Hydra* polyps in an unprecedented manner and at the same time provide insight into the evolution of host–microbe interactions.

5.3.1 Spatial localization of the bacteria in the *Hydra* host

The *Hydra* ectodermal epithelium can be considered as an inside-out gut, with the mucosal layer facing towards the outside (Schröder and Bosch 2016). A carbohydrate and glycoprotein-rich layer, called glycocalyx, covers the outer surface of the polyp (Figure 5.2). The glycocalyx is a multi-layered structure composed of at least five layers with first four layers anchored to the ectoderm. Interestingly, the presence of bacteria was detected only in the most distal layer of the glycocalyx (Fraune et al. 2015), which acts as the first barrier between the environmental bacteria and the host tissue. In *H. oligactis*, a few of the bacterial symbionts are endosymbiotic in the ectodermal epithelial cells (Fraune and Bosch 2007). Most of the functional studies were done in *H. vulgaris AEP* or *H. magnipapillata*, which do not harbor any endosymbiotic bacteria (Fraune and Bosch 2007; Fraune et al. 2015).

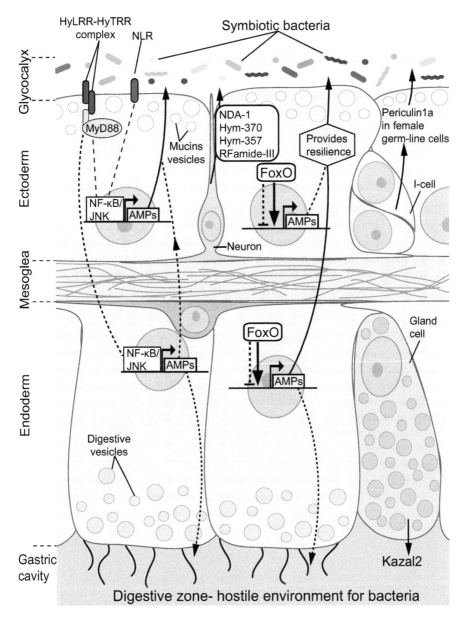

FIGURE 5.2 Summary of interactions between *Hydra* and symbiotic bacteria. Epithelial cells produce anti-microbial peptides (AMPs) belonging to Arminin/Hydramacin/ Periculin families. Gland cells in the endoderm express Kazal family AMPs. I-cells, after being differentiated into female germ-line cells express AMP- Periculin1a. Anti-bacterial neuropeptides NDA-1, Hym-370, Hym-357, and the RFamide III family are expressed in nerve cells. Stem-cell transcription factor FoxO plays a regulatory role in the expression of Hydramacin/Arminin/Kazal, without affecting Periculin and NDA-1. Mucin containing vesicles (in blue) in ectodermal epithelial cells contribute to the establishment of the glycocalyx.

5.3.2   Bacteria provide protection against fungal infection

Being able to separate the *Hydra* host and bacterial partners provides an efficient analytical framework to disentangle the contribution of each partner to the working of the holobiont. The germ-free *H. vulgaris AEP* hosts are highly susceptible to the infection by *Fusarium* sp. Fungus, suggesting a strong protective role of bacteria against a fungal pathogen. Colonization of hosts with single bacterial symbiont species failed to provide resistance against the fungi pointing to the importance of a complete and diverse bacterial community in maintaining the holobiont homeostasis.

5.3.3   The innate immune system shapes the host microbiome

Since the epithelial layer of *Hydra* is constantly in contact with a large variety of bacteria including pathogens, a wide array of competing forces seems to ensure the selection of specific bacterial species from the pool of bacterioplankton (Deines et al. 2017). But do *Hydra* polyps actively select their microbial symbionts? On testing neutrality in the microbial composition for *Hydra vulgaris AEP*, it was recently observed that the microbiome composition deviates from the theoretically predicted neutral community, indicating that under the prevailing environmental conditions a major part of the microbiota is affected by host-derived factors (Sieber et al. 2019).

These theoretical considerations are supported by experimental work, which over the last decade has shown that *Hydra*'s innate immune system plays a major role in shaping the species-specific microbiota (Bosch 2014; Klimovich and Bosch 2018). The recognition of the MAMPs occurs by Toll-like receptors (TLRs) and nucleotide binding and oligomerization domain (NOD) like receptors (NLRs) (Bosch et al. 2009). The structure of *Hydra* TLRs is unconventional: it is composed of leucine-rich repeat (LRR) domain containing protein (HyLRR) and a separate Toll/interleukin-1 receptor (TIR) domain containing protein (HyTRR). Both these proteins interact to form a functional TLR. The TIR domain of HyTRR interacts with the TIR domain of myeloid differentiation factor 88 (MyD88) protein. Upon stimulation by MAMPs, the signaling finally activates c-Jun N-terminal kinase (JNK) and *Hydra* orthologue of nuclear factor kappa-light-chain-enhancer of activated B cells (HyNF-κB). These transcription factors induce the expression of various anti-microbial peptides (AMPs) that play a crucial role in shaping the host microbiome (Augustin et al. 2009; Bosch 2014).

Endodermal epithelial cells express the AMP peptide family 'Hydramacin' (Bosch et al. 2009; Jung et al. 2009). Hydramacin expression is induced by MAMPs via the TLR signaling pathway and has antimicrobial activity against various gram-positive and gram-negative bacteria. The endothelial gland cells in the body column express a considerably high amount of Kazal2, a kazal-type serine protease inhibitor that targets the bacterial serine proteases (Augustin et al. 2009) (Figure 5.2), and provides protection against pathogenic bacteria in the gastric cavity. Although *Hydra* lacks any migratory phagocytic cells, it does possess phagocytic activity in the endodermal cell. It is through this property, that the aposymbiotic *H. viridissima* is able to establish a symbiosis with *Chlorella* algae. The same phagocytic property can also help to keep the endoderm free of bacteria by engulfing any bacteria entering with the food (Bosch et al. 2009).

Maternal protection of the developing progeny against pathogens or bacterial overgrowth is provided by the AMP Periculin1a which is expressed exclusively in maternal germ line cells (Fraune et al. 2010). This AMP persists during the initial phases of embryogenesis up to blastula stage, and controls the bacterial load as well as assists in selection of the colonizing bacteria. The function of maternal Periculin1a is later taken over by zygotic Periculin2b after midblastula transition. In an adult *H. vulgaris AEP* polyp, the endodermal knockdown of another family of AMP, the Arminin peptides, results in a reduced selection by the host for the colonizing bacterial partners. Since in the absence of Arminin peptides the bacteria from other donor species, *H. oligactis* and *H. viridissima*, can colonize the host, these AMPs are involved in maintaining a specific-specific microbiota.

The colonization of a newly hatched polyp follows a robust temporal trajectory (Franzenburg et al. 2013a). There is a high variability in the microbiota initially, followed by a transient increase in the relative abundance of the main colonizers of the adult microbiota. Following this, there is a drastic decrease in the microbial diversity before it stabilizes to adult microbiota at the end of four weeks.

5.3.4   Crosstalk between innate immunity and stem cell factors

In *Hydra*, and most likely in other organisms as well, the innate immune system is tightly linked to pathways controlling epithelial cell homeostasis. *Hydra*'s Forkhead box transcription factor (FoxO) is key to maintaining the stem cell identity of the ecto- and endodermal epithelial cells (Boehm et al. 2012) (Figure 5.2). The expression of FoxO is limited to the stem cell zone and absent in the terminally differentiated cells. Epithelial FoxO loss-of-function mutants revealed that a deficiency in FoxO signaling leads not only to malfunctions in cell cycle progression, but also to dysregulation of multiple families of genes encoding antimicrobial peptides (AMPs). FoxO loss-of-function polyps were more susceptible to colonization by foreign bacteria, and impaired in selection for bacteria resembling the native microbiome. FoxO-deficiency reduces the expression of AMPs, resulting in decreased selective pressure on colonizing microbial taxa and ultimately in reduced resilience of the microbiome (Boehm et al. 2012; Mortzfeld et al. 2018).

5.3.5   Crosstalk between the microbiota and the nervous system

In addition to the epithelial cells, the nervous system also plays an important role in shaping the microbial community of *Hydra*. First indication of crosstalk between both the systems came from the observation that loss of nerve-cell lineage resulted in an increased anti-bacterial activity of the *Hydra* polyps (Fraune et al. 2009). A recent study (Augustin et al. 2017) provided evidence that some sensory and ganglion neurons express a cationic neuropeptide called NDA-1, secrete it into the mucus layer and regulate the spatial distribution of the main colonizer, the gram-negative bacterium *Curvibacter*, along the *Hydra* trunk. The density of *Curvibacter* colonization is relatively low in the foot and tentacles of *Hydra*, where NDA-1 is strongly expressed, compared to the body column. Additionally, NDA-1 is highly potent in killing gram-positive bacteria. Strikingly, other neuropeptides, such as *Hydra* specific Hym-357 and Hym-370, and a member of the highly conserved

RFamide family, all previously characterized as classical neuromodulators eliciting motor activity, turned out to be also potent against gram-positive bacteria. Taken together, these findings indicate that distinct nerve cells contribute to the composition and spatial structure of *Hydra's* microbial community by expressing a variety of neuropeptides with distinct antimicrobial activities.

Based on these observations, we proposed (Klimovich and Bosch 2018) that during evolution the nervous system, in addition to its role in sensory input/motor output, plays a primordial role as part of the innate immune system.

All these observations, taken together, portray a rather complex network of molecular and cellular interactions controlling the establishment and maintenance of a stable microbiota in *Hydra* (Figure 5.2).

The level of complexity of the interspecies crosstalk in the holobiont *Hydra* is even larger since many of the bacterial partners harbour temperate phages bearing the capacity to cross infect other bacterial species. One such example is observed as competition between the top two main colonizers of *H. vulgaris AEP*, *Curvibacter sp.* and *Duganella sp.* (Li et al. 2015, 2017). Associated with the host, *Curvibacter* out-competes *Duganella*, while *in-vitro,* in monocultures the effect is reversed. However, the co-culture of both the bacteria *in-vitro* results in a non-linear reduction of growth of *Duganella* depending on the initial frequency of *Curvibacter*. When modelled mathematically, the interaction between both the bacteria in co-culture could only be explained by the presence of a third partner, a bacteriophage. Indeed, *Curvibacter* inhabits a pro-phage that can be activated to enter the lytic cycle and have lytic activity against *Duganella*. Thus, the *in-vivo* competition between the top two colonizers is likely to be mediated by a bacteriophage.

5.3.6   Effect of bacteria on host physiology

With the host factors have a profound effect on its microbiome, there also strong impacts of the bacterial partners on the host physiology and host behaviour. *Hydra* exhibits various behaviour like spontaneous contraction, feeding response to a stimulus and phototactic movement, to name a few. Spontaneous contractions are assumed to be regulated by the pacemaker activity of the neurons (Passano and Mccullough 1965). Interestingly, the frequency of the spontaneous contractions in germ-free *H. vulgaris AEP* animals is reduced to ~60% of the control animals (Murillo-Rincon et al. 2017). This effect can be restored, although not completely, by recolonization of these germ-free animals by normal microbiome. Moreover, the microbial extract from the *in-vitro* grown complete microbial community on a complex media, largely resulted in restoration of the contraction frequency. Bacteria, or bacteria-derived molecules, therefore, directly interfere with neuronal activity and function.

5.4   Conclusion: *Hydra*, an excellent model to understand inter-species interactions

Inter-species interactions in the freshwater polyp Hydra, between symbiotic algae and host cells, had been the subject of research since decades as they not only provide insights into the basic 'tool kit' necessary to establish symbiotic interactions, but are also of relevance in understanding the resulting evolutionary selection processes. A long-term persistence of symbiotic associations is prevalent not only in two-party interactions of *Hydra* and symbiotic algae, but also in more complex systems including stable associated bacteria and a species-specific virome (Franzenburg et al. 2013b; Grasis et al. 2014). Studying symbiotic inter-species interactions in *Hydra*, therefore, may be a paradigmatic example of a complex symbiotic community that influences the host's health and development. Our work has contributed to a paradigm shift in evolutionary immunology: components of the innate immune system, with its host-specific antimicrobial peptides, appear to have evolved in early branching metazoans because of the need to control the resident beneficial microbes, rather than because of invasive pathogens.

Acknowledgments

The work was supported in part by grants from the Deutsche Forschungsgemeinschaft (DFG) and the CRC 1182 ('Origin and Function of Metaorganisms'). T.C.G.B. appreciates support from the Canadian Institute for Advanced Research (CIFAR).

References

Allen, R. D., and A. K. Fok. 1983. "Nonlysosomal Vesicles (Acidosomes) Are Involved in Phagosome Acidification in Paramecium." *The Journal of Cell Biology* 97 (2). The Rockefeller University Press: 566–70. doi: 10.1083/jcb.97.2.566

Augustin, R., K. Schröder, A. P. Murillo Rincón, S. Fraune, F. Anton-Erxleben, E.-M. Herbst, J. Wittlieb et al. 2017. "A Secreted Antibacterial Neuropeptide Shapes the Microbiome of Hydra." *Nature Communications* 8 (1). Nature Publishing Group: 698. doi: 10.1038/s41467-017-00625-1

Augustin, R., S. Siebert, and T. C. G. Bosch. 2009. "Identification of a Kazal-Type Serine Protease Inhibitor with Potent Anti-Staphylococcal Activity as Part of Hydra's Innate Immune System." *Developmental & Comparative Immunology* 33 (7). Pergamon: 830–37. doi: 10.1016/J.DCI.2009.01.009

Baghdasarian, G., and L. Muscatine. 2000. "Preferential Expulsion of Dividing Algal Cells as a Mechanism for Regulating Algal-Cnidarian Symbiosis." *Biological Bulletin* 199 (3): 278–86. doi: 10.2307/1543184

Boehm, A.-M., K. Khalturin, F. Anton-Erxleben, G. Hemmrich, U. C. Klostermeier, J. A. Lopez-Quintero, H.-H. Oberg et al. 2012. "FoxO Is a Critical Regulator of Stem Cell Maintenance in Immortal Hydra." *Proceedings of the National Academy of Sciences of the United States of America* 109 (48). National Academy of Sciences: 19697–702. doi: 10.1073/pnas.1209714109

Bosch, T. C. G. 2013. "Cnidarian-Microbe Interactions and the Origin of Innate Immunity in Metazoans." *Annual Review of Microbiology* 67 (1). Annual Reviews: 499–518. doi: 10.1146/annurev-micro-092412-155626

Bosch, T. C. G. 2014. "Rethinking the Role of Immunity: Lessons from Hydra." *Trends in Immunology* 35 (10). Elsevier: 495–502. doi: 10.1016/j.it.2014.07.008

Bosch, T. C. G., R. Augustin, F. Anton-Erxleben, S. Fraune, G. Hemmrich, H. Zill, P. Rosenstiel et al. 2009. "Uncovering the Evolutionary History of Innate Immunity: The Simple Metazoan Hydra Uses Epithelial Cells for Host Defence." *Developmental & Comparative Immunology* 33 (4). Pergamon: 559–69. doi: 10.1016/J.DCI.2008.10.004

Bosch, T. C. G., and M. J. McFall-Ngai. 2011. "Metaorganisms as the New Frontier." *Zoology (Jena, Germany)* 114 (4). NIH Public Access: 185–90. doi: 10.1016/j.zool.2011.04.001

Campbell, R. D. 1990. "Transmission of Symbiotic Algae through Sexual Reproduction in Hydra: Movement of Algae into the Oocyte." *Tissue and Cell* 22 (2): 137–47. doi: 10.1016/0040-8166(90)90017-4

Cernichiari, E., L. Muscatine, and D. C. Smith. 1969. "Maltose Excretion by the Symbiotic Algae of Hydra Viridis." *Proceedings of the Royal Society B: Biological Sciences* 173 (1033): 557–76. doi: 10.1098/rspb.1969.0077

Davy, S. K. K., D. Allemand, and V. M. M. Weis. 2012. "Cell Biology of Cnidarian-Dinoflagellate Symbiosis." *Microbiology and Molecular Biology Reviews* 76 (2): 229–61. doi: 10.1128/MMBR.05014-11

Deines, P., T. Lachnit, and T. C. G. Bosch. 2017. "Competing Forces Maintain the *Hydra* Metaorganism." *Immunological Reviews* 279 (1). John Wiley & Sons, Ltd (10.1111): 123–36. doi: 10.1111/imr.12564

Dorling, M., P. J Mcauley, and H. Hodge. 1997. "Effect of PH on Growth and Carbon Metabolism of Maltose-Releasing Chlorella (Chlorophyta) Effect of pH on Growth and Carbon Metabolism of Maltose-Releasing Chlorella (Chlorophyta)." *European Journal of Phycology* 32 (1): 19–24. doi: 10.1080/09541449710001719335

Dunn, K. 1987. "Growth of Endosymbiotic Algae in the Green Hydra, Hydra Viridissima." http://jcs.biologists.org/content/joces/88/5/571.full.pdf

Franzenburg, S., S. Fraune, P. M. Altrock, S. Künzel, J. F. Baines, A. Traulsen, and T. C. G. Bosch. 2013a. "Bacterial Colonization of Hydra Hatchlings Follows a Robust Temporal Pattern." *The ISME Journal* 7 (4). Nature Publishing Group: 781–90. doi: 10.1038/ismej.2012.156

Franzenburg, S., J. Walter, S. Künzel, J. Wang, J. F. Baines, T. C. G. Bosch, and S. Fraune. 2013b. "Distinct Antimicrobial Peptide Expression Determines Host Species-Specific Bacterial Associations." *Proceedings of the National Academy of Sciences* 110 (39). National Academy of Sciences: E3730–38. doi: 10.1073/PNAS.1304960110

Fraune, S., Y. Abe, and T. C. G. Bosch. 2009. "Disturbing Epithelial Homeostasis in the Metazoan *Hydra* Leads to Drastic Changes in Associated Microbiota." *Environmental Microbiology* 11 (9). John Wiley & Sons, Ltd (10.1111): 2361–69. doi: 10.1111/j.1462-2920.2009.01963.x

Fraune, S., F. Anton-Erxleben, R. Augustin, S. Franzenburg, M. Knop, K. Schrö, D. Willoweit-Ohl, and T. C. Bosch. 2015. "Bacteria–Bacteria Interactions within the Microbiota of the Ancestral Metazoan Hydra Contribute to Fungal Resistance." *The ISME Journal* 9: 1543–56. doi: 10.1038/ismej.2014.239

Fraune, S., R. Augustin, F. Anton-Erxleben, J. Wittlieb, C. Gelhaus, V. B. Klimovich, M. P. Samoilovich, and T. C. G. G. Bosch. 2010. "In an Early Branching Metazoan, Bacterial Colonization of the Embryo Is Controlled by Maternal Antimicrobial Peptides." *Proceedings of the National Academy of Sciences of the United States of America* 107 (42). National Academy of Sciences: 18067–72. doi: 10.1073/pnas.1008573107

Fraune, S., and T. C. G. Bosch. 2007. "Long-Term Maintenance of Species-Specific Bacterial Microbiota in the Basal Metazoan Hydra." *Proceedings of the National Academy of Sciences* 104 (32). National Academy of Sciences: 13146–51. doi: 10.1073/PNAS.0703375104

Galliot, B. 2012. "Hydra, a Fruitful Model System for 270 Years." *The International Journal of Developmental Biology* 56 (6-7-8). UPV/EHU Press: 411–23. doi: 10.1387/ijdb.120086bg

Grasis, J. A., T. Lachnit, F. Anton-Erxleben, Y. W. Lim, R. Schmieder, S. Fraune, S. Franzenburg et al. 2014. "Species-Specific Viromes in the Ancestral Holobiont Hydra." Edited by John F. Rawls. *PLOS ONE* 9 (10). Public Library of Science: e109952. doi: 10.1371/journal.pone.0109952

Habetha, M., F. Anton-Erxleben, K. Neumann, and T. C. G. Bosch. 2003. "The Hydra Viridis/Chlorella Symbiosis. Growth and Sexual Differentiation in Polyps without Symbionts." *Zoology* 106 (2): 101–8. doi: 10.1078/0944-2006-00104

Hamada, M., K. Schröder, J. Bathia, U. Kürn, S. Fraune, M. Khalturina, K. Khalturin, C. Shinzato, N. Satoh, and T. C. G. Bosch. 2018. "Metabolic Co-Dependence Drives the Evolutionarily Ancient Hydra–Chlorella Symbiosis." *ELife* 7. doi: 10.7554/eLife.35122

Hohman, T. C., P. L. Mcneil, and L. Muscatine. 1982. "Phagosome-Lysosome Fusion Inhibited by Algal Symbionts of Hydra Viridis." *Journal of Cell Biology* 94 (1): 56–63. doi: 10.1083/jcb.94.1.56

Ishikawa, M., I. Yuyama, H. Shimizu, M. Nozawa, K. Ikeo, and T. Gojobori. 2016. "Different Endosymbiotic Interactions in Two Hydra Species Reflect the Evolutionary History of Endosymbiosis." *Genome Biology and Evolution* 8 (7): 2155–63. doi: 10.1093/gbe/evw142

Jolley, E., and D. C. Smith. 1978. "The Green Hydra Symbiosis. I. Isolation, Culture and Characteristics of the Chlorella Symbiont of 'European' Hydra Viridis." *New Phytologist* 81 (3). John Wiley & Sons, Ltd (10.1111): 637–45. doi: 10.1111/j.1469-8137.1978.tb01637.x

Jolley, E., and D. C. Smith. 1980. "The Green Hydra Symbiosis. II. The Biology of the Establishment of the Association." *Proceedings of the Royal Society of London. Series B. Biological Sciences* 207 (1168). The Royal Society: 311–33. doi: 10.1098/rspb.1980.0026

Jung, S., A. J. Dingley, R. Augustin, F. Anton-Erxleben, M. Stanisak, C. Gelhaus, T. Gutsmann et al. 2009. "Hydramacin-1, Structure and Antibacterial Activity of a Protein from the Basal Metazoan Hydra." *The Journal of Biological Chemistry* 284 (3). American Society for Biochemistry and Molecular Biology: 1896–905. doi: 10.1074/jbc.M804713200

Klimovich, A. V., and T. C. G. Bosch. 2018. "Rethinking the Role of the Nervous System: Lessons From the *Hydra* Holobiont." *BioEssays* 40 (9). John Wiley & Sons, Ltd: 1800060. doi: 10.1002/bies.201800060.

Klimovich, A., J. Wittlieb, and T. C. G. Bosch. 2019. "Transgenesis in Hydra to Characterize Gene Function and Visualize Cell Behavior." *Nature Protocols* 14 (7). Springer US: 2069–90. doi: 10.1038/s41596-019-0173-3

Kodama, Y., and M. Fujishima. 2015. "Differences in Infectivity between Endosymbiotic Chlorella Variabilis Cultivated Outside Host Paramecium Bursaria for 50 Years and Those Immediately Isolated from Host Cells after One Year of Reendosymbiosis." *Biology Open* 5 (1). The Company of Biologists Ltd: 55–61. doi: 10.1242/bio.013946

Li, X.-Y., T. Lachnit, S. Fraune, T. C. G. Bosch, A. Traulsen, M. Sieber, X.-Y. Li et al. 2017. "Temperate Phages as Self-Replicating Weapons in Bacterial Competition." *Journal of The Royal Society Interface* 14 (137). The Royal Society: 20170563. doi: 10.1098/rsif.2017.0563

Li, X.-Y., C. Pietschke, S. Fraune, P. M. Altrock, T. C. G. G. Bosch, and A. Traulsen. 2015. "Which Games Are Growing Bacterial Populations Playing?" *Journal of the Royal Society Interface* 12 (108). The Royal Society: 1–10. doi: 10.1098/rsif.2015.0121

Lu, Y., and T. D. Sharkey. 2006. "The Importance of Maltose in Transitory Starch Breakdown." *Plant, Cell and Environment* 29 (3): 353–66. doi: 10.1111/j.1365-3040.2005.01480.x

McAuley, P. J. 1986. "The Cell Cycle of Symbiotic Chlorella III Numbers of Algae in Green Hydra Digestive Cells Are Regulated at Digestive Cell Division." *Journal of Cell Science* 85: 63–71.

McNeil, P. L. 1981. "Mechanisms of Nutritive Endocytosis. I. Phagocytic Versatility and Cellular Recognition in Chlorohydra Digestive Cells, a Scanning Electron Microscope Study." *Journal of Cell Science* 49 (1).

McNeil, P. L., T. C. Hohman, and L. Muscatine. 1981. "Mechanisms of Nutritive Endocytosis. II. The Effect of Charged Agents on Phagocytic Recognition by Digestive Cells." *Journal of Cell Science* 52 (1).

Mortzfeld, B. M., J. Taubenheim, S. Fraune, A. V. Klimovich, and T. C. G. Bosch. 2018. "Stem Cell Transcription Factor FoxO Controls Microbiome Resilience in Hydra." *Frontiers in Microbiology* 9. Frontiers Media SA: 629. doi: 10.3389/fmicb.2018.00629

Murillo-Rincon, A. P., A. Klimovich, E. Pemöller, J. Taubenheim, B. Mortzfeld, R. Augustin, and T. C. G. Bosch. 2017. "Spontaneous Body Contractions Are Modulated by the Microbiome of Hydra." *Scientific Reports* 7 (1). Nature Publishing Group: 15937. doi: 10.1038/s41598-017-16191-x

Muscatine, L., and N. Neckelmann. 1981. "Regulation of Numbers of Algae in the Hydra-Chlorella Symbiosis." *Berichte Der Deutschen Botanischen Gesellschaft* 94 (1): 571–82. doi: 10.1111/j.1438-8677.1981.tb03428.x

Passano, L. M., and C. B. McCullough. 1965. "Co-Ordinating Systems and Behaviour in Hydra. I. Pacemaker system of the periodic contractions." *The Journal of Experimental Biology* 42: 205–31.

Schröder, K., and T. C. G. Bosch. 2016. "The Origin of Mucosal Immunity: Lessons from the Holobiont Hydra." *MBio* 7 (6). American Society for Microbiology (ASM). doi: 10.1128/mBio.01184-16

Sieber, M., L. Pita, N. W.-B. Uerid, P. Dirksen, J. Wang, B. Mortzfeldid, S. Franzenburgid et al. 2019. "Neutrality in the Metaorganism." *PLoS Biology* 17 (6). doi: 10.1371/journal.pbio.3000298

Wein, T., T. Dagan, S. Fraune, T. C. G. Bosch, T. B. H. Reusch, and N. F. Hülter. 2018. "Carrying Capacity and Colonization Dynamics of Curvibacter in the Hydra Host Habitat." *Frontiers in Microbiology* 9 (March). Frontiers: 443. doi: 10.3389/fmicb.2018.00443

Wittlieb, J., K. Khalturin, J. U. Lohmann, F. Anton-Erxleben, and T. C. G. Bosch. 2006. "Transgenic Hydra Allow in Vivo Tracking of Individual Stem Cells during Morphogenesis." *Proceedings of the National Academy of Sciences of the United States of America* 103 (16). National Academy of Sciences: 6208–11. doi: 10.1073/pnas.0510163103

Ye, S., K. N. Badhiwala, J. T. Robinson, W. H. Cho, and E. Siemann. 2019. "Thermal Plasticity of a Freshwater Cnidarian Holobiont: Detection of Trans-Generational Effects in Asexually Reproducing Hosts and Symbionts." *ISME Journal* 13 (8). Springer US: 2058–67. doi: 10.1038/s41396-019-0413-0

6 Hydra and Curvibacter
An intimate crosstalk at the epithelial interface

Timo Minten-Lange and Sebastian Fraune

Contents

6.1 Introduction

Hydra species represent a group of early branching metazoans in the animal tree of life, and are distributed in freshwater ecosystems worldwide. They belong to the phylum of Cnidaria, which is the sister group to all Bilateria. *Hydra* is a diploblastic organism with two true epithelial cell layers, the endo- and ectoderm. An acellular layer, the mesoglea, separates both epithelial layers. The ectoderm is covered by a multi-layer of glycoproteins called the glycocalyx (Fraune et al. 2015; Schröder and Bosch 2016). A characteristic feature of *Hydra* is its simple life cycle (Figure 6.1A). Under normal environmental conditions, it comprises only of a solitary polyp stage reproducing asexually by budding. Species-specific environmental cues, like temperature or food, can induce sexual reproduction (Figure 6.1A–C).

The innate immune system of *Hydra* relies on a rich repertoire of antimicrobial peptides and an evolutionary conserved set of pattern recognition receptors (Miller et al. 2007; Bosch et al. 2009; Franzenburg et al. 2012). This innate immune repertoire protects the metaorganism *Hydra* against pathogens (Bosch et al. 2009; Franzenburg et al. 2012) and maintains a homeostasis with beneficial microbes (Fraune et al. 2010; Franzenburg et al. 2013b; Augustin et al. 2017).

The bacterial community of adult *Hydra* polyps is relatively simple, with six bacterial species representing 90% of the total bacterial abundance (Fraune et al. 2015). Comparing the bacterial communities of different *Hydra* species maintained in the lab

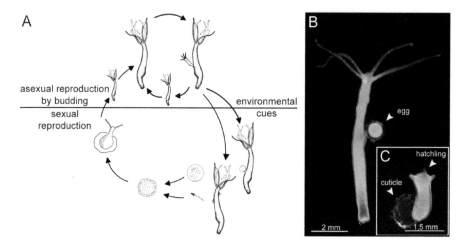

FIGURE 6.1 Life cycle and morphology of *Hydra*. (A) *Hydra* polyps mainly reproduce asexually via budding. Environmental cues can induce sexual reproduction. During sexual reproduction, juvenile polyps will hatch after a period of 2–5 weeks after fertilization. (B): Female polyp of *Hydra vulgaris* (AEP) with a developing embryo. (C) Juvenile *Hydra* hatching from the cuticle stage. (Figures B and C modified from Franzenburg, S. et al. 2013a. *The ISME Journal* 7 (4): 781–90.)

revealed a high degree of species-specificity and reflects the phylogenetic relationships of the *Hydra* host species (Franzenburg et al. 2013b). *Hydra* polyps sampled from the field associated with similar bacterial communities compared to laboratory animals support the aforementioned laboratory results of species-specific bacterial communities (Fraune and Bosch 2007). These findings strongly indicate distinct selective pressures within the *Hydra* epithelium, and that the host cells actively shape the composition of the colonizing microbiota.

6.2 *Hydra* and *Curvibacter*: The ideal duo to understand inter-kingdom communications

Curvibacter sp. represents the most abundant bacterial colonizer within bacterial communities associated with *Hydra* species of the vulgaris group (Franzenburg et al. 2013b; Fraune et al. 2015; Schwentner and Bosch 2015). At the same time, *Curvibacter* is detected in the laboratory as well as on *Hydra* collected from the wild (Fraune and Bosch 2007). Furthermore, *Curvibacter* is also detectable in most *Hydra* species outside of the vulgaris group, although in lower abundances (Franzenburg et al. 2013b).

Bacteria of the genus *Curvibacter* are common in a wide range of freshwater samples and are frequently found as colonizers of freshwater organisms (Ding and Yokota 2004; Fraune and Bosch 2007; McKenzie et al. 2012; Ma et al. 2016; Pittol et al. 2018). This close relationship of *Curvibacter* sp. with different host species makes the *Hydra–Curvibacter* interaction an ideal basis for a highly informative symbiosis model. *Hydra* and *Curvibacter* fulfil many requirements essential for cutting-edge symbiosis research (Koch and McFall-Ngai 2018): *Curvibacter* is easily cultivated in standard medium (Fraune et al. 2015) and *Hydra* has already been used as a model organism for centuries (Trembley 1744), with easy reproduction (asexual and sexual) in the lab. Another important factor for modern day science is the availability of genomic and transcriptomic resources. Both, genomic and transcriptomic data are available for *Hydra* and *Curvibacter* (Chapman et al. 2010; Hemmrich et al. 2012; Pietschke et al. 2017). In addition, molecular genetic methods are well established for both organisms, allowing functional studies in both symbiosis partners (Wittlieb et al. 2006; Wein et al. 2018; Klimovich et al. 2019).

Controlled recolonization experiments are an advantage for detailed studies of the interactions of *Hydra* and *Curvibacter*. It is possible to recolonize *Hydra* with single or multiple bacterial colonizers, after generation of germ-free polyps via antibiotic treatment (Rahat and Dimentman 1982; Franzenburg et al. 2013b; Fraune et al. 2015). Due to the transparent appearance of the animals and the ability of transgenic labelling of single host cells (Wittlieb et al. 2006) and of *Curvibacter* cells (Wein et al. 2018), microscopic analysis can be achieved in an *in vivo* context on single cell, as well as on whole tissue level in space and time. These possibilities turn *Hydra* and *Curvibacter* into a valuable symbiosis model to study and understand inter-kingdom communications in an *in vivo* context.

6.3   Spatial localization and transmission of *Curvibacter*

Hydra's ectodermal cells are covered by a multi-layered glycocalyx, which is separated into five distinct layers (Figure 6.2A) (Chapman et al. 2010; Böttger et al. 2012).

The four inner layers seem to be firmly attached to the ectodermal cell membrane. The outer layer consists of a loose meshwork of glycoproteins, which can be removed by a hypertonic salt wash (Schröder and Bosch 2016). Therefore, it was recently proposed to term the outer layer a mucus-like layer with analogies to the mucus in the mammalian colon (Johansson et al. 2011; Schröder and Bosch 2016). This mucus-like layer provides the habitat for *Curvibacter* cells (Figure 6.2B), whereas the four inner layers seem to be free of bacterial cells (Fraune et al. 2015). While *Curvibacter* homogenously colonizes the mucus-like layer of the head and the body column, tentacle tissue is colonized with higher densities (Augustin et al. 2017; Wein et al. 2018). The density gradient of *Curvibacter* cells between tentacles and head is controlled by an antimicrobial peptide named NDA-1, which is secreted by head specific neurons (Augustin et al. 2017). Interestingly, the foot region is less efficiently colonized by *Curvibacter* (Augustin et al. 2017). This phenomenon is correlated with a different mucus secreted by foot-specific cells (Philpott et al. 1966; Davis 1973).

To understand the transmission mode of *Curvibacter*, sexual and asexual reproduction of *Hydra* have to be considered separately. While asexual reproduction by budding is the main mode of reproduction in *Hydra*, sexual reproduction rarely occurs and is induced by species-specific environmental cues like temperature and food (Figure 6.1A) (Burnett 1973). During clonal reproduction, Hydra's tissue-transfer, from the mother polyp to the newly emerging bud, transmits *Curvibacter* cells passively, ensuring the vertical transmission of *Curvibacter* to the next clonal generation. In contrast, embryogenesis is occurring outside the mother polyp, but the

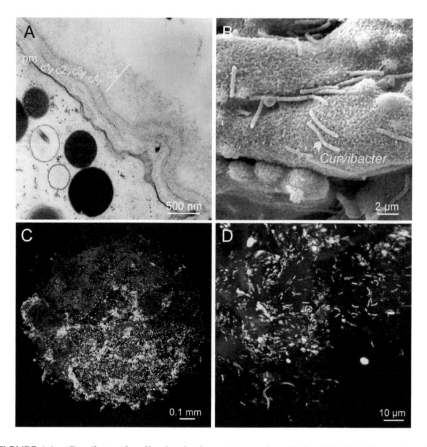

FIGURE 6.2 *Curvibacter* **localization in the metaorganism** *Hydra.* (A) *Hydra*'s ectodermal glycocalyx. Five distinct layers (c1–c5; pm, plasma membrane) of the glycocalyx can clearly be distinguished. (B) Raster electron micrograph (REM) of bacterial cells located on the surface of ectodermal cells. The yellow label marks a single *Curvibacter* cell. (C) Bacterial localization (blue) on the cuticle of 10-day-old embryos (D) FISH analysis of bacteria colonizing the cuticle of 10-day-old embryos, showing specifically labelled *Curvibacter* in yellow (yellow, overlay of a *Curvibacter*-specific probe in red and bacterial-specific probe in green). (Figures modified from Fraune, S. et al. 2010. *Proceedings of the National Academy of Sciences of the United States of America* 107 (42): 18067–72 and Fraune, S. et al. 2015. *The ISME Journal* 9 (7): 1543–56.)

embryos stay attached to the mother polyp (Figure 6.1B). During this time, secretion of maternal antimicrobial peptides of the periculin family protects the embryos (Fraune et al. 2010). After gastrulation, embryos start to secret zygotic periculin (Fraune et al. 2010) and produce a cuticle layer, protecting the embryo during the following developmental stages (Fröbius et al. 2003). *Curvibacter* cells are not detectable in early embryonic stages, most likely due to the secretion of periculin (Fraune et al. 2010). This suggests that *Curvibacter* is not directly transmitted to the oocyte. In contrast, a dense bacterial community colonizes later embryonic stages, like the cuticle stage (Figures 6.1B and 6.2C). Fluorescence *in situ* hybridization analysis specifically detects *Curvibacter* cells at the spike and cuticle stage (Figure 6.2D)

(Fraune et al. 2010). Considering the close proximity of mother tissue and embryo, it is likely that *Curvibacter* cells are vertically transmitted from the mother polyp to the cuticle layer of the embryo. During further development, the embryo detaches from the mother polyp and sinks to the ground. In the lab, juvenile polyps hatch after a period of 2–5 weeks (Figure 6.1C) (Klimovich et al. 2019). 16S rRNA analysis confirms the presence of *Curvibacter* cells attached to freshly hatched polyps (Franzenburg et al. 2013a). This suggests that polyps are colonized during hatching by *Curvibacter* cells attached to the cuticle, ensuring the vertical transmission of *Curvibacter* to the next sexual generation.

6.4 Establishment and carrying capacity of *Curvibacter* colonization

The establishment of *Curvibacter* colonization after hatching follows a highly dynamic process (Franzenburg et al. 2013a). Initial *Curvibacter* colonization is accompanied by a diverse colonization with other bacteria. During further succession, bacterial diversity decreases and *Curvibacter* starts to dominate the bacterial community after two weeks. This increase in *Curvibacter* abundance is followed by an intermediate decrease, followed by a final and stable increase after four weeks. Mathematical modelling suggests that host-derived factors and frequency-dependent bacteria-bacteria interactions are important aspects in the emergence of a stable *Curvibacter* colonization (Franzenburg et al. 2013a).

The importance of host factors for a stable *Curvibacter* colonization are also evident from experiments using the mutant Hydra SF-1 line (Fraune et al. 2009). Incubation of these polyps for a few hours at 28°C leads to apoptosis of the interstitial stem cells (Sugiyama and Fujisawa 1978; Terada et al. 1988). Consequently, these polyps lose their entire interstitial stem cell lineage. Over the course of the experiment, this loss, especially of nerve and secretory gland cells, leads to significant changes in the bacterial composition of the polyp's microbiome. Especially *Curvibacter's* abundance is drastically reduced (Fraune et al. 2009), indicating a strong dependence of *Curvibacter* fitness on host tissue homeostasis.

In adult polyps, *Curvibacter* reaches a relative abundance of 60%–80% (Franzenburg et al. 2013a; Fraune et al. 2015), with a total carrying capacity of 2×10^5 cells per polyp (Wein et al. 2018). Interestingly, recolonization experiments with fluorescence labelled *Curvibacter* cells revealed that a fully established colonization prevents colonization of newly incoming *Curvibacter* cells (Wein et al. 2018). These results suggest that a fully colonized *Hydra* polyp represents a closed system, with no exchange of bacterial cells between *Hydra* and environment. This enables a long-term association of both partners, a prerequisite of potential co-speciation of *Hydra* and *Curvibacter*.

6.5 *Curvibacter* function in the *Hydra* metaorganism

The influence of an intact microbiome on host fitness was shown in an experiment using germ-free *Hydra* (Fraune et al. 2015). When removing the microbiome by antibiotic treatment, germ-free *Hydra* cultures often suffer from infection by a fungus of the genus *Fusarium*. In vitro experiments showed that only a single bacterium from

Hydra's microbiome was able to inhibit growth of *Fusarium* sp. In mono-association with *Hydra* none of the colonizers alone could restore full resistance against fungal infection. Only recolonization with a complex bacterial community could protect against fungal infection, suggesting that bacteria–bacteria interactions are necessary for antifungal activity. Controlled recolonization experiments with the two most dominant bacterial colonizers of *Hydra*, *Curvibacter* sp. and *Duganella* sp., revealed synergistic activities restoring full antifungal activity, while in mono-association both had no protective capabilities (Fraune et al. 2015).

Another important function of *Curvibacter* is its ability to regulate and structure the microbiome colonizing *Hydra*. Co-culturing experiments of *Curvibacter* and *Duganella* revealed frequency-dependent, nonlinear growth rates, indicating strong interactions between these two bacteria (Li et al. 2015). Nevertheless, the observed growth rates of both bacteria are beyond the simple case of direct pairwise interactions, suggesting the presence of bacteriophages in one of them (Li et al. 2015). Genome analyses indeed proved the presence of an integrated and inducible prophage in the genome of *Curvibacter* (Pietschke et al. 2017). The isolated phage of *Curvibacter* is able to infect other bacteria of *Hydra*'s microbiome, like *Duganella*, and influences their abundances in culture and during *Hydra* colonization (Lange et al. 2019). These results highlight the importance of an intact microbiome for *Hydra*'s health, and *Curvibacter*'s role as an essential part of a protective and homeostatic microbiome.

6.6   Inter-kingdom communication between *Hydra* and *Curvibacter*

Since the beginning of metazoan evolution, multicellular organisms rely on the association with bacterial communities (King 2004). These associations are often mediated by communicating with each other through various chemical compounds, such as hormones and hormone-like small molecules produced by eukaryotes and bacteria (Pacheco and Sperandio 2009).

Bacteria use a communication system called quorum sensing (QS) to regulate a multitude of physiological processes (Schauder and Bassler 2001). Bacteria constantly produce small diffusible molecules, called autoinducers (AI). At a critical threshold, these AI bind to specific receptors, thereby activating expression of QS regulated genes. But bacterial AIs are not exclusively known for regulating the bacterial QS system. In some host-symbiont systems, the host also recognizes bacterial AIs. Some hosts are even able to interfere with or produce AIs or AI-like molecules, which in turn are recognized by the symbionts (Givskov et al. 1996; Manefield et al. 1999; Gao et al. 2003; Hughes and Sperandio 2008; Teplitski et al. 2011; Pérez-Montaño et al. 2013; Holm and Vikström 2014; Fetzner 2015; Palmer et al. 2016; Ismail et al. 2016; Mukherjee and Bassler 2019).

Inter-kingdom communication based on QS signalling molecules of the class acyl-homoserine lactones (AHL) also occurs in the interaction between *Hydra* and *Curvibacter*. A new eukaryotic mechanism was identified in *Hydra*, which enables the polyp to specifically modify long-chain AHLs by an oxidoreductase activity (Pietschke et al. 2017). In addition, *Curvibacter* expresses homologs of the *Vibrio* LuxI/LuxR system during host colonization, responsible for the production of long chain AHLs (Pietschke et al. 2017). Functional characterization of *Curvibacter*'s QS system showed that both, the host-modified and the non-modified AHL, are recognized by the same AHL receptor. Remarkably, even though both AHLs are recognized by the same receptor, gene expression profiles of *Curvibacter* differ, depending on the AHL version. Treatment with 3OHC12–homoserine lactone (HSL) leads to a strong activation of several metabolic pathways. Treatment with 3OC12-HSL leads to a strong upregulation of flagellar assembly and motility. Flagellin, the main structural component of the flagellum, is known as an important microbe-associated molecular pattern (MAMP) recognized by innate immune receptors. Investigating the impact of the different QS signals on metaorganism homeostasis *in vivo* showed that the host-modified signal promotes symbiont colonization, while the non-modified signal represses it. These results demonstrate that *Hydra* is able to alter quorum-sensing controlled behaviour of its bacterial colonizer *Curvibacter* to promote metaorganism assembly and resilience (Pietschke et al. 2017).

Other recent results indicate that *Curvibacter* is able to modify the Wnt-signaling pathway, a central signalling cascade in *Hydra* development (Taubenheim et al. 2019). The Wnt-signaling pathway most likely evolved in the common ancestor of multicellular animals (Holstein 2012). It is involved in several developmental processes in *Hydra*, like head formation (Hobmayer et al. 2000), control of bud formation (Lengfeld et al. 2009; Watanabe et al. 2014) and the differentiation of stem cells (Khalturin et al. 2007). Intriguingly, germ-free *Hydra* polyps react significantly more sensitive to the ectopic activation of Wnt-signaling compared to fully colonized polyps. Gene expression analyses led to the identification of two small secreted peptides, named Eco1 and Eco2, which are upregulated in the response to *Curvibacter* colonization. Functional analyses of the *eco* genes revealed that the corresponding peptides have an antagonistic function to Wnt-signaling in *Hydra* and influence stem cell differentiation (Taubenheim et al. 2019). These results suggest that *Curvibacter* produces signals recognized by *Hydra* and that their information is integrated into conserved developmental processes.

6.7   Outlook

Animal development has traditionally been viewed as an autonomous process directed by the host genome. In recent years, it became evident that bacterial cues provide a variety of signals that are integrated into the development and physiology of host organisms (McFall-Ngai et al. 2013). However, many questions remain to be answered: What is the nature of bacterial signals transmitted to the host? Which host receptors perceive these signals? How and why are bacterial signals integrated into the host signalling cascades controlling development and physiology? Have host receptors and bacterial signals co-evolved? Do host organisms modify bacterial signals for their own good?

The freshwater polyp *Hydra* and its bacterial symbiont *Curvibacter* are an ideal symbiosis model to study these questions of inter-kingdom communication in all of their details. Functional genetics and cell-specific reporter lines are available for both host and symbiont, allowing the investigation of gene function and *in vivo* imaging with spatial and temporal resolution. Temporal gene expression, in response to host colonization and interaction with other bacterial colonizers, can be analysed by using *Curvibacter* bioreporter systems. Further isolation and genomic sequencing of host-associated and pelagic *Curvibacter* strains will reveal the evolutionary adaptations of *Curvibacter* to its host and the underlying genetic mechanisms. Comparative genomics will lead to the identification of genes associated with an adaptation to a symbiotic lifestyle and to specific host species. Furthermore, *Curvibacter* knockout mutants can be used to test the identified genes for their requirement for successful host colonization, and their effects on host development and physiology.

Acknowledgments

We thank Hanna Domin and Jan Taubenheim for helpful comments on an earlier version of the manuscript. The work was supported by the Deutsche Forschungsgemeinschaft (DFG) (CRC1182 "Origin and function of Metaorganisms").

References

Augustin, R., K. Schröder, A.P. Murillo Rincón, S. Fraune, F. Anton-Erxleben, E.-M. Herbst, J. Wittlieb et al. 2017. "A Secreted Antibacterial Neuropeptide Shapes the Microbiome of Hydra." *Nature Communications* 8 (1). Springer US: 698.

Bosch, T.C.G., R. Augustin, F. Anton-Erxleben, S. Fraune, G. Hemmrich, H. Zill, P. Rosenstiel et al. 2009. "Uncovering the Evolutionary History of Innate Immunity: The Simple Metazoan Hydra Uses Epithelial Cells for Host Defence." *Developmental and Comparative Immunology* 33 (4): 559–69.

Böttger, A., A.C. Doxey, M.W. Hess, K. Pfaller, W. Salvenmoser, R. Deutzmann, A. Geissner et al. 2012. "Horizontal Gene Transfer Contributed to the Evolution of Extracellular Surface Structures: The Freshwater Polyp Hydra Is Covered by a Complex Fibrous Cuticle Containing Glycosaminoglycans and Proteins of the PPOD and SWT (Sweet Tooth) Families." *PLOS ONE* 7 (12): e52278.

Burnett, A.L. 1973. *Biology of Hydra.* Academic Press.

Chapman, J.A., E.F. Kirkness, O. Simakov, S.E. Hampson, T. Mitros, T. Weinmaier, T. Rattei et al. 2010. "The Dynamic Genome of Hydra." *Nature* 464 (7288): 592–96.

Davis, L.E. 1973. "Histological and Ultrastructural Studies of the Basal Disk of Hydra. I. The Glandulomuscular Cell." *Zeitschrift Fur Zellforschung Und Mikroskopische Anatomie (Vienna, Austria : 1948)* 139 (1): 1–27.

Ding, L., and A. Yokota. 2004. "Proposals of Curvibacter Gracilis Gen. Nov., Sp. Nov. and Herbaspirillum Putei Sp. Nov. for Bacterial Strains Isolated from Well Water and Reclassification of [Pseudomonas] Huttiensis, [Pseudomonas] Lanceolata, [Aquaspirillum] Delicatum and [Aquaspirillum]." *International Journal of Systematic and Evolutionary Microbiology* 54 (Pt 6): 2223–30.

Fetzner, S. 2015. "Quorum Quenching Enzymes." *Journal of Biotechnology* 201 (May). Elsevier B.V.: 2–14.

Franzenburg, S., S. Fraune, P.M. Altrock, S. Künzel, J.F. Baines, A. Traulsen, and T.C.G. Bosch. 2013a. "Bacterial Colonization of Hydra Hatchlings Follows a Robust Temporal Pattern." *The ISME Journal* 7 (4): 781–90.

Franzenburg, S., S. Fraune, S. Kunzel, J.F. Baines, T. Domazet-Loso, and T.C.G. Bosch. 2012. "MyD88-Deficient Hydra Reveal an Ancient Function of TLR Signaling in Sensing Bacterial Colonizers." *Proceedings of the National Academy of Sciences* 109 (47): 19374–79.

Franzenburg, S., J. Walter, S. Kunzel, J. Wang, J.F. Baines, T.C.G. Bosch, and S. Fraune. 2013b. "Distinct Antimicrobial Peptide Expression Determines Host Species-Specific Bacterial Associations." *Proceedings of the National Academy of Sciences* 110 (39): E3730–38.

Fraune, S., Y. Abe, and T.C.G. Bosch. 2009. "Disturbing Epithelial Homeostasis in the Metazoan Hydra Leads to Drastic Changes in Associated Microbiota." *Environmental Microbiology* 11 (9): 2361–69.

Fraune, S., F. Anton-Erxleben, R. Augustin, S. Franzenburg, M. Knop, K. Schröder, D. Willoweit-Ohl, and T.C.G. Bosch. 2015. "Bacteria–Bacteria Interactions within the Microbiota of the Ancestral Metazoan Hydra Contribute to Fungal Resistance." *The ISME Journal* 9 (7): 1543–56.

Fraune, S., R. Augustin, F. Anton-Erxleben, J. Wittlieb, C. Gelhaus, V.B. Klimovich, M.P. Samoilovich, and T.C.G. Bosch. 2010. "In an Early Branching Metazoan, Bacterial Colonization of the Embryo Is Controlled by Maternal Antimicrobial Peptides." *Proceedings of the National Academy of Sciences of the United States of America* 107 (42): 18067–72.

Fraune, S., and T.C.G. Bosch. 2007. "Long-Term Maintenance of Species-Specific Bacterial Microbiota in the Basal Metazoan Hydra." *Proceedings of the National Academy of Sciences of the United States of America* 104 (32): 13146–51.

Fröbius, A.C., G. Genikhovich, U. Kürn, F. Anton-Erxleben, and T.C.G. Bosch. 2003. "Expression of Developmental Genes during Early Embryogenesis of Hydra." *Development Genes and Evolution* 213 (9): 445–55.

Gao, M., M. Teplitski, J.B. Robinson, and W.D. Bauer. 2003. "Production of Substances by Medicago Truncatula That Affect Bacterial Quorum Sensing." *Molecular Plant-Microbe Interactions* 16 (9): 827–34.

Givskov, M., R. de Nys, M. Manefield, L. Gram, R. Maximilien, L. Eberl, S. Molin, P.D. Steinberg, and S. Kjelleberg. 1996. "Eukaryotic Interference with Homoserine Lactone-Mediated Prokaryotic Signalling." *Journal of Bacteriology* 178 (22): 6618–22.

Hemmrich, G., K. Khalturin, A.-M. Boehm, M. Puchert, F. Anton-Erxleben, J. Wittlieb, U.C. Klostermeier et al. 2012. "Molecular Signatures of the Three Stem Cell Lineages in Hydra and the Emergence of Stem Cell Function at the Base of Multicellularity." *Molecular Biology and Evolution* 29 (11): 3267–80.

Hobmayer, B., F. Rentzsch, K. Kuhn, C.M. Happel, C.C. von Laue, P. Snyder, U. Rothbächer, and T.W. Holstein. 2000. "WNT Signalling Molecules Act in Axis Formation in the Diploblastic Metazoan Hydra." *Nature* 407 (6801): 186–89.

Holm, A., and E. Vikström. 2014. "Quorum Sensing Communication between Bacteria and Human Cells: Signals, Targets, and Functions." *Frontiers in Plant Science* 5 (June): 309.

Holstein, T.W. 2012. "The Evolution of the Wnt Pathway." *Cold Spring Harbor Perspectives in Biology* 4 (7): a007922.

Hughes, D.T., and V. Sperandio. 2008. "Inter-Kingdom Signalling: Communication between Bacteria and Their Hosts." *Nature Reviews. Microbiology* 6 (2): 111–20.

Ismail, A.S., J.S. Valastyan, and B.L. Bassler. 2016. "A Host-Produced Autoinducer-2 Mimic Activates Bacterial Quorum Sensing." *Cell Host & Microbe* 19 (4). Elsevier Inc.: 470–80.

Johansson, M.E.V., D. Ambort, T. Pelaseyed, A. Schütte, J.K. Gustafsson, A. Ermund, D.B. Subramani et al. 2011. "Composition and Functional Role of the Mucus Layers in the Intestine." *Cellular and Molecular Life Sciences: CMLS* 68 (22): 3635–41.

Khalturin, K., F. Anton-Erxleben, S. Milde, C. Plötz, J. Wittlieb, G. Hemmrich, and T.C.G. Bosch. 2007. "Transgenic Stem Cells in Hydra Reveal an Early Evolutionary Origin for Key Elements Controlling Self-Renewal and Differentiation." *Developmental Biology* 309 (1): 32–44.

King, N. 2004. "The Unicellular Ancestry of Animal Development." *Developmental Cell* 7 (3): 313–25.

Klimovich, A., J. Wittlieb, and T.C.G. Bosch. 2019. "Transgenesis in Hydra to Characterize Gene Function and Visualize Cell Behavior." *Nature Protocols* 14 (7). Springer US: 2069–90.

Koch, E.J., and M. McFall-Ngai. 2018. "Model Systems for the Study of How Symbiotic Associations between Animals and Extracellular Bacterial Partners Are Established and Maintained." *Drug Discovery Today: Disease Models* 28. Elsevier Ltd: 3–12.

Lange, J., S. Fraune, T.C.G. Bosch, and T. Lachnit. 2019. "The Neglected Part of the Microbiome: Prophage TJ1 Regulates the Bacterial Community of the Metaorganism Hydra." *BioRxiv* 607325.

Lengfeld, T., H. Watanabe, O. Simakov, D. Lindgens, L. Gee, L. Law, H.A. Schmidt, S. Ozbek, H. Bode, and T.W. Holstein. 2009. "Multiple Wnts Are Involved in Hydra Organizer Formation and Regeneration." *Developmental Biology* 330 (1): 186–99.

Li, X.-Y., C. Pietschke, S. Fraune, P.M. Altrock, T.C.G. Bosch, and A. Traulsen. 2015. "Which Games Are Growing Bacterial Populations Playing?" *Journal of the Royal Society, Interface* 12 (108): 20150121.

Ma, D., Z. Hao, R. Sun, M. Bartlam, and Y. Wang. 2016. "Genome Sequence of a Typical Ultramicrobacterium, Curvibacter Sp. Strain PAE-UM, Capable of Phthalate Ester Degradation." *Genome Announcements* 4 (1): 1–2.

Manefield, M., R. de Nys, K. Naresh, R. Roger, M. Givskov, S. Peter, and S. Kjelleberg. 1999. "Evidence That Halogenated Furanones from Delisea Pulchra Inhibit Acylated Homoserine Lactone (AHL)-Mediated Gene Expression by Displacing the AHL Signal from Its Receptor Protein." *Microbiology* 145 (2): 283–91.

McFall-Ngai, M., M.G. Hadfield, T.C.G. Bosch, H.V. Carey, T. Domazet-Lošo, A.E. Douglas, N. Dubilier et al. 2013. "Animals in a Bacterial World, a New Imperative for the Life Sciences." *Proceedings of the National Academy of Sciences of the United States of America* 110 (9): 3229–36.

McKenzie, V.J., R.M. Bowers, N. Fierer, R. Knight, and C.L. Lauber. 2012. "Co-Habiting Amphibian Species Harbor Unique Skin Bacterial Communities in Wild Populations." *The ISME Journal* 6 (3). Nature Publishing Group: 588–96.

Miller, D.J., G. Hemmrich, E.E. Ball, D.C. Hayward, K. Khalturin, N. Funayama, K. Agata, and T.C.G. Bosch. 2007. "The Innate Immune Repertoire in Cnidaria—Ancestral Complexity and Stochastic Gene Loss." *Genome Biology* 8 (4): R59.

Mukherjee, S., and B.L. Bassler. 2019. "Bacterial Quorum Sensing in Complex and Dynamically Changing Environments." *Nature Reviews. Microbiology* 17 (6): 371–82.

Pacheco, A.R., and V. Sperandio. 2009. "Inter-Kingdom Signaling: Chemical Language between Bacteria and Host." *Current Opinion in Microbiology* 12 (2): 192–98.

Palmer, A.G., A. Mukherjee, D.M. Stacy, S. Lazar, J.-M. Ané, and H.E. Blackwell. 2016. "Interkingdom Responses to Bacterial Quorum Sensing Signals Regulate Frequency and Rate of Nodulation in Legume-Rhizobia Symbiosis." *ChemBioChem* 17 (22): 2199–205.

Pérez-Montaño, F., I. Jiménez-Guerrero, R. Contreras Sánchez-Matamoros, F.J. López-Baena, F.J. Ollero, M.A. Rodríguez-Carvajal, R.A. Bellogín, and M.R. Espuny. 2013. "Rice and Bean AHL-Mimic Quorum-Sensing Signals Specifically Interfere with the Capacity to Form Biofilms by Plant-Associated Bacteria." *Research in Microbiology* 164 (7): 749–60.

Philpott, D.E., A.B. Chaet, and A.L. Burnett. 1966. "A Study of the Secretory Granules of the Basal Disk of Hydra." *Journal of Ultrastructure Research* 14 (1–2): 74–84.

Pietschke, C., C. Treitz, S. Forêt, A. Schultze, S. Künzel, A. Tholey, T.C.G. Bosch, and S. Fraune. 2017. "Host Modification of a Bacterial Quorum-Sensing Signal Induces a Phenotypic Switch in Bacterial Symbionts." *Proceedings of the National Academy of Sciences of the United States of America* 114 (40): E8488–97.

Pittol, M., E. Scully, D. Miller, L. Durso, L. Mariana Fiuza, and V.H. Valiati. 2018. "Bacterial Community of the Rice Floodwater Using Cultivation-Independent Approaches." *International Journal of Microbiology* 2018. Hindawi: 1–13.

Rahat, M., and C. Dimentman. 1982. "Cultivation of Bacteria-Free Hydra Viridis: Missing Budding Factor in Nonsymbiotic Hydra." *Science (New York, N.Y.)* 216 (4541): 67–68.

Schauder, S., and B.L. Bassler. 2001. "The Languages of Bacteria." *Genes & Development* 15 (12): 1468–80.

Schröder, K., and T.C.G. Bosch. 2016. "The Origin of Mucosal Immunity: Lessons from the Holobiont Hydra." *MBio* 7 (6): 1–9.

Schwentner, M., and T.C.G. Bosch. 2015. "Revisiting the Age, Evolutionary History and Species Level Diversity of the Genus Hydra (Cnidaria: Hydrozoa)." *Molecular Phylogenetics and Evolution* 91 (October). Elsevier Inc.: 41–55.

Sugiyama, T., and T. Fujisawa. 1978. "Genetic Analysis of Developmental Mechanisms in Hydra. II. Isolation and Characterization of an Interstitial Cell-Deficient Strain." *Journal of Cell Science* 29 (February): 35–52.

Taubenheim, J., D. Willoweit-Ohl, M. Knop, S. Franzenburg, B. Tcg, and S. Fraune. 2019. "Bacteria- and Temperature-Regulated Peptides Modulate Beta-Catenin Signaling in Hydra." *BioRxiv*, 747303. https://doi.org/10.1101/747303

Teplitski, M., U. Mathesius, and K.P. Rumbaugh. 2011. "Perception and Degradation of N-Acyl Homoserine Lactone Quorum Sensing Signals by Mammalian and Plant Cells." *Chemical Reviews* 111 (1): 100–16.

Terada, H., T. Sugiyama, and Y. Shigenaka. 1988. "Genetic Analysis of Developmental Mechanisms in Hydra. XVIII. Mechanism for Elimination of the Interstitial Cell Lineage in the Mutant Strain Sf-1." *Developmental Biology* 126 (2): 263–69.

Trembley, A. 1744. *Mémoires Pour l'histoire Des Polypes d'eau Douce. 1744.*

Watanabe, H., H.A. Schmidt, A. Kuhn, S.K. Höger, Y. Kocagöz, N. Laumann-Lipp, S. Ozbek, and T.W. Holstein. 2014. "Nodal Signalling Determines Biradial Asymmetry in Hydra." *Nature* 515 (7525): 112–15.

Wein, T., T. Dagan, S. Fraune, T.C.G. Bosch, T.B.H. Reusch, and N.F. Hülter. 2018. "Carrying Capacity and Colonization Dynamics of Curvibacter in the Hydra Host Habitat." *Frontiers in Microbiology* 9 (MAR): 443.

Wittlieb, J., K. Khalturin, J.U. Lohmann, F. Anton-Erxleben, and T.C.G. Bosch. 2006. "Transgenic Hydra Allow in Vivo Tracking of Individual Stem Cells during Morphogenesis." *Proceedings of the National Academy of Sciences of the United States of America* 103 (16): 6208–11.

7 The coral holobiont highlights the dependence of cnidarian animal hosts on their associated microbes

Claudia Pogoreutz, Christian R. Voolstra,
Nils Rädecker, Virginia Weis,
Anny Cardenas, and Jean-Baptiste Raina

Contents

7.1   Introduction: The coral holobiont as an ecosystem engineer and its reliance on associated microbes

The productivity and biodiversity of coral reefs are unmatched in the marine environment (Hatcher 1990). Surrounded by oligotrophic oceans, coral reefs are buzzing oases of life in a marine desert (Darwin 1842). Tropical coral reefs cover just 0.1% of the seafloor but provide habitat for ~32% of all marine multicellular species (Fisher et al. 2015) and contribute to the livelihoods of more than 600 million people (Moberg and Folke 1999; Wilkinson 2008; Spalding et al. 2017). This entire ecosystem is supported by its foundation species: reef-building corals (Figure 7.1). These organisms sustain the immense productivity of coral reefs (Wild et al. 2004), contribute to the reef food web, and their calcareous skeletons form the structural basis of the reef framework.

As early as the 19th century, scientists identified the cohabitation of benthic cnidarians with intracellular photosynthetic dinoflagellates, initially termed "zooxanthellae" (Krueger 2017) (Figure 7.2). Building on more than a century of research, the coral–dinoflagellate symbiosis is one of the best characterized eukaryotic endosymbioses, and a powerful model to understanding the functioning of symbioses in general. Corals also harbor a diverse array of other microbes comprised of protists, fungi, bacteria, archaea, and viruses. This collective is called the coral holobiont (Rohwer et al. 2002). The coral metaorganism, by comparison, is a more restricted definition and typically only describes the coral host and associated microbes for which a function has been proposed or is known (Jaspers et al. 2019). This suite of organisms forms a complex network of symbiotic interactions that extend the metabolic repertoire, immunity, and

FIGURE 7.1   Scleractinian corals are the foundation species of coral reef ecosystems. Complex symbiotic interactions facilitate nutrient uptake and recycling, thereby enabling corals to thrive in highly oligotrophic environments and to build the structural framework of coral reefs. Image credit: Anna Roik.

FIGURE 7.2 Overview and diversity of cnidarian–dinoflagellate symbioses. (a) Reef-building or stony coral (Anthozoa, Scleractinia: *Acropora humilis*); (b) Fire coral, a calcifying hydrocoral (Hydrozoa: *Millepora platyphylla*); (c) Blue Coral, a calcifying "soft" coral (Anthozoa, Octocorallia: *Heliopora coerulea*); (d) upside-down jellyfish (Scyphozoa: *Cassiopea* sp.). All of these marine cnidarian holobiont systems are intimately engaged in a mutualistic relationship with dinoflagellates of the family Symbiodiniaceae.

environmental adaptation of the coral host (Muscatine 1990; Rosenberg et al. 2007; Ritchie 2012; Rädecker et al. 2015; Ziegler et al. 2017, 2019; Robbins et al. 2019). Microbes can therefore be considered fundamental to the ecological success of corals and the reefs they build (Bang et al. 2018).

Reef ecosystems have existed for ~240 million years, but face unprecedented and accelerating decline: the 2015–2017 global coral bleaching event affected 74% of reefs worldwide, and up to half of the coral cover was lost on the Great Barrier Reef alone, the largest reef system in the world (Hughes et al. 2018b). Future predictions for coral reefs are dramatic: even under a 1.5°C warming scenario, it is expected that coral reefs will decline by a further 70%–90%, with larger losses of up to 99% projected to be highly likely under a 2.0°C warming scenario (IPCC 2018). Such trajectories make the understanding of coral holobiont functioning and the contribution of its various microbes critical, not only to comprehend how symbiotic interactions have shaped the most biodiverse marine ecosystem on Earth, but also to help conserve and protect corals and the reefs they build for future generations. In this chapter, we discuss the state of knowledge of coral–microbe interactions and how they shape the ecology, resilience, and adaptation of the coral holobiont.

7.2   The coral–Symbiodiniaceae relationship

7.2.1   Symbiodiniaceae: Micro-algal engines of the coral holobiont machinery

The ecological success of the coral–Symbiodiniaceae symbiosis is based on efficient nutrient recycling between host and symbiont (Figure 7.3). The driving force of this bidirectional nutrient exchange ultimately lies in the complementary nutrient limitations of the two symbiotic partners (Shantz et al. 2016; Bang et al. 2018). While the heterotrophic coral host is limited by organic nutrient availability (e.g., glucose), the autotrophic intracellular dinoflagellates are limited by inorganic nutrients (e.g., carbon dioxide or ammonium) (Muscatine et al. 1989; Falkowski et al. 1993; Rädecker et al. 2015). These reciprocal metabolic exchanges are governed by the uptake of limiting nutrients and release of excess nutrients by each symbiotic partner (Muscatine 1990; Cunning et al. 2017). Symbiodiniaceae translocate high rates of excess photosynthetically-fixed carbon to the coral host. The host metabolism, in turn, produce waste compounds such as carbon dioxide through respiration available to the Symbiodiniaceae (Falkowski et al. 1993). This nutrient exchange is so efficient that the translocation of photosynthates can fully meet or even exceed the respiratory requirements of the coral host (Muscatine and Porter 1977; Muscatine 1990; Falkowski et al. 1993), and hence constitutes its primary energy source (Tremblay et al. 2012). As a consequence, symbiotic coral hosts may overcome their carbon limitation and shift instead towards a nitrogen-limited state (Cunning et al. 2017; Rädecker et al. 2018).

In a stable symbiosis, host and symbionts are nitrogen-limited and compete for available environmental ammonium (Pernice et al. 2012). Increasing evidence suggests that this resource competition is critical for maintaining the functioning of the coral-Symbiodiniaceae symbiosis (Cui et al. 2019). Coral hosts use the translocated carbon for ammonium assimilation required for amino acid synthesis (Cui et al. 2019). Consequently, carbon translocation reduces nitrogen availability for the dinoflagellates. The nutrient cycling in the intact symbiosis is thus stabilized by a positive feedback loop: because symbionts translocate carbon, they are nitrogen-limited; and because they are nitrogen-limited, a substantial fraction of photosynthetically fixed carbon cannot be channeled towards their biomass and growth, and is hence released to the host. While this "selfish" interaction between host and symbiont is central to the ecological success of the coral–Symbiodiniaceae relationship, it also renders the symbiosis highly susceptible to environmental disturbance.

Initially considered to be a single species, *Symbiodinium microadriaticum* (Freudenthal 1962), the recently established family Symbiodiniaceae currently encompasses seven distinct genera (*Symbiodinium*—formerly Clade A, *Breviolum* (B), *Cladocopium* (C), *Durusdinium* (D), *Effrenium* (E), *Fugacium* (F), and *Gerakladium* (G)) that have originated and diversified alongside reef-building corals approximately 160 mya ago (LaJeunesse et al. 2018). Coral reefs of the Indo-Pacific are almost exclusively dominated by *Cladocopium* and *Durusdinium* (LaJeunesse et al. 2003, 2004), with at least 50–100 possible "species" currently identified around Australia alone (LaJeunesse et al. 2003; Silverstein et al., 2011). The remarkable diversity of this family influences their host's susceptibility to environmental fluctuations, such as thermal stress and salinity (Berkelmans and van Oppen 2006; Sampayo et al.

2008). In the world's warmest reefs of the Persian/Arabian Gulf, corals predominantly harbor *Cladocopium thermophilum*, an association that is central to the thermo-tolerance of these coral communities (Hume et al. 2016). Similarly, the stress-tolerant *Durusdinium trenchii* has rapidly taken over Caribbean corals following repeated anthropogenic disturbances and increasing seawater temperatures (Pettay et al. 2015). However, thermal adaptation of reef communities represents a significant physiological tradeoff: corals harboring *D. trenchii* typically grow slower (Little et al. 2004) and incorporate only half the amount of photosynthates compared to those associated with *Cladocopium* (Cantin et al. 2009). Recently, multiple Symbiodiniaceae genomes have become available (Shoguchi et al. 2013; Lin et al. 2015; Aranda et al. 2016), and comparative analyses have revealed that these organisms possess an extensive transporter repertoire for carbon and nitrogen metabolites, which is unique among dinoflagellates and likely underpins their symbiotic lifestyle (Aranda et al. 2016).

7.2.2 Innate immunity, symbiosis sensing, and cell signaling

The regulation of the coral–Symbiodiniaceae partnership is complex and only partially understood. There is strong evidence that the host innate immune system plays a large role in mechanisms of recognition, maintenance, and dysbiosis of the association (Weis 2008). In the majority of coral species, the symbiosis is established anew with each host generation. The algae are acquired via phagocytosis by nutritive phagocytes that comprise the host gastrodermal tissue (endoderm) (Fadlallah 1983). However, instead of being digested, the algae specific to a particular host persist and proliferate within host vacuoles (termed symbiosomes) (Colley and Trench 1983; Schwarz et al. 1999). This process is mediated by the host innate immune system (Palmer 2018). The process of phagocytosis is a complex part of innate immunity that is highly conserved across the Metazoa (Underhill and Ozinsky 2002).

Microbes, including Symbiodiniaceae, are arrayed with a variety of microbe-associated molecular patterns (MAMPs), including glycans, that are recognized by host pattern recognition receptors (PRRs) on phagocyte cell surfaces (Weis 2008). These MAMP–PRR interactions launch a variety of signaling cascades that determine the fate of the phagocytized microbe. The model in corals is that tolerogenic pathways allow for the persistence and proliferation of algal symbionts inside symbiosomes, while resistant pathways are launched during dysbiosis and bleaching (see next section) that reject and remove the symbiont. There is overwhelming evidence for the presence of MAMP–PRR interactions and downstream innate immune signaling in corals. The majority of evidence comes from now extensive -omics studies that repeatedly point to elaboration, enhancement, and overexpression of innate immunity genes in corals and other symbiotic cnidarians (Rodriguez-Lanetty et al. 2004; Shinzato et al. 2011; Baumgarten et al. 2015; Mohamed et al. 2016; Voolstra et al. 2017; Cunning et al. 2018; Shumaker et al. 2019). There are also a variety of studies, often in sea anemone model systems, that provide evidence of innate immune pathway function including: host lectin–symbiont glycan interactions (Wood-Charlson et al. 2006; Bay et al. 2011; Parkinson et al. 2018), scavenger receptors (Neubauer et al. 2016), thrombospondin type 1 repeat proteins (Wolfowicz et al. 2016; Neubauer et al. 2017), complement system (Poole et al. 2016), the master immunity gatekeeper NF-κB (Mansfield et al.

2017), tolerogenic TGFβ pathway (Detournay et al. 2012; Berthelier et al. 2017), sphingolipid signaling (Kitchen and Weis 2017; Kitchen et al. 2017), and Rab protein signaling and other evidence of endosomal trafficking (Chen et al. 2003, 2004).

Also critical to symbiosis regulation is the maintenance of host and symbiont biomass ratios through cell cycle regulation. Host and symbiont biomass ratios reach a dynamic homeostasis after symbionts fully colonize a host (Davy et al. 2012). Symbiont populations *in hospite* grow much more slowly than those in culture and are arrested at the G1/S transition (Smith and Muscatine 1999). Mechanisms that coordinate the two cell cycles and an understanding of how this coregulation becomes decoupled during dysbiosis have yet to be revealed.

7.2.3 Coral bleaching: The breakdown of the coral–Symbiodiniaceae relationship

Stressful environmental conditions such as temperature anomalies, nutrient enrichment, or pollution can result in so-called coral bleaching, a (general) stress response characterized by whitening of the coral tissue caused primarily by the loss of Symbiodiniaceae endosymbionts from the tissue via expulsion, host cell apoptosis/detachment, digestion, or exocytosis of the symbiont cells (Gates et al. 1992; Douglas 2003; Dunn et al. 2007; Davy et al. 2012). Coral bleaching can also occur via the loss of photosynthetic pigment from the symbionts *in hospite* (Jones et al. 1998). Several triggers that induce the coral bleaching cascade have been proposed, including oxidative stress (Lesser 1997) and changes in nutrient stoichiometry (Wiedenmann et al. 2012; Morris et al. 2019). However, the underlying cellular mechanisms are still not fully understood.

Corals may recover from a bleaching event by repopulating their tissues with Symbiodiniaceae (Jones et al. 2008; Silverstein et al. 2015). However, as bleaching effectively results in the disruption of carbon fixation by Symbiodiniaceae and subsequent loss of photosynthate translocation to the coral host (Ezzat et al. 2015), coral host starvation may eventually result in mortality as a consequence of prolonged bleaching. The availability of heterotrophic food sources and the heterotrophic capacity of the host, thus, may determine the resilience of corals during heat stress (Grottoli et al. 2006). Mass coral bleaching events at the ecosystem scale have resulted in the loss of entire reefs, or reef systems (Hoegh-Guldberg 2011; Hughes et al. 2018a) and are projected to increase in the future due to climate change driving ocean warming (IPCC 2018).

The majority of contemporary observations of ecosystem-scale coral bleaching coincide with high-temperature anomalies, such as the El Niño Southern Oscillation (ENSO), or marine heatwaves, which can push corals beyond their thermal limits (Hoegh-Guldberg 1999; Hughes et al. 2003, 2017, 2018a). The first observations of coral bleaching were made in 1983 in the Eastern Pacific near the Panama-Costa Rica border (Glynn 1983). Due to the increasing severity and frequency of high-temperature anomalies attributed to global warming, coral bleaching events have become more common in the past decades. Since then, three mass bleaching events have occurred at the global scale, subsequently named the First, Second, and Third Global Bleaching Event, recorded in 1997/1998, 2009/2010, and 2015–2016, respectively (Hughes et al.

2017). A recent analysis of bleaching records of 100 reefs from 1980–2016 has shown that the average turnaround time between bleaching events has halved in the past thirty years and is now only six years (Hughes et al. 2018a). This concerning trend means that the likelihood of recurring annual mass bleaching events in the coming decade is increasing, and the amount of time in between these bleaching events is no longer sufficient to allow full reef recovery (Hughes et al. 2018a).

7.3   Symbiodiniaceae–bacteria relationships

Interactions between bacteria and Symbiodiniaceae are challenging to study *in hospite* because of the complex nature of the coral holobiont. The physiology, functional diversity, and stress tolerance of Symbiodiniaceae have been studied in cultured strains for decades to disentangle their contribution to the health and functioning of the holobiont. Symbiodiniaceae cultures harbor very diverse bacterial communities with abundances exceeding those of the algal cells by almost two orders of magnitude (Ritchie 2012; Lawson et al. 2018). Recurring bacterial taxa such as *Marinobacter* (Gammaproteobacteria), *Labrenzia*, and other *Roseobacter* (Alphaproteobacteria), have been identified in association with a wide diversity of Symbiodiniaceae genera (Ritchie 2012; Lawson et al. 2018). Strikingly, the same bacterial taxa are also known to positively influence the growth of many phytoplankton species (Seymour et al. 2017). Future studies on Symbiodiniaceae–bacteria interactions should use axenic cultures to characterize the effect of specific bacteria on the growth and physiology of the dinoflagellates. Protocols to set up axenic microalgal cultures exist and have successfully been applied to multiple phytoplankton species (Shishlyannikov et al. 2011; Cho et al. 2013). Establishing axenic Symbiodiniaceae cultures is an important first step towards identifying the functional roles and reciprocal exchanges occurring between Symbiodiniaceae and their bacterial partners.

7.4   Diversity and function of microbes associated with the coral host

7.4.1   The host as a habitat

Coral holobionts associate with a range of bacteria, archaea, and viruses, among other microorganisms (Rohwer et al. 2002; Wegley et al. 2007) (Figure 7.3). Coral holobionts can be separated into three distinct compartments: the surface mucus layer (SML), the coral tissue, and the underlying aragonite skeleton. The three compartments are governed by distinct physicochemical properties and environmental gradients (Ferrer and Szmant 1988; Wangpraseurt et al. 2016; Pernice et al. 2019) as well as distinct associated microbial communities (Sweet et al. 2011; Pollock et al. 2018). Most coral-associated prokaryotes have not yet been cultivated and the functional contributions of specific taxa are largely unknown. Nonetheless, prokaryotic functions are crucial to understanding microbial contribution to coral health and adaptation. Molecular tools have allowed for the in-depth characterization of bacterial communities associated with corals and revealed specific assemblages of bacteria differing between the SML, tissue, and skeleton. SML-associated bacteria are largely dominated by

FIGURE 7.3   The coral holobiont consists of the animal host and a diversity of eukaryotic and prokaryotic microorganisms, and their potentially numerous metabolic interactions. Microbial community composition and structure within the holobiont readily respond to environmental fluctuations (e.g., of temperature, nutrients, or light availability), or stress. The eukaryotic **Symbiodiniaceae** are intracellular tissue-associated microalgal symbionts that engage in an intimate and efficient nutrient-exchange mutualism with their coral host. Through their high carbon acquisition, fixation, and translocation rates to the host, Symbiodiniaceae can be considered true unicellular "engines" of the coral holobiont, which are able to provide organic carbon at rates that can fully meet or even exceed the respiratory requirements of the coral animal host. Among the prokaryotic associates of the coral holobiont, **bacteria** are the best-studied members. Coral-associated bacterial communities are highly diverse, uneven, and exhibit distinct compositions in the coral surface mucus layer, the coral tissues, and the coral skeleton. In the coral skeleton, **endolithic algae** (including the abundant and diverse *Ostreobium* spp.) and **fungi** form dense bands. These endoliths are hypothesized to metabolically interact with the coral host tissue. The metabolic functions of prokaryotes, endolithic algae, and endolithic fungi are potentially highly diverse; whether and how they may contribute to holobiont functioning is an area of active exploration.

Gammaprotebacteria, commonly represented by members of the genus *Vibrio* (Koren and Rosenberg 2006) and by a broad range of taxa presumed to be commensals that can inhibit the growth of pathogens (Ritchie 2006). Shifts in SML-associated bacteria communities towards a dominance of Alphaproteobacteria, Verrucomicrobia, and Cyanobacteria have been reported in mucus sheets of *Porites* corals, in particular related to aging mucus (Glasl et al. 2016), variation in nutrient composition, and heat stress (Lee et al. 2015, 2016), although studies from bleached and healthy coral mucus from the Persian/Arabian Gulf showed no difference between bacterial communities (Hadaidi et al. 2017).

Bacterial assemblages associated with coral tissues have been widely studied. Tissue-associated bacterial communities are strongly driven by coral and Symbiodiniaceae genotypes (Rohwer et al. 2002). *Endozoicomonas*, Alteromonadaceae, and *Ralstonia* are examples of tissue-associated taxa ubiquitously found across many coral species from widely separated locations (Bayer et al. 2013; Ainsworth et al. 2015; Roder et al. 2015; Hernandez-Agreda et al. 2016; Neave et al. 2016a,b; Certner and Vollmer 2018). Endosymbiotic and episymbiotic tissue-associated bacterial communities have been characterized by coupling laser microdissection and next-generation sequencing. The orders Rhizobiales, Caulobacterales, and Burkholderiales have been identified in association with Symbiodiniaceae, while members of the family Endozoicomonadaceae, and orders Rickettisales and Rhodobacterales have been identified in association with coral host cells (Ainsworth et al. 2015; Neave et al. 2016b, 2017). To date, several genomes of coral-associated *Endozoicomonas* are available (Neave et al. 2014, 2017; Ding et al. 2016), as well as protocols for isolation and culture (Pogoreutz and Voolstra 2018).

The coral skeleton contains a rich microbial community, which is often dominated by a diversity of cyanobacteria (Yamazaki et al. 2008). Some of these cyanobacteria have been cultured and characterized, such as *Plectonema terebrans, Mastigocoleus testarum,* and *Halomicronema excentricum.* Recent studies revealed that while skeleton-associated cyanobacteria occur at low relative abundances, they can be highly diverse (Marcelino and Verbruggen 2016). Other bacterial functional groups, such as anoxygenic phototrophic bacteria and anaerobic green sulphur bacteria, have also been identified in coral skeletons by spectral signatures of bacteriochlorophylls and via next-generation sequencing (Magnusson et al. 2007; Ralph et al. 2007; Yang et al. 2016).

Besides such broad compartmentalization, localization of microbes within a host can provide information on putative functions. Fluorescence *in situ* hybridization (FISH) approaches have shown that in visibly healthy hosts, bacteria tend to form aggregates (Ainsworth et al. 2006; Bayer et al. 2013; Wada et al. 2016, 2019; Neave et al. 2017), which can contain one or more cell morphologies (Wada et al. 2019). In contrast, diseased and lesioned coral tissues appear "overgrown" and crowded by bacteria (Ainsworth et al. 2006; Wada et al. 2016). Two bacterial genera previously identified and imaged in coral tissues are particularly worthwhile mentioning: (1) the abundant *Endozoicomonas*, which aggregate in the gastroderm of *Stylophora pistillata* (Bayer et al. 2013; Neave et al. 2017) and *Pocillopora verrucosa* (Neave et al. 2017); and (2) the putative parasite *Candidatus* Aquarickettsia rohweri, which is present in the mucocytes of the coral ectoderm (Klinges et al. 2019). This genus lacks genes for nitrogen metabolism and the synthesis of most sugars but maintains the genetic machinery for sensing and responding to extracellular nitrogen, as well as a complete type IV secretion system, collectively suggesting that *Ca*. A. rohweri may be a true coral parasite (Klinges et al. 2019).

7.4.2   Diversity of coral-associated bacteria and interspecies interactions

Studies on coral-associated microbes have mostly focused on bacterial diversity associated with coral tissues. Coral bacterial communities are typically diverse, uneven, and consist of hundreds to thousands of bacterial taxa (Rohwer et al. 2002; Bourne et al. 2008; Ziegler et al. 2016). Structure and composition of bacterial communities readily respond to the host's environment, that is, reflect environmental gradients, fluctuations, and habitat suitability (Thurber et al. 2009; Morrow et al. 2015; Roder et al. 2015; Ziegler et al. 2017). In adult corals, bacterial community shifts due to environmental changes can occur within a few days (Thurber et al. 2009) and in some cases within hours (Ziegler et al. 2017), with older colonies exhibiting delayed microbiome shifts, possibly due to differences in microbial composition, bacteria–bacteria interactions, or host energetics (Sweet et al. 2017).

The tissue-associated *Endozoicomonas* commonly display substantial changes in relative abundance depending on coral or reef health (Bayer et al. 2013; Roder et al. 2015; Neave et al. 2016a). The relative abundance of this genus is usually high in healthy corals and on reefs with high coral cover, but low on degraded reefs and in stressed, bleached, or diseased corals (Bourne et al. 2008; Meyer et al. 2014; Morrow et al. 2015). It was hence suggested that *Endozoicomonas* may be important for coral holobiont health, but its potential roles have yet to be identified. Increases in abundance of opportunistic bacteria and putative pathogens, such as Vibrionaceae, Rhodobacteraceae, or Flavobacteriaceae, often occur in chronically or severely stressed corals (Cárdenas et al. 2012; Roder et al. 2014a,b; Ziegler et al. 2016; Gignoux-Wolfsohn et al. 2017; Certner and Vollmer 2018). Stress-associated bacterial community shifts in corals are also reflected in the functional gene repertoire of the microbiome, showing increases in abundance of genes involved in virulence, stress resistance, or sulfur and nitrogen metabolism (Thurber et al. 2009).

Interactions between bacteria exert essential selective forces that sculpt coral microbial assemblages. For instance, the production of antibacterial molecules is one of the most important mechanisms used by native commensal bacteria for shaping the diversity of coral-associated prokaryotes and controlling the presence of pathogens (Sweet et al. 2011). Antibacterial activity has been demonstrated in numerous bacterial taxa (Ritchie 2006; Nissimov et al. 2009; Rypien et al. 2010; Kvennefors et al. 2012; Pereira et al. 2017) and coral extracts (Kelman et al. 2006; Gochfeld and Aeby 2008), but only a few compounds have been identified to date (Raina et al. 2016). Other bacteria–bacteria interactions also occur through quorum sensing (QS), which allows cells to communicate and synchronize gene expression in a concerted density-dependent manner to coordinate population behaviors. A broad spectrum of coral-associated bacteria, including some known coral pathogens, rely on QS to control colonization, virulence, and extracellular enzyme production (Tait et al. 2010; Alagely et al. 2011; de O Santos et al. 2011; Golberg et al. 2011). Given the importance of QS in the induction of virulence-related traits in opportunistic and pathogenic bacteria, many coral-associated commensal bacteria are capable of disrupting QS circuits as a strategy to prevent and mitigate pathogen invasion (Golberg et al. 2013), and some of these taxa might be critical for our understanding of QS in the context of coral diseases (Zimmer et al. 2014; Meyer et al. 2016; Certner and Vollmer 2018).

7.4.3   Acquisition of bacterial associates and their roles in early coral life-stages

Coral sexual reproduction occurs either through spawning where gametes are released into the water column, or through internally fertilizing gametes and brooding larvae inside the coral polyp (Fadlallah 1983). In spawning corals, the establishment of coral–bacteria symbioses is widely believed to happen horizontally (i.e., through acquisition of symbionts from the environment) during the coral pelagic larval phase, or even after settlement and metamorphosis (Apprill et al. 2009; Sharp et al. 2010). Conversely, in brooders, bacteria can be vertically transmitted (i.e., directly passed from the parent colony to the planula before its release). Bacterial communities exhibit dynamic changes between the different ontogenetic stages of coral development (Damjanovic et al. 2019; Epstein et al. 2019), likely reflecting a succession of microbial functions relevant for the holobiont (Bernasconi et al. 2019) and potentially following a winnowing process (Epstein et al. 2019).

A broad phylogenetic range of bacteria can be vertically or horizontally acquired. The most common associates of gametes, embryos, and larvae of brooders and spawners include the genera *Roseobacter*, *Marinobacter*, *Alteromonas, Vibrio, Bradyrhizobia,* and *Endozoicomonas. Roseobacter-* and *Alteromonas*-affiliated sequences are consistently found in very early developmental stages of several coral species (Apprill et al. 2009; Ceh et al. 2012, 2013; Sharp et al. 2012). These taxa are metabolically diverse and have been linked with the production of antibiotic compounds to counter coral pathogens (Piekarski et al. 2009). Furthermore, *Alteromonas* spp., *Vibrio* spp., and the diazotroph *Bradyrhizobia* might provide biologically available nitrogen to coral larvae (Ceh et al. 2013; Lema et al. 2014). The transmission of potentially beneficial bacteria from parent coral colonies to gametes or early ontogenetic stages might be a mechanism to ensure microbial inheritance across generations. In addition, bacterial behavior is likely to play a role in microbiome establishment. Indeed, many environmentally acquired bacterial symbionts use chemotaxis—the ability to direct their movement towards or away from specific chemicals—to locate their hosts (Raina et al. 2019). Chemotaxis may be a particularly prevalent mechanism employed on reefs because the coral surface is characterized by strong gradients of organic compounds that can act as cues for microorganisms (Ochsenkühn et al. 2018). Several coral-associated bacterial families, such as Endozoicomonadaceae, Rhodobacteraceae, or Oceanospirillaceae, exhibit chemoattraction towards constituents of coral mucus (Tout et al. 2015). Chemotaxis and motility, however, can also enable coral pathogens, such as *Vibrio shiloi* and *V. coralliilyticus*, to locate and infect their hosts (Banin et al. 2001;

Meron et al. 2009) using chemical cues such as dimethylsulfoniopropionate (DMSP) present in the coral mucus (Garren et al. 2014).

Bacteria are also directly involved in the transition between pelagic and benthic life stages in the coral life cycle. Indeed, after a pelagic phase (typically ranging from a week to more than 100 days), coral larvae must attach to a suitable reef structure to metamorphose into colony-forming juveniles (Connolly and Baird 2010). Habitat-specific environmental cues, mainly produced by specific bacteria associated with crustose coralline algae (CCA), contribute to coral larval recruitment and metamorphosis (Harrington et al. 2004; Webster et al. 2004; Tebben et al. 2015). Clear shifts in the bacterial community structure of CCA occur following thermal stress, resulting in significant reduction of coral larvae recruitment (Webster et al. 2011). Similarly, antibiotic treatment of larval cultures reduced settlement rates, suggesting that the presence of certain bacteria is essential for settlement induction (Vermeij et al. 2009).

7.4.4 Coral probiotics

Understanding the dynamic interplay of coral-associated microbes in relation to the prevailing environment is deeply interconnected with the setup, maintenance, and inheritance of microbial symbiotic relationships. The underlying premise is that microbial associates can adapt quickly to the surrounding environment and contribute functions that support coral holobiont health and resilience (Reshef et al. 2006; Ziegler et al. 2017, 2019; Bang et al. 2018). In recent years, efforts have been channeled into coral "probiotics" applications. Their ultimate goal is to fast-track ecological adaptation to global climate change by designing physiologically augmented coral holobionts (Peixoto et al. 2017). The research field of coral probiotics includes the isolation and screening of native bacterial associates for functional genes beneficial to coral health, and subsequent physiological assays to determine holobiont performance after inoculation with putatively beneficial bacterial isolates (Rosado et al. 2018). Experimental inoculation with mixed consortia of native coral bacterial isolates harboring dinitrogen fixation (*nifH*), denitrification (*nirK*), and DMSP-degrading (*dmdA*) genes resulted in partial mitigation of coral bleaching compared to controls or corals challenged with the temperature-dependent pathogen *Vibrio coralliilyticus* (Rosado et al. 2018). Open questions to this line of research are the temporal stability of the observed beneficial effects, the underlying mechanistic nature, and the potential for application of coral probiotics at the reef scale.

7.4.5 Contribution of bacteria to holobiont nutrient cycling

The ubiquitous coral symbionts *Endozoicomonas* harbor large numbers of genes involved in amino acid synthesis and carbohydrate cycling, suggesting its involvement in holobiont nutrient cycling (Neave et al. 2017), with different strains potentially exhibiting a different genetic and metabolic makeup (Neave et al. 2017; Pogoreutz et al. 2018). Taxonomy-based functional inference was used recently and suggested a role of Endozoicomonadaceae in processes related to nitrate reduction in giant clams (Rossbach et al. 2019). As such, Endozoicomonadaceae may provide otherwise inaccessible nitrogen sources, including ammonia, to the coral host.

Nitrogen (N) cycling is a critical component of holobiont health (Cardini et al. 2014; Rädecker et al. 2015; Pogoreutz et al. 2017a). Most research on nitrogen cycling in the coral holobiont has focused on prokaryotic dinitrogen (N_2) fixation (Shashar et al. 1994; Lesser et al. 2007; Rädecker et al. 2014; Bednarz et al. 2017), while the assessment of other major nitrogen cycling pathways such as nitrification, denitrification, and ANAMMOX has only received marginal attention to date (Wafar et al. 1990; Tilstra et al. 2019), and is currently limited to describing the presence of functional genes in sequencing datasets (Wegley et al. 2007; Siboni et al. 2008; Neave et al. 2017). Prokaryotic N_2 fixation commonly occurs in reef-building corals, helping supply the holobiont with "new" bioavailable nitrogen (Lesser et al. 2007; Cardini et al. 2015; Benavides et al. 2017). The biologically fixed nitrogen is then assimilated by both, coral (Benavides et al. 2016; Bednarz et al. 2017) and Symbiodiniaceae (Lesser et al. 2007; Cardini et al. 2015; Pogoreutz et al. 2017a). Nitrogen assimilation rates, however, appear to depend on environmental nitrogen availability, highlighting the importance of integrating environmental context into the study of coral holobiont function. The particular role of N_2 fixation to coral heat stress (bleaching) remains to be determined. Elevated temperatures rapidly cause an increase in the relative abundance and activity of coral-associated N_2 fixers (Santos et al. 2014; Cardini et al. 2016). It was previously concluded that excess nitrogen supply has the potential to ameliorate the effects of heat stress caused by global warming in corals (Santos et al. 2014; Cardini et al. 2016). An increase in holobiont N_2 fixation, however, can shift the N:P ratio of dinoflagellate symbionts in Red Sea corals (Pogoreutz et al. 2017a), thereby destabilizing the coral–algae symbiosis, resulting in coral bleaching (Wiedenmann et al. 2012). Ultimately, whether increases in N_2 fixation during heat stress have beneficial or detrimental effects on coral holobiont health may likely be determined by the environmental context (e.g., ambient nutrient regime), host nutritional state, or heterotrophic capacity (Bednarz et al. 2017; Pogoreutz et al. 2017b), and will require further mechanistic studies considering all major nitrogen cycling pathways.

In addition, symbiotic interactions between corals and bacteria might involve the cycling of essential compounds, for instance vitamins. Indeed, the cnidarian host may rely on bacterial symbionts for the provision of cobalamin, which is required for methionine synthesis by both corals and Symbiodiniaceae (Robbins et al. 2019). Concentrations of cobalamin in the coral gastrovascular cavity (coelenteron) are up to 35 times higher than in surrounding reef waters (Agostini et al. 2009), strongly suggesting that the dense bacterial communities harbored in the gastrovascular cavity are producing this essential molecule. In addition, some species from the genus *Acropora* are lacking an essential enzyme to synthesize the amino acid cysteine (Shinzato et al. 2011), and likely rely on their associated microbes for its provision.

7.4.6   Archaea associated with the coral holobiont

Corals associate with a diversity of archaea including representatives from the Crenarchaeota and Euryarchaeota. Crenarchaeota of the class Thermoprotei often dominate the archaeal community, while most abundant euryarchaeal members are affiliated to the Marine Group II and Thermoplasma (Kellogg 2004; Siboni et al.

2008). Interestingly, SML-associated archaeal sequences are most similar to obligate and facultative anaerobic and uncultivated archaea from anoxic environments, suggesting anaerobic microniches within the SML (Kellogg 2004). In terms of absolute abundance, archaeal cell numbers can comprise up to half of the prokaryotic community with an average of $>10^7$cells/cm^2 on the surface of *Porites astreoides* colonies (Wegley et al. 2004).

The diversity of ammonia-oxidizing archaea (AOA) has also been evaluated in coral tissues by amplifying the *amoA* gene encoding the alpha-subunit of the ammonia monooxygenase (Beman et al. 2007; Siboni et al. 2008). In addition, AOA are suggested to be less host-specific and more geographically dependent (Siboni et al. 2012). The presence of *amo* genes in coral-associated archaea was also supported by an integrated genomic approach of Thaumarchaeota genomes assembled from *Porites lutea* metagenomes (Robbins et al. 2019). This study revealed the presence of other relevant key metabolic pathways, including the reductive tricarboxylic acid cycle, cobalamin synthesis, and taurine dioxygenase in the Thaumarchaeota genomes, suggesting these symbionts may contribute to the host's demand for essential vitamins and carbon metabolism (Robbins et al. 2019).

7.4.7   Protists and fungi associated with the coral holobiont

Two photosynthetic alveolates, *Chromera velia* (Moore et al. 2008) and *Vitrella brassicaformis* (Oborník et al. 2012), which are the closest free-living relatives of the large parasitic phylum Apicomplexa, are protists commonly associated with corals worldwide (Janouškovec et al. 2013). A recent transcriptomic study revealed that the coral host response to *C. velia* inoculation was similar to that of a parasite or pathogen infection in vertebrates (Mohamed et al. 2018). This suggests that *C. velia*, despite its photoautotrophic capabilities, is not involved in mutualistic interactions with corals, but rather parasitic or commensal (Mohamed et al. 2018). In addition to these two alveolates, the presence of apicomplexans in coral tissues has been reported for the past 30 years (Upton and Peters 1986; Toller et al. 2002; Šlapeta and Linares 2013; Clerissi et al. 2018) and it was recently revealed that a single apicomplexan lineage is ubiquitously associated with corals, and might be the second most abundant microeukaryote group (after Symbiodiniaceae) associated with coral tissues (Janouškovec et al. 2012; Kwong et al. 2019). Although the nature of the association between these "corallicolids" and the coral host remains unknown, their genomes lack all genes encoding for photosystem proteins, but retained the four ancestral genes involved in chlorophyll biosynthesis (Kwong et al. 2019).

Deeper in the skeleton, endolithic protist algae can form dense bands visible to the unaided eye below the tissues of many coral species and are often dominated by the filamentous green algae *Ostreobium* spp. (Siphonales, Chlorophyta) (Kornmann and Sahling 1980). Recent molecular studies have revealed the astonishing genetic diversity of this group, with up to 80 taxonomic units at the near-species level (Marcelino and Verbruggen 2016; Marcelino et al. 2017, 2018; Verbruggen et al. 2017). These filamentous algae colonize the skeleton of coral juveniles early in their development (Massé et al. 2018) and can interact with the coral tissue through transfers of photosynthates (Schlichter et al. 1995; Fine and Loya 2002; Pernice et al.

2019). High-throughput amplicon sequencing has also revealed the presence of other, less abundant, boring green microalgae closely related to *Phaeophila*, *Bryopsis*, *Chlorodesmis*, *Cladophora*, *Pseudulvella*, and red algae from the Bangiales order in coral skeletons (Marcelino and Verbruggen 2016).

Fungi are prevalent in corals and have been well studied in the coral skeleton where they penetrate the calcium carbonate microstructures and ultimately interact with *Ostreobium* cells (Le Campion-Alsumard et al. 1995). Along with endolithic algae, fungi are present in the newly deposited coral skeleton (Bentis et al. 2000; Golubic et al. 2005), where they exhibit rapid growth to match skeletal accretion (Le Campion-Alsumard et al. 1995). Fungi were the most abundant microorganisms in the *Porites astreoides* metagenome, contributing to 38% of the microbial sequences (Wegley et al. 2007). These fungi belonged mainly to the phylum Ascomycota, but also members of Basidiomycota and Chytridiomycota were detected. Based on their genomic potential, these organisms might play a role in nitrogen recycling through the reduction of nitrate and nitrite to ammonia and subsequent ammonia assimilation (Wegley et al. 2007).

7.5   Summary and Outlook

The coral–dinoflagellate symbiosis is a well-studied system with high potential for understanding mechanisms of host–microbe interactions, controls, and co-evolution. The functional importance of bacteria, archaea, and other protists in this complex system has only recently started to emerge and the resulting picture is that of a holobiont where all partners depend on and interact with each other. These complex interdomain relationships are orchestrated and regulated by immune systems. It is therefore important to adopt a two-pronged approach, elucidating the immune responses of each individual partner (i.e., coral host, Symbiodiniaceae, bacteria, and other microbes), but also gaining a greater understanding of immunity at the holobiont scale. To elucidate this complex network of interactions, the Aiptasia model system may help to functionally interrogate the mechanistic underpinnings of the ecology, stress resilience, and environmental adaptation of cnidarian holobionts. For instance, the ability of genetic and microbiome manipulation of Aiptasia allows for detailed studies into the contribution of specific genes or microbes.

It is understood that productivity, structural complexity, and biodiversity of coral reef ecosystems critically depend on healthy coral holobionts, which in turn are linked to the diversity and identity of the associated microbes. This implies the need for a radically different approach to address the key questions of coral health and resilience in the face of climate change: one must examine the complexity of the coral holobiont in its entirety, that is, considering the diversity and function of all associated organisms. Gaining a holistic understanding of the biology and functioning of coral holobionts is of prime importance, given that coral reefs are at the brink of ecological collapse due to the effects of climate change and local stressors. New strategies to mitigate reef loss are now more important than ever. One such strategy may lie in the manipulation of holobiont immunity or associated microbes to enable ecological adaptation at a rate and scale that matches the pace of environmental change.

References

Agostini, S., Suzuki, Y., Casareto, B.E., Nakano, Y., Michio, H., and N. Badrun. 2009. Coral symbiotic complex: Hypothesis through vitamin B12 for a new evaluation. *Galaxea, J. Coral Reef Stud.* 11:1–11. https://doi.org/10.3755/galaxea.11.1.

Ainsworth, T., Krause, L., Bridge, T., Torda, G., Raina, J-B., Zakrzewski, M. et al. 2015. The coral core microbiome identifies rare bacterial taxa as ubiquitous endosymbionts. *ISME J.* 9:2261–2274. https://doi.org/10.1038/ismej.2015.39.

Ainsworth, T.D., Fine, M., Blackall, L.L., and O. Hoegh-Guldberg. 2006. Fluorescence *In Situ* Hybridization and spectral imaging of coral-associated bacterial communities. *Appl. Environ. Microbiol.* 72:3016–3020. https://doi.org/10.1128/AEM.72.4.3016-3020.2006.

Alagely, A., Krediet, C.J., Ritchie, K.B., and M. Teplitski. 2011. Signaling-mediated cross-talk modulates swarming and biofilm formation in a coral pathogen *Serratia marcescens*. *ISME J.* 5:1609–1620. https://doi.org/10.1038/ismej.2011.45.

Apprill, A., Marlow, H.Q., Martindale, M.Q., and M.S. Rappé. 2009. The onset of microbial associations in the coral *Pocillopora meandrina*. *ISME J.* 3:685–699. https://doi.org/10.1038/ismej.2009.3.

Aranda, M., Li, Y., Liew, Y.J., Baumgarten, S., Simakov, O., Wilson, M. et al. 2016. Genomes of coral dinoflagellate symbionts highlight evolutionary adaptations conducive to a symbiotic lifestyle. *Sci. Rep.* 6:39734. https://doi.org/10.1038/srep39734.

Bang, C., Dagan, T., Deines, P., Dubilier, N., Duschl, W.J., Fraune, S. et al. 2018. Metaorganisms in extreme environments: do microbes play a role in organismal adaptation? *Zoology* 127:1–19. https://doi.org/10.1016/j.zool.2018.02.004.

Banin, E., Israely, T., Fine, M., Loya, Y., and E. Rosenberg. 2001. Role of endosymbiotic zooxanthellae and coral mucus in the adhesion of the coral-bleaching pathogen *Vibrio shiloi* to its host. *FEMS Microbiol. Lett.* 199:33–37. https://doi.org/10.1111/j.1574-6968.2001. tb10647.x.

Baumgarten, S., Simakov, O., Esherick, L.Y., Liew, Y.J., Lehnert, E.M., Michell, C.T. et al. 2015. The genome of *Aiptasia*, a sea anemone model for coral biology. *Proc. Natl. Acad. Sci.* 112:11893–11898. https://doi.org/10.1073/pnas.1513318112.

Bay, L.K., Cumbo, V.R., Abrego, D., Kool, J.T., Ainsworth, T.D., and B.L. Willis. 2011. Infection dynamics vary between *Symbiodinium* types and cell surface treatments during establishment of endosymbiosis with coral larvae. *Diversity* 3:356–374. https://doi.org/10.3390/d3030356.

Bayer, T., Neave, M.J., Alsheikh-Hussain, A., Aranda, M., Yum, L.K., Mincer, T. et al. 2013. The microbiome of the Red Sea coral *Stylophora pistillata* is dominated by tissue-associated *Endozoicomonas* bacteria. *Appl. Environ. Microbiol.* 79:4759–4762. https://doi.org/10.1128/AEM.00695-13.

Bednarz, V.N., Grover, R., Maguer, J.F., Fine, M., and C. Ferrier-Pagès. 2017. The assimilation of diazotroph-derived nitrogen by scleractinian corals depends on their metabolic status. *mBio* 8:e02058–e02016. https://doi.org/10.1128/mBio.02058-16.

Beman, J.M., Roberts, K.J., Wegley, L., Rohwer, F., and C.A. Francis. 2007. Distribution and diversity of archaeal ammonia monooxygenase genes associated with corals. *Appl. Environ. Microbiol.* 73:5642–5647. https://doi.org/10.1128/AEM.00461-07.

Benavides, M., Bednarz, V.N., and C. Ferrier-Pagès. 2017. Diazotrophs: Overlooked key players within the coral symbiosis and tropical reef ecosystems? *Front. Mar. Sci.* 4:10. https://doi.org/10.3389/fmars.2017.00010.

Benavides, M., Houlbreque, F., Camps, M., Lorrain, A., Grosso, O., and S. Bonnet. 2016. Diazotrophs: A non-negligible source of nitrogen for the tropical coral *Stylophora pistillata*. *J. Exp. Biol.* 219:2608–2612. https://doi.org/10.1242/jeb.139451.

Bentis, C.J., Kaufman, L., and S. Golubic. 2000. Endolithic fungi in reef-building corals (Order: Scleractinia) are common, cosmopolitan, and potentially pathogenic. *Biol. Bull.* 198:254–260. https://doi.org/10.2307/1542528.

Berkelmans, R., and M.J.H. van Oppen. 2006. The role of zooxanthellae in the thermal tolerance of corals: A 'nugget of hope' for coral reefs in an era of climate change. *Proc. R. Soc. B. Biol. Sci.* 273:2305–2312. https://doi.org/10.1098/rspb.2006.3567.

Bernasconi, R., Stat, M., Koenders, A., Paparini, A., Bunce, M., and M.J. Huggett. 2019. IEstablishment of coral-bacteria symbioses reveal changes in the core bacterial community with host ontogeny. *Front. Microbiol.* 10:1529. https://doi.org/10.3389/fmicb.2019.01529.

Berthelier, J., Schnitzler, C.E., Wood-Charlson, E.M., Poole, A.Z., Weis, V.M., and O. Detournay. 2017. Implication of the host TGFβ pathway in the onset of symbiosis between larvae of the coral *Fungia scutaria* and the dinoflagellate *Symbiodinium* sp. (clade C1f). *Coral Reefs* 36:1263–1268. http://dx.doi.org/10.1007/s00338-017-1621-6.

Bourne, D., Iida, Y., Uthicke, S., and C. Smith-Keune. 2008. Changes in coral-associated microbial communities during a bleaching event. *ISME J.* 2:350–363. https://doi.org/10.1038/ismej.2007.112.

Cantin, N.E., Van Oppen, M.J.H., Willis, B.L., Mieog, J.C., and A.P. Negri. 2009. Juvenile corals can acquire more carbon from high-performance algal symbionts. *Coral Reefs* 28:405–414. https://doi.org/10.1007/s00338-009-0478-8.

Cárdenas, A., Rodriguez-R, L.M., Pizarro, V., Cadavid, L.F., and C. Arévalo-Ferro. 2012. Shifts in bacterial communities of two Caribbean reef-building coral species affected by white plague disease. *ISME J.* 6:502–512. https://doi.org/10.1038/ismej.2011.123.

Cardini, U., Bednarz, V.N., Foster, R.A., and C. Wild. 2014. Benthic N_2 fixation in coral reefs and the potential effects of human-induced environmental change. *Ecol. Evol.* 4:1706–1727. https://doi.org/10.1002/ece3.1050.

Cardini, U., Bednarz, V.N., Naumann, M.S., Van Hoytema, N., Rix, L., Foster, R.A. et al. 2015. Functional significance of dinitrogen fixation in sustaining coral productivity under oligotrophic conditions. *Proc. R. Soc. B. Biol. Sci.* 282:20152257. https://doi.org/10.1098/rspb.2015.2257.

Cardini, U., Van Hoytema, N., Bednarz, V.N., Rix, L., Foster, R.A., Al-Rshaidat, M.M.D. et al. 2016. Microbial dinitrogen fixation in coral holobionts exposed to thermal stress and bleaching. *Environ. Microbiol.* 18:2620–2633. https://doi.org/10.1111/1462-2920.13385.

Ceh, J., Raina, J-B., Soo, R.M., van Keulen, M., and D.G. Bourne. 2012. Coral-bacterial communities before and after a coral mass spawning event on Ningaloo Reef. *PLOS ONE* 7:3–9. https://doi.org/10.1371/journal.pone.0036920.

Ceh, J., van Keulen, M., and D.G. Bourne. 2013. Intergenerational transfer of specific bacteria in corals and possible implications for offspring fitness. *Microb. Ecol.* 65:227–231. https://doi.org/10.1007/s00248-012-0105-z.

Certner, R.H., and S.V. Vollmer. 2018. Inhibiting bacterial quorum sensing arrests coral disease development and disease-associated microbes. *Environ. Microbiol.* 20:645–657. https://doi.org/10.1111/1462-2920.13991.

Chen, M., Cheng, Y., Hong, M., and L. Fang. 2004. Molecular cloning of Rab5 (ApRab5) in *Aiptasia pulchella* and its retention in phagosomes harboring live zooxanthellae. *Biochem. Biophys. Res. Comm.* 324:1024–1033. https://doi.org/10.1016/j.bbrc.2004.09.151.

Chen, M., Cheng, Y., Sung, P., Kuo, C., and L. Fang. 2003. Molecular identification of Rab7 (ApRab7) in *Aiptasia pulchella* and its exclusion from phagosomes harboring zooxanthellae. *Biochem. Biophys. Res. Comm.* 308:586–595. https://doi.org/10.1016/s0006-291x(03)01428-1.

Cho, D.H., Ramanan, R., Kim, B.H., Lee, J., Kim, S., Yoo, C. et al. 2013. Novel approach for the development of axenic microalgal cultures from environmental samples. *J. Phycol.* 49:802–810. https://doi.org/10.1111/jpy.12091.

Clerissi, C., Brunet, S., Vidal-Dupiol, J., Adjeroud, M., Lepage, P., Guillou, L. et al. 2018. Protists within corals: The hidden diversity. *Front. Microbiol.* 9:2043. https://dx.doi.org /10.3389%2Ffmicb.2018.02043.

Colley, N., and R. Trench. 1983. Selectivity in phagocytosis and persistence of symbiotic algae by the scyphistoma stage of the jellyfish *Cassiopeia xamachana*. *Proc. R. Soc. B. Biol. Sci.* 219d:61–82. https://doi.org/10.1098/rspb.1983.0059.

Connolly, S.R., and A.H. Baird. 2010. Estimating dispersal potential for marine larvae: Dynamic models applied to scleractinian corals. *Ecology* 91:3572–3583. https://doi. org/10.1890/10-0143.1.

Cui, G., Liew, Y.J., Li, Y., Kharbatia, N., Zahran, N.I., Emwas, A-H. et al. 2019. Host-dependent nitrogen recycling as a mechanism of symbiont control in Aiptasia. *PLOS Genet.* 15:e1008189. https://doi.org/10.1371/journal.pgen.1008189.

Cunning, R., Bay, R.A., Gillette, P., Baker, A.C., and N. Traylor-Knowles. 2018. Comparative analysis of the *Pocillopora damicornis* genome highlights role of immune system in coral evolution. *Sci. Rep.* 8:16134. https://doi.org/10.1038/s41598-018-34459-8.

Cunning, R., Muller, E.B., Gates, R.D., and R.M. Nisbet 2017. A dynamic bioenergetic model for coral-*Symbiodinium* symbioses and coral bleaching as an alternate stable state. *J. Theor. Biol.* 431:49–62. https://doi.org/10.1016/j.jtbi.2017.08.003.

Damjanovic, K., van Oppen, M.J.H., Menéndez, P., and L.L. Blackall. 2019. Experimental inoculation of coral recruits with marine bacteria indicates scope for microbiome manipulation in *Acropora tenuis* and *Platygyra daedalea*. *Front. Microbiol.* 10:1702. https://doi.org/10.3389/fmicb.2019.01702.

Darwin, C. 1842. *The Structure and Distribution of Coral Reefs.* Smith, Elder and Co.

Davy, S.K., Allemand, D., and V.M. Weis. 2012. Cell biology of cnidarian-dinoflagellate symbiosis. *Microbiol. Mol. Biol. Rev.* 76:229–261. https://doi.org/10.1128/MMBR.05014-11.

de O Santos, E., Alves, N.. Jr., Dias, G.M., Mazotto, A.M., Vermelho, A., Vora, G.J. et al. 2011. Genomic and proteomic analyses of the coral pathogen *Vibrio coralliilyticus* reveal a diverse virulence repertoire. *ISME J.* 5:1471–1483. https://doi.org/10.1038/ismej.2011.19.

Detournay, O., Schnitzler, C.E., Poole, A., and V.M. Weis. 2012. Regulation of cnidarian-dinoflagellate mutualisms: Evidence that activation of a host TGFβ innate immune pathway promotes tolerance of the symbiont. *Dev. Comp. Immunol.* 38:525–537. https:// doi.org/10.1016/j.dci.2012.08.008.

Ding, J.Y., Shiu, J.H., Chen, W.M., Chiang, Y.R., and S.L. Tang. 2016. Genomic insight into the host-endosymbiont relationship of *Endozoicomonas montiporae* CL-33 T with its coral host. *Front. Microbiol.* 7:251. https://doi.org/10.3389/fmicb.2016.00251.

Douglas, A.E. 2003. Coral bleaching - How and why? *Mar. Pollut. Bull.* 46:385–392. https:// doi.org/10.1016/S0025-326X(03)00037-7.

Dunn, S.R., Schnitzler, C.E., and V.M. Weis. 2007. Apoptosis and autophagy as mechanisms of dinoflagellate symbiont release during cnidarian bleaching: Every which way you lose. *Proc. R. Soc. B. Biol. Sci.* 274:3079–3085. https://doi.org/10.1098/rspb.2007.0711.

Epstein, H.E., Torda, G., Munday, P.L., and M.J.H. Van Oppen. 2019. Parental and early life stage environments drive establishment of bacterial and dinoflagellate communities in a common coral. *ISME J.* 13:1635–1638. https://doi.org/10.1038/s41396-019-0358-3.

Ezzat, L., Maguer, J-F., Grover, R., and C. Ferrier-Pagès. 2015. New insights into carbon acquisition and exchanges within the coral – dinoflagellate symbiosis under NH_4^+ and NO_3^- supply. *Proc. R. Soc. B. Biol. Sci.* 282:20150610. https://doi.org/10.1098/rspb.2015.0610.

Fadlallah, Y.H. 1983. Coral reefs sexual reproduction, development and larval biology in scleractinian corals: A review. *Coral Reefs* 2:129–150. https://doi.org/10.1007/ BF00336720.

Falkowski, P.G., Dubinsky, Z., Muscatine, L., and L. McCloskey. 1993. Population control in symbiotic corals. *Bioscience* 43:606–611. https://doi.org/10.2307/1312147.

Ferrer, L.M., and A.M. Szmant. 1988. Nutrient regeneration by the endolithic community in coral skeletons. *Proceedings of the 6th International Coral Reef Symposium* 2:1–4.

Fine, M., and Y. Loya. 2002. Endolithic algae: an alternative source of photoassimilates during coral bleaching. *Proc. R. Soc. B. Biol. Sci.* 269:1205–1210. https://doi.org/10.1098/rspb.2002.1983.

Fisher, R., O'Leary, R.A., Low-Choy, S., Mengersen, K., Knowlton, N., Brainard, R.E. et al. 2015. Species richness on coral reefs and the pursuit of convergent global estimates. *Curr. Biol.* 25:500–505. https://doi.org/10.1016/j.cub.2014.12.022.

Freudenthal, H.D. 1962. *Symbiodinium* gen. nov. and *Symbiodinium microadriaticum* sp. nov., a zooxanthella: Taxonomy, life cycle, and morphology. *J. Protozool.* 9:45–52. https://doi.org/10.1111/j.1550-7408.1962.tb02579.x.

Garren, M., Son, K., Raina, J-B., Rusconi, R., Menolascina, F., Shapiro, O.H. et al. 2014. A bacterial pathogen uses dimethylsulfoniopropionate as a cue to target heat-stressed corals. *ISME J.* 8:999–1007. https://doi.org/10.1038/ismej.2013.210.

Gates, R.D., Baghdasarian, G., and L. Muscatine. 1992. Temperature stress causes host cell detachment in symbiotic cnidarians: implications for coral bleaching. *Biol. Bull.* 182:324–332. https://doi.org/10.2307/1542252.

Gignoux-Wolfsohn, S.A., Aronson, F.M., and S.V. Vollmer. 2017. Complex interactions between potentially pathogenic, opportunistic, and resident bacteria emerge during infection on a reef-building coral. *FEMS Microbiol. Ecol.* 93:fix080. https://doi.org/10.1093/femsec/fix080.

Glasl, B., Herndl, G.J., and P.R. Frade. 2016. The microbiome of coral surface mucus has a key role in mediating holobiont health and survival upon disturbance. *ISME J.* 10:2280–2292. https://doi.org/10.1038/ismej.2016.9.

Glynn, P.W. 1983. Extensive "bleaching" and death of reef corals on the Pacific coast of Panama. *Environ. Conserv.* 10:149–154. https://doi.org/10.1017/S0376892900012248.

Gochfeld, D.J., and G.S. Aeby. 2008. Antibacterial chemical defenses in Hawaiian corals provide possible protection from disease. *Mar. Ecol. Prog. Ser.* 362:119–128. https://doi.org/10.3354/meps07418.

Golberg, K., Eltzov, E., Shnit-Orland, M., Marks, R.S., and A. Kushmaro. 2011. Characterization of quorum sensing signals in coral-associated bacteria. *Microb. Ecol.* 61:783–792. https://doi.org/10.1007/s00248-011-9848-1.

Golberg, K., Pavlov, V., Marks, R.S., and A. Kushmaro. 2013. Coral-associated bacteria, quorum sensing disrupters, and the regulation of biofouling. *Biofouling* 29:669–682. https://doi.org/10.1080/08927014.2013.796939.

Golubic, S., Radtke, G., and T. Le Campion-Alsumard. 2005. Endolithic fungi in marine ecosystems. *Trends Microbiol.* 13:229–235. https://doi.org/10.1016/j.tim.2005.03.007.

Grottoli, A.G., Rodrigues, L.J., and J. E. Palardy. 2006. Heterotrophic plasticity and resilience in bleached corals. *Nature* 440:1186–1189. https://doi.org/10.1038/nature04565.

Hadaidi, G., Röthig, T., Yum, L.K., Ziegler, M., Arif, C., Roder, C. et al. 2017. Stable mucus-associated bacterial communities in bleached and healthy corals of *Porites lobata* from the Arabian Seas. *Sci. Rep.* 7:45362. https://doi.org/10.1038/srep45362.

Harrington, L., Fabricius, K., De'Ath, G., and A. Negri. 2004. Recognition and selection of settlement substrata determine post-settlement survival in corals. *Ecology* 85:3428–3437. https://doi.org/10.1890/04-0298.

Hatcher, B.G. 1990. Coral reef primary productivity. A hierarchy of pattern and process. *Trends Ecol. Evol.* 5:149–155. https://doi.org/10.1016/0169-5347(90)90221-X.

Hernandez-Agreda, A., Leggat, W., Bongaerts, P., and T. Ainsworth. 2016. The microbial signature provides insight into the mechanistic basis of coral success across reef habitats. *mBio* 7:e00560–e00516. https://doi.org/10.1128/mBio.00560-16.

Hoegh-Guldberg, O. 1999. Climate Change, coral bleaching and the future of the world's coral reefs. *Mar. Freshw. Res.* 50:839–866. https://doi.org/10.1071/MF99078.

Hoegh-Guldberg, O. 2011. Coral reef ecosystems and anthropogenic climate change. *Reg. Environ. Chang.* 11:215–227. https://doi.org/10.1007/s10113-010-0189-2.

Hughes, T.P., Anderson, K.D., Connolly, S.R., Heron, S.F., Kerry, J.T., Lough, J.M. et al. 2018a. Spatial and temporal patterns of mass bleaching of corals in the Anthropocene. *Science* 359:80–83. https://doi.org/10.1126/science.aan8048.

Hughes, T.P., Baird, A.H., Bellwood, D.R., Card, M., Connolly, S.R., Folke, C. et al. 2003. Climate change, human impacts, and the resilience of coral reefs. *Science* 301:929–933. https://doi.org/10.1126/science.1085046.

Hughes, T.P., Kerry, J.T., Álvarez-Noriega, M., Álvarez-Romero, J.G., Anderson, K.D., Baird, A.H. et al. 2017. Global warming and recurrent mass bleaching of corals. *Nature* 543:373–377. https://doi.org/10.1038/nature21707.

Hughes, T.P., Kerry, J.T., and T. Simpson. 2018b. Large-scale bleaching of corals on the Great Barrier Reef. *Ecology* 99:501. https://doi.org/10.1002/ecy.2092.

Hume, B.C.C., Voolstra, C.R., Arif, C., D'Angelo, C., Burt. J.A., Eyal, G. et al. 2016. Ancestral genetic diversity associated with the rapid spread of stress-tolerant coral symbionts in response to holocene climate change. *Proc. Natl. Acad. Sci.* 113:4416–4421. https://doi.org/10.1073/pnas.1601910113.

IPCC. 2018. *Summary for Policymakers*. In: Global warming of 1.5°C. An IPCC Special Report on the impacts of global warming of 1.5°C above pre-industrial levels and related global greenhouse gas emission pathways, in the context of strengthening the global response to the threat of climate change, sustainable development, and efforts to eradicate poverty. https://www. ipcc. ch/sr15/. Accessed 10/2019.

Janouškovec, J., Horák, A., Barott, K.L., Rohwer, F.L., and P.J. Keeling. 2012. Global analysis of plastid diversity reveals apicomplexan-related lineages in coral reefs. *Curr. Biol.* 22:R518–R519. https://doi.org/10.1016/j.cub.2012.04.047.

Janouškovec, J., Horák, A., Barott, K.L., Rohwer, F.L., and P.J. Keeling. 2013. Environmental distribution of coral-associated relatives of apicomplexan parasites. *ISME J.* 7:444–447. https://doi.org/10.1038/ismej.2012.129.

Jaspers, C., Fraune, S., Arnold, A.E., Miller, D.J., Bosch, T.C.G., and C.R. Voolstra. 2019. Resolving structure and function of metaorganisms through a holistic framework combining reductionist and integrative approaches. *Zoology* 133:81–87. https://doi.org/10.1016/j.zool.2019.02.007.

Jones, A.M., Berkelmans, R., Van Oppen, M.J.H., Mieog, J.C., and W. Sinclair. 2008. A community change in the algal endosymbionts of a scleractinian coral following a natural bleaching event: Field evidence of acclimatization. *Proc. R. Soc. B. Biol. Sci.* 275:1359–1365. https://doi.org/10.1098/rspb.2008.0069.

Jones, R.J., Hoegh-Guldberg, O., Larkum, A.W.D., and U. Schreiber. 1998. Temperature-induced bleaching of corals begins with impairment of the CO_2 fixation mechanism in zooxanthellae. *Plant, Cell Environ.* 21:1219–1230. https://doi.org/10.1046/j.1365-3040.1998.00345.x.

Kellogg, C.A. 2004. Tropical Archaea: Diversity associated with the surface microlayer of corals. *Mar. Ecol. Prog. Ser.* 273:81–88. https://ui.adsabs.harvard.edu/link_gateway/2004MEPS..273...81K/doi:10.3354/meps273081.

Kelman, D., Kashman, Y., Rosenberg, E., Kushmaro, A., and Y. Loya. 2006. Antimicrobial activity of Red Sea corals. *Mar. Biol.* 149:357–363. https://doi.org/10.1007/s00227-005-0218-8.

Kitchen, S.A., Poole, A.Z., and V.M. Weis. 2017. Sphingolipid metabolism of a sea anemone is altered by the presence of dinoflagellate symbionts. *Biol. Bull.* 233:242–254. https://doi.org/10.1086/695846.

Kitchen, S.A., and V.M. Weis. 2017. The sphingosine rheostat is involved in the cnidarian heat stress response but not necessarily in bleaching. *J. Exp. Biol.* 220:1709–1720. https://doi.org/10.1242/jeb.153858.

Klinges, J.G., Rosales, S.M., McMinds, R., Shaver, E.C., Shantz, A.A., Peters, E.C. et al. 2019. Phylogenetic, genomic, and biogeographic characterization of a novel and ubiquitous marine invertebrate-associated Rickettsiales parasite, *Candidatus* Aquarickettsia rohweri, gen. nov., sp. nov. *ISME J.* 13:2938–2953. https://doi.org/10.1038/s41396-019-0482-0.

Koren, O., and E. Rosenberg. 2006. Bacteria associated with mucus and tissues of the coral *Oculina patagonica* in summer and winter. *Appl. Environ. Microbiol.* 72:5254–5259. https://doi.org/10.1128/AEM.00554-06.

Kornmann, P., and P. Sahling. 1980. *Ostreobium quekettii* (Codiales, Chlorophyta). *Helgoländer Meeresuntersuchungen* 34:115–122.

Krueger, T. 2017. Concerning the cohabitation of animals and algae – an English translation of K. Brandt's 1881 presentation "Ueber das Zusammenleben von Thieren und Algen". *Symbiosis* 71:167–174. https://doi.org/10.1007/s13199-016-0439-2

Kvennefors, E.C.E., Sampayo, E., Kerr, C., Vieira, G., Roff, G., and A.C. Barnes. 2012. Regulation of bacterial communities through antimicrobial activity by the coral holobiont. *Microb. Ecol.* 63:605–618. https://doi.org/10.1007/s00248-011-9946-0.

Kwong, W.K., del Campo, J., Mathur, V., Vermeij, M.J.A., and P.J. Keeling. 2019. A widespread coral-infecting apicomplexan with chlorophyll biosynthesis genes. *Nature* 568:103–107. https://doi.org/10.1038/s41586-019-1072-z.

LaJeunesse, T.C., Bhagooli, R., Hidaka, M., DeVantier, L., Done, T., Schmidt, G.W. et al. 2004. Closely related *Symbiodinium* spp. differ in relative dominance in coral reef host communities across environmental, latitudinal and biogeographic gradients. *Mar. Ecol. Prog. Ser.* 284:147–161.

LaJeunesse, T.C., Loh, W.K.W., Van Woesik, R., Hoegh-Guldberg, O., Schmidt, G.W., and W.K. Fitt. 2003. Low symbiont diversity in southern Great Barrier Reef corals, relative to those of the Caribbean. *Limnol. Oceanogr.* 48:2046–2054. https://doi.org/10.4319/lo.2003.48.5.2046.

LaJeunesse, T.C., Parkinson, J.E., Gabrielson, P.W., Jeong, H.J., Reimer, J.D., Voolstra, C.R. et al. 2018. Systematic revision of Symbiodiniaceae highlights the antiquity and diversity of coral endosymbionts. *Curr. Biol.* 28:2570–2580. https://doi.org/10.1016/j.cub.2018.07.008.

Lawson, C.A., Raina, J.B., Kahlke, T., Seymour, J.R., and D.J. Suggett. 2018. Defining the core microbiome of the symbiotic dinoflagellate, *Symbiodinium*. *Environ. Microbiol. Rep.* 10:7–11. https://doi.org/10.1111/1758-2229.12599.

Le Campion-Alsumard, T., Golubic, S., and K. Priess. 1995. Fungi in corals: symbiosis or disease? Interaction between polyps and fungi causes pearl-like skeleton biomineralization. *Mar. Ecol. Prog. Ser.* 117:137–147. https://ui.adsabs.harvard.edu/link_gateway/1995MEPS..117..137L/doi:10.3354/meps117137.

Lee, S.T.M., Davy, S.K., Tang, S.L., Fan, T.Y., and P.S. Kench. 2015. Successive shifts in the microbial community of the surface mucus layer and tissues of the coral *Acropora muricata* under thermal stress. *FEMS Microbiol. Ecol.* 91:fiv142. https://doi.org/10.1093/femsec/fiv142.

Lee, S.T.M., Davy, S.K., Tang, S.L., and P.S. Kench. 2016. Mucus sugar content shapes the bacterial community structure in thermally stressed *Acropora muricata*. *Front. Microbiol.* 7:371. https://doi.org/10.3389/fmicb.2016.00371.

Lema, K.A., Bourne, D.G., and B.L. Willis. 2014. Onset and establishment of diazotrophs and other bacterial associates in the early life history stages of the coral *Acropora millepora*. *Mol. Ecol.* 23:4682–4695. https://doi.org/10.1111/mec.12899.

Lesser, M.P. 1997. Oxidative stress causes coral bleaching during exposure to elevated temperatures. *Coral Reefs* 16:187–192. https://doi.org/10.1007/s003380050073.

Lesser, M.P., Falcón, L.I., Rodríguez-Román, A., Enríquez, S., Hoegh-Guldberg, O., and R. Iglesias-Prieto. 2007. Nitrogen fixation by symbiotic cyanobacteria provides a source of nitrogen for the scleractinian coral *Montastraea cavernosa*. *Mar. Ecol. Prog. Ser.* 346:143–152. https://doi.org/10.3354/meps07008.

Lin, S., Cheng, S., Song, B., Zhong, X., Lin, X., Li, W. et al. 2015. The *Symbiodinium kawagutii* genome illuminates dinoflagellate gene expression and coral symbiosis. *Science* 350:691–694. https://doi.org/10.1126/science.aad0408.

Little, A.F., Van Oppen, M.J.H., and B.L. Willis. 2004. Flexibility in algal endosymbioses shapes growth in reef corals. *Science* 304:1492–1494. https://doi.org/10.1126/science.1095733.

Magnusson, S.H., Fine, M., and M. Kühl. 2007. Light microclimate of endolithic phototrophs in the scleractinian corals *Montipora monasteriata* and *Porites cylindrica*. *Mar. Ecol. Prog. Ser.* 332:119–128. https://doi.org/10.3354/meps332119.

Mansfield, K.M., Carter, N.M., Nguyen, L., Cleves, P.A., Alshanbayeva, A., Williams, L.M. et al. 2017. Transcription factor NF-κB is modulated by symbiotic status in a sea anemone model of cnidarian bleaching. *Sci. Rep.* 7: https://doi.org/10.1038/s41598-017-16168-w.

Marcelino, V.R., Morrow, K.M., van Oppen, M.J.H., Bourne, D.G., and H. Verbruggen. 2017. Diversity and stability of coral endolithic microbial communities at a naturally high pCO$_2$ reef. *Mol. Ecol.* 26:5344–5357. https://doi.org/10.1111/mec.14268.

Marcelino, V.R., Van Oppen, M.J.H., and H. Verbruggen. 2018. Highly structured prokaryote communities exist within the skeleton of coral colonies. *ISME J.* 12:300–303. https://doi.org/10.1038/ismej.2017.164.

Marcelino, V.R., and H. Verbruggen. 2016. Multi-marker metabarcoding of coral skeletons reveals a rich microbiome and diverse evolutionary origins of endolithic algae. *Sci. Rep.* 6:31508. https://doi.org/10.1038/srep31508.

Massé, A., Domart-Coulon, I., Golubic, S., Duché, D., and A. Tribollet. 2018. Early skeletal colonization of the coral holobiont by the microboring Ulvophyceae *Ostreobium* sp. *Sci. Rep.* 8:2293. https://doi.org/10.1038/s41598-018-20196-5.

Meron, D., Efrony, R., Johnson, W.R., Schaefer, A.L., Morris, P.J., Rosenberg, E. et al. 2009. Role of flagella in virulence of the coral pathogen *Vibrio coralliilyticus*. *Appl. Environ. Microbiol.* 75:5704–5707. https://doi.org/10.1128/AEM.00198-09.

Meyer, J.L., Gunasekera, S.P., Scott, R.M., Paul, V.J., and M. Teplitski. 2016. Microbiome shifts and the inhibition of quorum sensing by Black Band Disease cyanobacteria. *ISME J.* 10:1204–1216. https://doi.org/10.1038/ismej.2015.184.

Meyer, J.L., Paul, V.J., and M. Teplitski. 2014. Community shifts in the surface microbiomes of the coral *Porites astreoides* with unusual lesions. *PLOS ONE* 9: e100316. https://doi.org/10.1371/journal.pone.0100316.

Moberg, F., and C. Folke. 1999. Ecological goods and services of coral reef ecosystems. *Ecol. Econ.* 29:215–233. https://doi.org/10.1016/S0921-8009(99)00009-9.

Mohamed, A.R., Cumbo, V., Harii, S., Shinzato, C., Chan, C.X., Ragan, M.A. et al. 2016. The transcriptomic response of the coral *Acropora digitifera* to a competent *Symbiodinium* strain: the symbiosome as an arrested early phagosome. *Mol. Ecol.* 5:3127–3141.

Mohamed, A.R., Cumbo, V.R., Harii, S., Shinzato, C., Chan, C.X., Ragan, M.A. et al. 2018. Deciphering the nature of the coral-*Chromera* association. *ISME J.* 12:776–790. https://doi.org/10.1038/s41396-017-0005-9.

Moore, R.B., Oborník, M., Janouškovec, J., Chrudimský, T., Vancová, M., Green, D.H. et al. 2008. A photosynthetic alveolate closely related to apicomplexan parasites. *Nature* 451:959–963. https://doi.org/10.1038/nature06635.

Morris, L.A., Voolstra, C.R., Quigley, K.M., Bourne, D.G., and L.K. Bay 2019. Nutrient availability and metabolism affect the stability of coral–Symbiodiniaceae symbioses. *Trends Microbiol.* 27:678–689. https://doi.org/10.1016/j.tim.2019.03.004.

Morrow, K.M., Bourne, D.G., Humphrey, C., Botté, E.S., Laffy, P., Zaneveld, J. et al. 2015. Natural volcanic CO$_2$ seeps reveal future trajectories for host-microbial associations in corals and sponges. *ISME J.* 9:894–908. https://doi.org/10.1038/ismej.2014.188.

Muscatine, L. 1990. The role of symbiotic algae in carbon and energy flux in reef corals. In: Dubinsky Z. (eds) *Ecosystems of the World: Coral Reefs*. Elsevier, Amsterdam, pp. 75–87.

Muscatine, L., Falkowski, P.G., Dubinsky, P.A., Cook, C.A., and L.R.R. McCloskey. 1989. The effect of external nutrient resources on the population dynamics of zooxanthellae in a reef coral. *Proc. R. Soc. London Ser B Biol. Sci.* 236:311–324. https://doi.org/10.1098/rspb.1989.0025.

Muscatine, L., and J.W. Porter. 1977. Reef corals: Mutualistic symbioses adapted to nutrient-poor environments. *Bioscience* 27:454–460. https://doi.org/10.2307/1297526.

Neave, M.J., Apprill, A., Ferrier-Pagès, C., and C.R. Voolstra. 2016a. Diversity and function of prevalent symbiotic marine bacteria in the genus *Endozoicomonas*. *Appl. Microbiol. Biotechnol* 100: 8315–8324. https://doi.org/10.1007/s00253-016-7777-0.

Neave, M.J., Michell, C.T., Apprill, A., and C.R. Voolstra. 2014. Whole-genome sequences of three symbiotic *Endozoicomonas* strains. *Genome Announc.* 2:e00802–e00814. https://doi.org/10.1128/genomeA.00802-14.

Neave, M.J., Michell, C.T., Apprill, A., and C.R. Voolstra. 2017. *Endozoicomonas* genomes reveal functional adaptation and plasticity in bacterial strains symbiotically associated with diverse marine hosts. *Sci. Rep.* 7:40579. https://doi.org/10.1038/srep40579.

Neave, M.J., Rachmawati, R., Xun, L., Michell, C.T., Bourne, D.G., Apprill, A. et al. 2016b. Differential specificity between closely related corals and abundant *Endozoicomonas* endosymbionts across global scales. *ISME J.* 11:186–200. https://doi.org/10.1038/ismej.2016.95.

Neubauer, E-F., Poole, A.Z., Detournay, O., Weis, V.M., and S.K. Davy. 2016. The scavenger receptor repertoire in six cnidarian species and its putative role in cnidarian-dinoflagellate symbiosis. *PeerJ.* 4:e2692. https://doi.org/10.7717/peerj.2692.

Neubauer, E-F., Poole, A.Z., Neubauer, P., Detournay, O., Tan, K., Davy, S.K. et al. 2017. A diverse host thrombospondin-type-1 repeat protein repertoire promotes symbiont colonization during establishment of cnidarian-dinoflagellate symbiosis. *Elife* 6:e24494. https://doi.org/10.7554/eLife.24494

Nissimov, J., Rosenberg, E., and C.B. Munn. 2009. Antimicrobial properties of resident coral mucus bacteria of *Oculina patagonica*. *FEMS Microbiol. Lett.* 292:210–215. https://doi.org/10.1111/j.1574-6968.2009.01490.x.

Oborník, M., Modrý, D., Lukeš, M., Černotíková-Stříbrná, E., Cihlář, J., Tesařová, M. et al. 2012. Morphology, ultrastructure and life cycle of *Vitrella brassicaformis* n. sp., n. gen., a novel chromerid from the Great Barrier Reef. *Protist* 163:306–323. https://doi.org/10.1016/j.protis.2011.09.001.

Ochsenkühn, M.A., Schmitt-Kopplin, P., Harir, M., and S.A. Amin. 2018. Coral metabolite gradients affect microbial community structures and act as a disease cue. *Commun Biol.* 1:184. https://doi.org/10.1038/s42003-018-0189-1.

Palmer, C.V. 2018. Immunity and the coral crisis. *Commun Biol.* 1:91. https://doi.org/10.1038/s42003-018-0097-4.

Parkinson, J.E., Tivey, T.R., Mandelare, P.E., Adpressa, D.A., Loesgen, S., and V.M. Weis 2018. Subtle differences in symbiont cell surface glycan profiles do not explain species-specific colonization rates in a model cnidarian-algal symbiosis. *Front. Microbiol.* 9:842. https://doi.org/10.3389/fmicb.2018.00842.

Peixoto, R.S., Rosado, P.M., Leite, D.C. de A., Rosado, A.S., and D.G. Bourne. 2017. Beneficial microorganisms for corals (BMC): Proposed mechanisms for coral health and resilience. *Front. Microbiol.* 8:341. https://doi.org/10.3389/fmicb.2017.00341.

Pereira, L.B., Palermo, B.R.Z., Carlos, C., and L.M.M Ottoboni. 2017. Diversity and antimicrobial activity of bacteria isolated from different Brazilian coral species. *FEMS Microbiol. Lett.* 364:fnx164. https://doi.org/10.1093/femsle/fnx164,

Pernice, M., Meibom, A., Van Den Heuvel, A., Kopp, C., Domart-Coulon, I., Hoegh-Guldberg, O. et al. 2012. A single-cell view of ammonium assimilation in coral–dinoflagellate symbiosis. *ISME J.* 6:1314–1324. https://doi.org/10.1038/ismej.2011.196.

Pernice, M., Raina, J-B., Rädecker, N., Cárdenas, A., Pogoreutz, C., and Voolstra, C.R. 2019. Down to the Bone: the role of overlooked endolithic microbiomes in reef coral health. *ISME J.* 14: 325–334. https://doi.org/10.1038/s41396-019-0548-z.

Pettay, D.T., Wham, D.C., Smith, R.T., Iglesias-Prieto, R., and T.C. LaJeunesse. 2015. Microbial invasion of the Caribbean by an Indo-Pacific coral zooxanthella. *Proc. Natl. Acad. Sci.* 112:7513–7518. https://doi.org/10.1073/pnas.1502283112.

Piekarski, T., Buchholz, I., Drepper, T., Schobert, M., Wagner-Doebler, I., Tielen, P. et al. 2009. Genetic tools for the investigation of *Roseobacter* clade bacteria. *BMC Microbiol.* 9:265. https://doi.org/10.1186/1471-2180-9-265.

Pogoreutz, C., Rädecker, N., Cárdenas, A., Gärdes, A., Voolstra, C.R., and C. Wild. 2017a. Sugar enrichment provides evidence for a role of nitrogen fixation in coral bleaching. *Glob. Chang. Biol.* 23:3838–3848. https://doi.org/10.1111/gcb.13695.

Pogoreutz, C., Rädecker, N., Cárdenas, A., Gärdes, A., Wild, C., and C.R. Voolstra. 2017b. Nitrogen fixation aligns with *nifH* abundance and expression in two coral trophic functional groups. *Front. Microbiol.* 8:1187. https://doi.org/10.3389/fmicb.2017.01187.

Pogoreutz, C., Rädecker, N., Cárdenas, A., Gärdes, A., Wild, C., and C.R. Voolstra. 2018. Dominance of *Endozoicomonas* bacteria throughout coral bleaching and mortality suggests structural inflexibility of the *Pocillopora verrucosa* microbiome. *Ecol. Evol.* 8:2240–2252. https://doi.org/10.1002/ece3.3830.

Pogoreutz, C., and C.R. Voolstra. 2018. Isolation, culturing, and cryopreservation of *Endozoicomonas* (Gammaproteobacteria: Oceanospirillales: Endozoicomonadaceae) from reef-building corals. *Protocols.io* https://doi.org/10.17504/protocols.io.t2aeqae

Pollock, F.J., McMinds, R., Smith, S., Bourne, D.G., Willis, B.L., Medina, M. et al. 2018. Coral-associated bacteria demonstrate phylosymbiosis and cophylogeny. *Nat. Comm.* 9:4921. https://doi.org/10.1038/s41467-018-07275-x.

Poole, A.Z., Kitchen, S.A., and V.M. Weis. 2016. The role of complement in cnidarian-dinoflagellate symbiosis and immune challenge in the sea anemone *Aiptasia pallida*. *Front. Microbiol.* 7:519. https://doi.org/10.3389/fmicb.2016.00519.

Rädecker, N., Meyer, F.W., Bednarz, V.N., Cardini, U., and C. Wild. 2014. Ocean acidification rapidly reduces dinitrogen fixation associated with the hermatypic coral *Seriatopora hystrix*. *Mar. Ecol. Prog. Ser.* 511:297–302. https://doi.org/10.3354/meps10912.

Rädecker, N., Pogoreutz, C., Voolstra, C.R., Wiedenmann, J., and C. Wild. 2015. Nitrogen cycling in corals: the key to understanding holobiont functioning? *Trends Microbiol.* 23:490–497. https://doi.org/10.1016/j.tim.2015.03.008.

Rädecker, N., Raina, J-B., Pernice, M., Perna, G., Guagliardo, P., Kilburn, M.R. et al. 2018. Using Aiptasia as a model to study metabolic interactions in cnidarian-*Symbiodinium* symbioses. *Front. Physiol* 9:214. https://doi.org/10.3389/fphys.2018.00214.

Raina, J-B., Fernandez, V., Lambert, B., Stocker, R., and J.R. Seymour. 2019. The role of microbial motility and chemotaxis in symbiosis. *Nat. Rev. Microbiol.* 17:284–294. https://doi.org/10.1038/s41579-019-0182-9.

Raina, J-B., Tapiolas, D., Motti, C.A., Foret, S., Seemann, T., Tebben, J. et al. 2016. Isolation of an antimicrobial compound produced by bacteria associated with reef-building corals. *PeerJ.* 4:e2275. https://doi.org/10.7717/peerj.2275.

Ralph, P.J., Larkum, A.W.D., and M. Kühl. 2007. Photobiology of endolithic microorganisms in living coral skeletons: 1. Pigmentation, spectral reflectance and variable chlorophyll fluorescence analysis of endoliths in the massive corals *Cyphastrea serailia*, *Porites lutea* and *Goniastrea australensis*. *Mar. Biol.* 152:395–404. https://doi.org/10.1007/s00227-007-0694-0.

Reshef, L., Koren, O., Loya, Y., Zilber-Rosenberg, I., and E. Rosenberg. 2006. The Coral Probiotic Hypothesis. *Environ. Microbiol.* 8:2068–2073. https://doi.org/10.1111/j.1462-2920.2006.01148.x.

Ritchie, K.B. 2006. Regulation of microbial populations by coral surface mucus and mucus-associated bacteria. *Mar. Ecol. Prog. Ser.* 322:1–14. https://doi.org/10.3354/meps322001.

Ritchie, K.B. 2012. Bacterial symbionts of corals and Symbiodinium. In: Rosenberg E., Gophna U. (eds) *Beneficial Microorganisms in Multicellular Life Forms.* Springer, Berlin, Heidelberg, pp. 139–150. https://doi.org/10.1007/978-3-642-21680-0_9.

Robbins, S.J., Singleton, C.M., Chan, C.X., Messer, L.F., Geers, A.U., Ying, H. et al. 2019. A genomic view of the reef-building coral *Porites lutea* and its microbial symbionts. *Nat. Microbiol.* 4:2090–2100. https://doi.org/10.1038/s41564-019-0532-4.

Roder, C., Arif, C., Bayer, T., Aranda, M., Daniels, C., Shibl, A. et al. 2014a. Bacterial profiling of White Plague Disease in a comparative coral species framework. *ISME J.* 8:31–39. https://doi.org/10.1038/ismej.2013.127.

Roder, C., Arif, C., Daniels, C., Weil, E., and C.R. Voolstra. 2014b. Bacterial profiling of White Plague Disease across corals and oceans indicates a conserved and distinct disease microbiome. *Mol. Ecol.* 23:965–974. https://doi.org/10.1111/mec.12638.

Roder, C., Bayer, T., Aranda, M., Kruse, M., and C.R. Voolstra. 2015. Microbiome structure of the fungid coral *Ctenactis echinata* aligns with environmental differences. *Mol. Ecol.* 24:3501–3511. https://doi.org/10.1111/mec.13251.

Rodriguez-Lanetty, M., Krupp, D.A., and V.M. Weis. 2004. Distinct ITS types of *Symbiodinium* in Clade C correlate with cnidarian/dinoflagellate specificity during onset of symbiosis. *Mar. Ecol. Prog. Ser.* 275:97–102. https://doi:10.3354/meps275097

Rohwer, F., Seguritan, V., Azam, F., and N. Knowlton. 2002. Diversity and distribution of coral-associated bacteria. *Mar. Ecol. Prog. Ser.* 243:1–10. https://doi:10.3354/meps243001.

Rosado, P.M., Leite, D.C.A., Duarte, G.A.S., Chaloub, R.M., Jospin, G., Nunes da Rocha, U. et al. 2018. Marine probiotics: increasing coral resistance to bleaching through microbiome manipulation. *ISME J.* 921–936. https://doi.org/10.1038/s41396-018-0323-6.

Rosenberg, E., Koren, O., Reshef, L., Efrony, R., and I. Zilber-Rosenberg. 2007. The role of microorganisms in coral health, disease and evolution. *Nat. Rev. Microbiol.* 5:355–362. https://doi.org/10.1038/nrmicro1635.

Rossbach, S., Cárdenas, A., Perna, G.H., Duarte, C.M., and C.R. Voolstra. 2019. Tissue specific microbiomes of the Red Sea giant clam *Tridacna maxima* highlight differential abundance of Endozoicomonadaceae. *Front. Microbiol.* 10:2661. https://doi.org/10.3389/fmicb.2019.02661.

Rypien, K.L., Ward, J.R., and F. Azam. 2010. Antagonistic interactions among coral-associated bacteria. *Environ. Microbiol.* 12:28–39. https://doi.org/10.1111/j.1462-2920.2009.02027.x.

Sampayo, E.M., Ridgway, T., Bongaerts, P., and O. Hoegh-Guldberg. 2008. Bleaching susceptibility and mortality of corals are determined by fine-scale differences in symbiont type. *Proc. Natl. Acad. Sci.* 105:10444–10449. https://doi.org/10.1073/pnas.0708049105.

Santos, H.F., Carmo, F.L., Duarte, G., Dini-Andreote, F., Castro, C.B., Rosado, A.S. et al. 2014. Climate change affects key nitrogen-fixing bacterial populations on coral reefs. *ISME J.* 8:2272–2279. https://doi.org/10.1038/ismej.2014.70.

Schlichter, D., Zscharnack, B., and H. Krisch. 1995. Transfer of photoassimilates from endolithic algae to coral tissue. *Naturwissenschaften* 82:561–564.

Schwarz, J., Krupp, D., and V. Weis. 1999. Late larval development and onset of symbiosis in the scleractinian coral *Fungia scutaria. Biol. Bull.* 196:70–79. https://doi.org/10.2307/1543169.

Seymour, J.R., Amin, S.A., Raina, J-B., and R. Stocker. 2017. Zooming in on the phycosphere: The ecological interface for phytoplankton-bacteria relationships. *Nat. Microbiol.* 2:17065. https://doi.org/10.1038/nmicrobiol.2017.65.

Shantz, A.A., Lemoine, N.P., and D.E. Burkepile. 2016. Nutrient loading alters the performance of key nutrient exchange mutualisms. *Ecol. Lett.* 19:20–28. https://doi.org/10.1111/ ele.12538.

Sharp, K.H., Distel, D., and V.J. Paul. 2012. Diversity and dynamics of bacterial communities in early life stages of the Caribbean coral *Porites astreoides*. *ISME J.* 6:790–801. https:// doi.org/10.1038/ismej.2011.144.

Sharp, K.H., Ritchie, K.B., Schupp, P.J., Ritson-Williams, R., and V.J. Paul. 2010. Bacterial acquisition in juveniles of several broadcast spawning coral species. *PLOS ONE* 5:e10898. 10.1371/journal.pone.0010898.

Shashar, N., Cohen, Y., Loya, Y., and N. Sar. 1994. Nitrogen fixation (acetylene reduction) in stony corals: evidence for coral-bacteria interactions. *Mar. Ecol. Prog. Ser.* 111:259–264. https://www.jstor.org/stable/24849564.

Shinzato, C., Shoguchi, E., Kawashima, T., Hamada, M., Hisata, K., Tanaka, M. et al. 2011. Using the *Acropora digitifera* genome to understand coral responses to environmental change. *Nature* 476:320–323. https://doi.org/10.1038/nature10249.

Shishlyannikov, S.M., Zakharova, Y.R., Volokitina, N.A., Mikhailov, I.S., Petrova, D.P., and Y.V. Likhoshway. 2011. A procedure for establishing an axenic culture of the diatom *Synedra acus* subsp. *radians* (Kütz.) Skabibitsch from Lake Baikal. *Limnol. Oceanogr. Methods* 9:478–484. https://doi.org/10.4319/lom.2011.9.478.

Shoguchi, E., Shinzato, C., Kawashima, T., Gyoja, F., Mungpakdee, S., Koyanagi, R. et al. 2013. Draft assembly of the *Symbiodinium minutum* nuclear genome reveals dinoflagellate gene structure. *Curr. Biol.* 23:1399–1408. https://doi.org/10.1016/j.cub.2013.05.062.

Shumaker, A., Putnam, H.M., Qiu, H., Price, D.C., Zelzion, E., Harel, A. et al. 2019. Genome analysis of the rice coral *Montipora capitata*. *Sci. Rep.* 9:2571. https://doi.org/10.1038/ s41598-019-39274-3.

Siboni, N., Ben-Dov, E., and A. Kushmaro. 2012. Geographic specific coral-associated ammonia-oxidizing archaea in the northern Gulf of Eilat (Red Sea). *Microb. Ecol.* 64:18–24. https://doi.org/10.1007/s00248-011-0006-6.

Siboni, N., Ben-Dov, E., Sivan, A., and A. Kushmaro. 2008. Global distribution and diversity of coral-associated Archaea and their possible role in the coral holobiont nitrogen cycle. *Environ. Microbiol.* 10:2979–2990. https://doi.org/10. 1111/j.1462-2920.2008.01718.x.

Silverstein, R.N., Correa, A.M.S., LaJeunesse, T.C., and A.C. Baker. 2011. Novel algal symbiont (*Symbiodinium* spp.) diversity in reef corals of Western Australia. *Mar. Ecol. Prog. Ser.* 422:63–75. https://doi.org/10.3354/meps08934.

Silverstein, R.N., Cunning, R., and A.C. Baker. 2015. Change in algal symbiont communities after bleaching, not prior heat exposure, increases heat tolerance of reef corals. *Glob. Chang. Biol.* 21:236–249. https://doi.org/10.1111/gcb.12706.

Šlapeta, J., and M.C. Linares. 2013. Combined amplicon pyrosequencing assays reveal presence of the apicomplexan "type-N" (cf. *Gemmocystis cylindrus*) and *Chromera velia* on the Great Barrier Reef, Australia. *PLOS ONE* 8:e76095. https://doi.org/10.1371/ journal.pone.0076095.

Smith, G.J., and L. Muscatine. 1999. Cell cycle of symbiotic dinoflagellates: Variation in G1 phase-duration with anemone nutritional status and macronutrient supply in the *Aiptasia pulchella-Symbiodinium pulchrorum* symbiosis. *Mar. Biol.* 134:405–418. https://doi. org/10.1007/s002270050557.

Spalding, M., Burke, L., Wood, S.A., Ashpole, J., Hutchison, J., and P. zu Ermgassen. 2017. Mapping the global value and distribution of coral reef tourism. *Mar. Policy* 82:104–113. https://doi.org/10.1016/j.marpol.2017.05.014.

Sweet, M.J., Brown, B.E., Dunne, R.P., Singleton, I., and M. Bulling. 2017. Evidence for rapid, tide-related shifts in the microbiome of the coral *Coelastrea aspera*. *Coral Reefs* 36:815–828. https://doi.org/10.1007/s00338-017-1572-y.

Sweet, M.J., Croquer, A., and J.C. Bythell. 2011. Bacterial assemblages differ between compartments within the coral holobiont. *Coral Reefs* 30:39–52. https://doi.org/10.1007/s00338-010-0695-1.

Tait, K., Hutchison, Z., Thompson, F.L., and C.B. Munn. 2010. Quorum sensing signal production and inhibition by coral-associated *Vibrios*. *Environ. Microbiol. Rep.* 2:145–150. https://doi.org/10.1111/j.1758-2229.2009.00122.x.

Tebben, J., Motti, C.A., Siboni, N., Tapiolas, D.M., Negri, A.P., Schupp, P.J. et al. 2015. Chemical mediation of coral larval settlement by crustose coralline algae. *Sci. Rep.* 5:10803. https://doi.org/10.1038/srep10803.

Thurber, R.V., Willner-Hall, D., Rodriguez-Mueller, B., Desnues, C., Edwards, R.A., Angly, F. et al. 2009. Metagenomic analysis of stressed coral holobionts. *Environ. Microbiol.* 11:2148–2163. https://doi.org/10.1111/j.1462-2920.2009.01935.x.

Tilstra, A., El-Khaled, Y., Roth, F., Rädecker, N., Pogoreutz, C., Voolstra, C.R. et al. 2019. Denitrification aligns with N_2 fixation in Red Sea corals. *Sci. Rep.* 9:19460. https://doi.org/10.1038/s41598-019-55408-z.

Toller, W.W., Rowan, R., and N. Knowlton. 2002. Genetic evidence for a protozoan (phylum Apicomplexa) associated with corals of the *Montastraea annularis* species complex. *Coral Reefs* 21:143–146. https://doi.org.10.1007/s00338-002-0220-2.

Tout, J., Jeffries, T.C., Petrou, K., Tyson, G.W., Webster, N.S., Garren, M. et al. 2015. Chemotaxis by natural populations of coral reef bacteria. *ISME J.* 9:1764–1777. https://doi.org/10.1038/ismej.2014.261.

Tremblay, P., Grover, R., Maguer, J.F., Legendre, L., and C. Ferrier-Pages. 2012. Autotrophic carbon budget in coral tissue: a new ^{13}C-based model of photosynthate translocation. *J. Exp. Biol.* 215:1384–1393. https://doi: 10.1242/jeb.065201.

Underhill, D.M., and A. Ozinsky. 2002. Phagocytosis of microbes: complexity in action. *Annu. Rev. Immunol.* 20:825–852. https://doi.org/10.1146/annurev.immunol.20.103001.114744.

Upton, S.J., and E.C. Peters. 1986. A new and unusual species of *Coccidium* (Apicomplexa: Agamococcidiorida) from Caribbean scleractinian corals. *J. Invertebr. Pathol.* 47:184–193.

Verbruggen, H., Marcelino, V.R., Guiry, M.D., Cremen, M.C.M., and C.J. Jackson 2017. Phylogenetic position of the coral symbiont *Ostreobium* (Ulvophyceae) inferred from chloroplast genome data. *J. Phycol.* 53:790–803. https://doi.org/10.1111/jpy.12540.

Vermeij, M.J.A., Smith, J.E., Smith, C.M., Vega Thurber, R., and S.A. Sandin. 2009. Survival and settlement success of coral planulae: Independent and synergistic effects of macroalgae and microbes. *Oecologia* 159:325–336. https://doi.org/10.1007/s00442-008-1223-7.

Voolstra, C.R., Li, Y., Liew, Y.J., Baumgarten, S., Zoccola, D., Flot, J.F. et al. 2017. Comparative analysis of the genomes of *Stylophora pistillata* and *Acropora digitifera* provides evidence for extensive differences between species of corals. *Sci. Rep.* 7:17583. https://doi.org/10.1038/s41598-017-17484-x.

Wada, N., Ishimochi, M., Matsui, T., Pollock, F.J., Tang, S.-L., Ainsworth, T.D. et al. 2019. Characterization of coral-associated microbial aggregates (CAMAs) within tissues of the coral *Acropora hyacinthus*. *Sci. Rep.* 9:14662. https://doi.org/10.1038/s41598-019-49651-7.

Wada, N., Pollock, F.J., Willis, B.L., Ainsworth, T.D., Mano, N., and D.G. Bourne. 2016. *In situ* visualization of bacterial populations in coral tissues: pitfalls and solutions. *PeerJ.* 4:e2424. https://doi.org/10.7717/peerj.2424.

Wafar, M.M., Wafar, S., and David, J.J. 1990. Nitrification in reef corals. *Limnol. Oceanogr.* 35:725–730. https://doi.org/10.4319/lo.1990.35.3.0725.

Wangpraseurt, D., Pernice, M., Guagliardo, P., Kilburn, M.R., Clode, P.L., Polerecky, L. et al. 2016. Light microenvironment and single-cell gradients of carbon fixation in tissues of symbiont-bearing corals. *ISME J.* 10:788–792. https://doi.org/10.1038/ismej.2015.133.

Webster, N.S., Smith, L.D., Heyward, A.J., Watts, E.M., Webb, R.I., Blackall, L.L. et al. 2004. Metamorphosis of a scleractinian coral in response to microbial biofilms. *Appl. Environ. Microbiol.* 70:1213–1221. https://doi.org.10.1128/AEM.70.2.1213-1221.2004.

Webster, N.S., Soo, R., Cobb, R., and A.P. Negri. 2011. Elevated seawater temperature causes a microbial shift on crustose coralline algae with implications for the recruitment of coral larvae. *ISME J.* 5:759–770. https://doi.org/10.1038/ismej.2010.152.

Wegley, L., Edwards, R., Rodriguez-Brito, B., Liu, H., and F. Rohwer. 2007. Metagenomic analysis of the microbial community associated with the coral *Porites astreoides*. *Environ. Microbiol.* 9:2707–2719. https://doi:10.1111/j.1462-2920-2007.01383.x.

Wegley, L., Yu, Y., Breitbart, M., Casas, V., Kline, D.I., and F. Rohwer. 2004. Coral-associated archaea. *Mar. Ecol. Prog. Ser.* 273:89–96. https://doi:10.3354/meps273089.

Weis, V.M. 2008. Cellular mechanisms of Cnidarian bleaching: stress causes the collapse of symbiosis. *J. Exp. Biol.* 211:3059–3066. https://doi: 10.1242/jeb.009597.

Wiedenmann, J., D'Angelo, C., Smith, E.G., Hunt, A.N., Legiret, F.-E., Postle A.D. et al. 2012. Nutrient enrichment can increase the susceptibility of reef corals to bleaching. *Nat. Clim. Chang.* 3:160–164. https://doi.org/10.1038/nclimate1661.

Wild, C., Huettel, M., Klueter, A., Kremb, S.G., Rasheed, M.Y.M., and B.B. Jørgensen. 2004. Coral mucus functions as an energy carrier and particle trap in the reef ecosystem. *Nature* 428:66–70. https://doi.org/10.1038/nature02344.

Wilkinson, C. 2008. *Status of Coral Reefs of the World: Global Coral Reef Monitoring Network and Reef and Rainforest Research Centre.* Townsville, Australia. 296 pp.

Wolfowicz, I., Baumgarten, S., Voss, P.A., Hambleton, E.A., Voolstra, C.R., Hatta, M. et al. 2016 *Aiptasia* sp. larvae as a model to reveal mechanisms of symbiont selection in cnidarians. *Sci. Rep.* 6:32366. https://doi.org/10.1038/srep32366.

Wood-Charlson, E.M., Hollingsworth, L.L., Krupp, D.A., and V.M. Weis. 2006. Lectin/glycan interactions play a role in recognition in a coral/dinoflagellate symbiosis. *Cell Microbiol.* 8:1985–1993. https://doi.org/10.1111/j.1462-5822.2006.00765.x.

Yamazaki, S.S., Nakamura, T., Yuen, Y.S., and H. Yamasaki. 2008. Reef-building coral *Goniastrea aspera* harbor a novel filamentous cyanobacterium in their skeleton. *Proceedings of the 11th International Coral Reef Symposium* 1:265–268.

Yang, S.H., Lee, S.T.M., Huang, C.R., Tseng, C.H., Chiang, P.W., Chen, C.P. et al. 2016. Prevalence of potential nitrogen-fixing, green sulfur bacteria in the skeleton of reef-building coral *Isopora palifera*. *Limnol. Oceanogr.* 61:1078–1086. https://doi.org/10.1002/lno.10277.

Ziegler, M., Grupstra, C.G.B., Barreto, M.M., Eaton, M., BaOmar, J., Zubier, K. et al. 2019. Coral bacterial community structure responds to environmental change in a host-specific manner. *Nat. Comm.* 10:1–11. https://doi.org/10.1038/s41467-019-10969-5.

Ziegler, M., Roik, A., Porter, A., Zubier, K., Mudarris M.S., Ormond R. et al. 2016. Coral microbial community dynamics in response to anthropogenic impacts near a major city in the central Red Sea. *Mar. Pollut. Bull.* 105:629–640. https://doi.org/10.1016/j.marpolbul.2015.12.045.

Ziegler, M., Seneca, F.O., Yum, L.K., Palumbi, S.R., and C.R. Voolstra 2017. Bacterial community dynamics are linked to patterns of coral heat tolerance. *Nat. Comm.* 8:14213. https://doi.org/10.1038/ncomms14213.

Zimmer, B.L., May, A.L., Bhedi, C.D., Dearth, S.P., Prevatte, C.W., Pratte, Z. et al. 2014. Quorum sensing signal production and microbial interactions in a polymicrobial disease of corals and the coral surface mucopolysaccharide layer. *PLOS ONE* 9:e108541. https://doi.org/10.1371/journal.pone.0108541.

8 Extra-intestinal regulation of the gut microbiome

The case of C. elegans *TGFβ/SMA signaling*

Rebecca Choi, Dan Kim, Stacy Li, Meril Massot, Vivek Narayan, Samuel Slowinski, Hinrich Schulenburg, and Michael Shapira

All authors contributed equally to the writing of this chapter.

Contents

8.1 Introduction: *Caenorhabditis elegans* as a model for studying the holobiont

C. elegans is a bacterivore. Reproductive populations are typically found in environments rich in rotting organic matter, such as compost or fallen rotting fruit, which are also rich in fast growing bacteria (Felix and Braendle 2010). This co-existence suggests evolutionary interactions between *C. elegans* and various bacteria, some as food, others as persistent gut inhabitants. However, decades of culturing worms in the lab with *E. coli* as a sole food source—an easily cultivable and nourishing bacterium, yet poor colonizer—left a gap in our understanding of the interactions between worms and their microbes. This is now being corrected with studies characterizing the *C. elegans* gut microbiome, in wild isolates or in wildtype worms raised in the lab in natural-like microcosms (Berg et al. 2016a; Dirksen et al. 2016). The same advantages making *C. elegans* a generally useful model organism, such as its simple body plan, transparency, fast development, and short lifespan, also make it useful for microbiome research; several studies have thus demonstrated bacterial acceleration of worm development, modulation of host metabolism, and enhancement of pathogen resistance (Montalvo-Katz et al. 2013; Watson et al. 2014; Berg et al. 2016b; Dirksen et al. 2016; Samuel et al. 2016). Moreover, the ability to work with genetically homogenous populations of self-fertilizing worms facilitates the identification of gene effects on microbiome composition and function, by minimizing interindividual variation. This is starting to yield insights into the role of host genetics in shaping gut microbiome structure and function, as discussed in this chapter (Berg et al. 2016b, 2019). Furthermore, the genetic tractability of *C. elegans* enables facile generation of transgenic worms expressing transgenes from tissue-specific promoters, enabling the study of the role of intertissue communication in shaping and responding to the gut microbiome. In the following sections, we will examine what can be achieved in this model and how it may serve to shed light on the factors and principles that shape the gut microbiome.

8.2   The *C. elegans* gut microbiome and the factors that shape it

Work in the past few years has shown that *C. elegans* harbors a diverse yet characteristic gut microbiome, similar in worms raised in different environments (both in the wild and in natural-like microcosms), and significantly different from the respective environmental communities (Berg et al. 2016a; Dirksen et al. 2016; Zhang et al. 2017). A core gut microbiome was described, consisting of bacterial families known for their flexible metabolism and fast growth, and dominated by bacteria of the *Enterobacteriaceae* and *Pseudomonadaceae* families (Berg et al. 2016a; Zhang et al. 2017). As a rule, members of the *C. elegans* gut microbiome are beneficial, and those examined demonstrated contributions to infection resistance and acceleration of development (Montalvo-Katz et al. 2013; Berg et al. 2016b; Dirksen et al. 2016; Samuel et al. 2016). In some cases, underlying mechanisms for beneficial contributions are known, such as the priming of p38 MAPK-dependent immune gene expression, enabled by the commensal *Pseudomonas mendocina* (Montalvo-Katz et al. 2013), or the production of antifungal and antibacterial compounds by several other *Pseudomonas* isolates (Dirksen et al. 2016; Kissoyan et al. 2019). Other beneficial contributions may be attributed to metabolic interactions between the worm and its commensals. Whole genome sequence analysis of a representative selection of bacteria from the *C. elegans* gut microbiome suggested that the community as a whole can provide all the nutrients required for *C. elegans* growth, reproduction, and survival (Zimmermann et al. 2019). Among those, members of the *Pseudomonas* and *Ochrobactrum* genera were found to be capable of providing the full range of metabolites necessary to support worm growth, including vitamin B12, which is essential for nematode development and fertility, and could not be produced by any of the other sequenced microbiome members (Watson et al. 2014, 2016; Zimmermann et al. 2019). Metabolic network analysis based on bacterial genomic data revealed specific metabolic modules that were significantly associated with bacterial ability to colonize worms and affected population growth. For example, fermentation of pyruvate to (S)-acetoin was positively correlated with bacterial ability to colonize. This fermentation pathway produces diacetyl as an intermediate, which is known to attract worms and promote feeding activity (Ryan et al. 2014). Thus, prolonged time spent on lawns producing diacetyl could increase the likelihood of colonization. Additional associations were revealed, forming hypotheses to be tested. Similar metabolic analyses performed on transcriptomic and proteomic data from worms colonized with *Ochrobactrum* species indicated a role for host amino acids, carbohydrates, vitamin, and folate metabolism, as well as N-glycan and serotonin/octopamine biosynthesis, in shaping the worm interactions with microbes (Yang et al. 2019). Together, these genomic analyses suggest that metabolic interactions between the host and the microbiome are fundamental for determining microbiome composition, extending an emerging theme describing metabolic control over microbial community assembly (Goldford et al. 2018).

Other studies, utilizing compost microcosms, have shown that additional factors are important for shaping the worm gut microbiome. First, temperature-dependent host-associated processes were found to uncouple temperature-dependent changes in the worm gut microbiome from temperature-dependent changes in the environmental microbiome. This was taken to represent effects of host physiology on microbiome composition (Berg et al. 2016a). In addition, network analysis comparing gut microbiomes to the respective environmental communities indicated that interspecies interactions between microbiome members also played a role in shaping microbiome composition, with bacteria of core gut families identified as hubs for competitive interactions (Berg et al. 2016a). What underlies such physiological or competitive interactions could still be metabolic interactions, but alternatively, such interactions could be attributed to host immunity, bacterial toxins, or all of the above together.

Potentially underlying all interactions between the host and its microbiome is genetics, both of the host and of its microbes. Supporting this, a study based on 16S sequencing, which compared the composition of gut microbiomes in different *C. elegans* strains and related species raised in identical microcosm environments, identified significant contributions of host genetics to microbiome composition (Berg et al. 2016b). Constrained multivariate analysis estimated host genetics as responsible for 12% of the overall variation in microbiome composition. This is likely an underestimate, due to the limited phylogenetic resolution offered by 16S sequence data. Indeed, subsequent functional evaluation of gut isolates, most of which of the *Enterbacter cloacae* clade, demonstrated that host genetics determined the function (or lack thereof) of commensals of the same clade, otherwise indistinguishable by sequence.

8.3 The intestinal niche

Host genetics contributes to defining the intestinal niche—the structure and surface proteins of the intestinal epithelium, chemical composition, nutrient availability, and the presence of antimicrobial proteins. It can further define host behavior, affecting its interactions with the environment. The *C. elegans* gut is slightly acidic; luminal pH cycles between 4.4 and about 6 in a period of 45 seconds, linked to the defecation cycle, with acidification enabled by the VHA-6 proton transporter (Allman et al. 2009; Bender et al. 2013). Oxygen levels in the worm gut have not been measured, but considering the small size of the intestine, lack of apparent compartmentalization, and its frequent opening to the outside world, it is likely to be aerobic. Composition of the worm gut microbiome supports this, with more than a few obligate aerobes (Berg et al. 2016a). Mucin secretion and glycosylation are also important characteristics of the intestinal niche. In humans, heavily glycosylated mucin proteins serve as food for a subset of gut bacteria, chief among them is *Akkermansia muciniphila*, which makes up to 3% of the human colon microbiome. By liberating mucin-associated oligosaccharides, *Akkermansia* cross-feeds additional bacteria (Belzer et al. 2017; Umu et al. 2017; Van Herreweghen et al. 2018). Glycosylation patterns of mucins further diversify environments, with sugars that can be digested only by bacteria with the suitable enzymes (Etienne-Mesmin et al. 2019). Highlighting the importance of such glycosylation patterns, disruption of the fucosyltransferase gene *fut2* causes extensive changes in gut microbiome composition in mice (Kashyap et al. 2013). Mucus also serves as a selective substrate for adhesion of bacteria with suitable surface proteins, potentially enhancing gut colonization by slow-growing taxa (Juge 2012; McLoughlin et al. 2016). In *C. elegans* electron micrographs of the intestine show what appears to be a mucus layer overlaying intestinal microvilli (Figure 8.1B). However, the proteins that make this layer are not known (although several mucin homologs are identified in the *C. elegans* genome). Nevertheless, glycosylation is prevalent, and glycosylation patterns, determined by different enzymes, affect localization of secreted proteins, including lectins (Maduzia et al. 2011). Lectins, further discussed below, are sugar and lipid binding proteins, which play conserved roles in host innate immunity (Vaishnava et al. 2011; Hoving et al. 2014; Casals et al. 2018), and were shown to contribute to immune protection also in *C. elegans* (O'Rourke et al. 2006; Irazoqui et al. 2010; Simonsen et al. 2011; Dierking et al. 2016).

8.4   Host immunity and its role in shaping the intestinal niche

Host immunity is a dominant factor among those which could shape gut microbiome structure and function. In fact, some researchers have gone as far as suggesting that immunity evolved first and foremost under selection to recognize and manage complex communities of beneficial microbes (Hedrick 2004; McFall-Ngai 2007). Innate immunity is the more conserved branch of the immune system, shared between vertebrates and invertebrates alike. Lectins represent a significant part of *C. elegans* innate immunity, making a large gene family (283 members), of which many are known to be expressed in the gut, and most encode secreted proteins (Pees et al. 2016). Lectins, in particular C-type lectins, are also among the genes most strongly and reproducibly induced during *C. elegans* responses to pathogens, but different subsets are induced by different pathogens, as well as by nonpathogenic commensals (Shapira et al. 2006; Wong et al. 2007; Pees et al. 2016; Berg et al. 2019; Yang et al. 2019). Interestingly, studies using fluorescent reporters for lectin gene expression demonstrated differential expression, with some C-type lectins preferentially localized to the posterior gut (e.g. *clec-52* and *clec-60*) and others (*clec-66*) localized to both the anterior, as well as the posterior portions of the gut (Irazoqui et al. 2010;

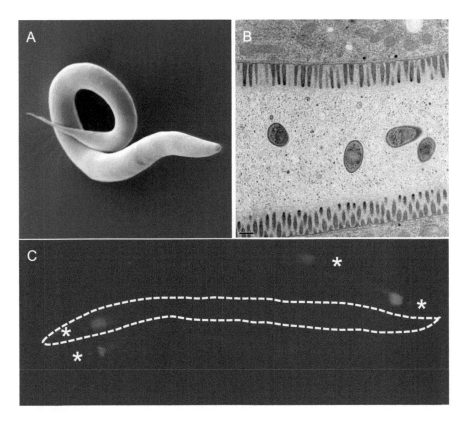

FIGURE 8.1 *C. elegans* **colonization by its** *Enterobacter* commensal. (A) A scanning electron micrograph of a mature worm. (B) A transmission electron micrograph showing initial colonization of the worm intestine by *E. cloacae* (C) A fluorescent image showing tdTomato-expressing *E. cloacae* colonizing the anterior gut. One worm is outlined. Asterisks mark the head region. Images are not to scale.

Pees et al. 2016). Considering the simple cylindrical structure of the intestine and its small size, such local specializations are surprising, and their functional significance is still unknown. Lysozymes, enzymes capable of hydrolyzing the bacterial cell wall, are another conserved family of immune effectors shown to play important roles in *C. elegans* immune responses. At least one member of this family (*ilys-3*) also demonstrated differentially localized intestinal expression (Irazoqui et al. 2010). In addition, worms express antimicrobial peptides (AMPs) divided between several families. Some are conserved, such as members of the defensin gene family (named *abf* genes), while others are specific to worms. Caenopores are among the latter, and encode saposin-like peptides with demonstrated pore-forming antibacterial activities (Roeder et al. 2010). Most members of this family are expressed in the intestine, and several (namely *spp-3*, *spp-5*, *spp-8* and *spp-18*) were further shown to be upregulated in worms raised in the presence of their normal commensals, as compared to standard *E. coli* food bacteria (Berg et al. 2019; Yang et al. 2019). In mammals, AMPs are among the fastest evolving gene families, and are thought to be at the forefront of an arms race with bacterial resistance (Peschel and Sahl 2006). As an outcome of AMP diversification in this arms race, different members of AMP families tend to have selective efficiency against different microbes (typically shown for pathogens), and different host strains demonstrate selective susceptibility to bacterial pathogens hinging on their particular AMP cocktail (Salger et al. 2016; Schmitt et al. 2016; Romoli et al. 2017). Such selectivity may affect also beneficial microbes or commensals.

The innate immune system enables rapid responses to nonself, typically based on the identification of molecular patterns associated with pathogens (PAMPs), or just microbes (MAMPs), as well as molecular patterns associated with downstream damage (DAMPs). Recognition of such molecular patterns is achieved by pattern recognition receptors (PRR), such as Toll-like, RIG-I-like, or NOD-like receptors, or the inflammasome complex. The innate immune system is suited to quickly respond to changes in microbial composition, enabling it to dynamically shape the intestinal niche, affecting both nutrient availability and antimicrobial agents (Thaiss et al. 2016). For example, TLR signaling was shown to regulate intestinal fucosylation (Pickard et al. 2014); also, the inflammasome complex, in different configurations, was shown to regulate epithelial AMP production, and its disruption caused intestinal dysbiosis (Hu et al. 2015; Levy et al. 2015). Whereas the function of the innate immune system is phylogenetically conserved, its activation mechanisms in different organisms vary, suggesting that its interactions with gut microbes may also follow distinct trajectories. For example, the first PRR to be identified, drosophila's *Toll*, is activated by a peptide generated downstream to soluble PRRs (Michel et al. 2001; Gottar et al. 2002); in contrast, vertebrate Toll-like receptors (or similar PRRs) often bind MAMPs directly (Botos et al. 2011). Furthermore, work in *C. elegans* so far provided only circumstantial evidence for PAMP/MAMP recognition (Twumasi-Boateng and Shapira 2012), and the one Toll homolog, TOL-1, is largely dispensable for immune responses (Pujol et al. 2001; Tenor and Aballay 2008). Instead, several lines of evidence suggest that responses to DAMPs, detection of pathogens by chemosensory neurons, and a surveillance mechanism guarding conserved targets of bacterial toxins (e.g. translation, cellular respiration) may play more dominant roles in *C. elegans* immunity, both against pathogens and potentially also in its interactions with gut commensals (Pujol et al. 2008; Dunbar et al. 2012; Melo and Ruvkun 2012; Meisel et al. 2014; Zugasti et al. 2014; Zhang et al. 2015).

While microbial recognition in *C. elegans* is not fully understood, much more is known about subsequent activation of signaling cascades that coordinate intestinal immune responses to different microbes. These include the p38 MAP kinase pathway, insulin/insulin-like growth factor signaling (IIS), and TGF-β signaling, and their downstream transcription factor mediators, which include SKN-1, ATF-7, DAF-16, as well as ZIP-2, HLH-30, and ELT-2 (Mallo et al. 2002; Shivers et al. 2010; Hoeven et al. 2011; Visvikis et al. 2014; Block et al. 2015; Reddy et al. 2016; Tjahjono and Kirienko 2017; Lee et al. 2018). Whereas more is known about the involvement of these pathways in antipathogen responses, new research has begun to describe their roles in host–microbiome interactions. The role of TGFβ signaling is the focus of the subsequent sections. In addition, recent work has unraveled new roles for the central intestinal regulator ELT-2 in regulating immune gene expression (as well as affecting reproduction) during interactions with nonpathogenic *Ochrobactrum* commensals, and further suggested involvement of the IIS pathway (Yang et al. 2016; Yang et al. 2019).

8.5   Multitissue contributions of TGFβ signaling control anterior gut commensal abundance and function

Enrichment for immune genes, including lectins, AMPs, and hydrolytic enzymes is a recurring theme in gene expression studies where *C. elegans* is exposed to complex environmental microbiomes (Berg et al. 2019; Yang et al. 2019). Enrichment for known targets of immune regulators can further suggest candidates that may be central in shaping the microbiome, and the availability of mutant strains in the *C. elegans* research community makes the testing of such hypotheses straightforward. Recent work focusing on mutants for central immune regulators employed a synthetic environmental community—prepared by mixing equal portions of cultures from 30 previously isolated *C. elegans* gut bacteria representing most of the worm core gut microbiome families (Berg et al. 2019). Initially germ-free larvae, wildtype, or mutant, were raised to adulthood on the synthetic community. Following washing and surface sterilization, worms were ground and their gut microbiome analyzed using calibrated quantitative PCR with universal eubacterial primers (evaluating overall microbiome size), or family-level taxa-specific primers, providing measurements for evaluating relative abundance. The most significant changes in microbiome size and composition were observed in mutants for *dbl-1*, which encodes a TGFβ ligand most similar to the vertebrate BMP-1. Disruption of *dbl-1* caused a threefold increase in bacterial load, due to a bloom specifically of *Enterobacter* species; other members of the microbiome were unable to take advantage of *dbl-1* disruption, even in the absence of *Enterobacter*. In contrast to *dbl-1* mutants, *tnt-3* mutants, which are defective in grinding, showed an increase in microbiome size without any significant change in its composition, indicating that the effects of the *dbl-1* disruption on microbiome composition were downstream to intestinal intake, and were associated with changes in the intestinal niche. Importantly, TGFβ disruption caused not only a change in *Enterobacter* abundance, but also turned an otherwise beneficial commensal, which protects worms from subsequent infection (Berg et al. 2016b), into an opportunistic pathogen. These findings highlight the importance of proper microbiome control, where more of a good thing is not necessarily better, and demonstrate the inherent dangers of bacterial imbalances, or dysbiosis.

An *Enterobacter cloacae* derivative expressing a fluorescent protein was used to examine additional TGFβ/BMP mutants (all of which are named *sma* for their small body size), showing that it was their immune functions, rather than developmental, that affected *Enterobacter* commensals. Interestingly, overexpression of the DBL-1 ligand, instead of simply bringing back control over *Enterobacter* colonization to wildtype levels (as in Figure 8.1C), reduced colonization specifically in the anterior intestine, suggesting that this is where TGFβ signaling exerted its control, and possibly that this is where *Enterobacter* species initiated colonization. Considering the small size and simple structure of the worm gut, spatial differentiation is not assumed. However, the localized effects of DBL-1 overexpression suggest localized functional differentiation along the gut longitudinal axis. Adding this observation to the earlier anecdotal reports of localized intestinal expression of certain lectins (see above) supports the existence of local intestinal subregions in *C. elegans*, differing in concentrations of AMPs and perhaps of surface proteins, giving rise to compartmentalization of colonization, similarly to the situation in organisms with more complex body plans.

DBL-1 is primarily expressed in neurons (Suzuki et al. 1999). Its heterodimeric receptor subunits (SMA-6 and DAF-4) and downstream transcriptional regulators (SMA-2, 3 and 4) are expressed more ubiquitously, including the pharynx, epidermis, and intestine (Wang et al. 2002). The simplest explanation for the importance of TGFβ signaling for *Enterobacter* intestinal abundance would involve binding of DBL-1 to intestinal receptors and activation of localized gene expression, for example of AMPs, which directly controls *Enterobacter*. However, experiments attempting to restore TGFβ control using tissue specific expression of the downstream TGFβ mediator SMA-3/SMAD ruled this out: SMA-3 expression in the intestine was unable to reduce intestinal *Enterobacter* to wildtype levels. On the other hand, SMA-3 expression in the pharynx, and even more so in the epidermis, effectively restored wildtype-level control. This suggests that the effects of TGFβ signaling, while focused on the anterior gut, are indirect, originating in extraintestinal tissues and depending on downstream signal(s) to regulate intestinal function.

8.6 TGFβ signaling and cell nonautonomous regulation of intestinal function

Neuronal DBL-1 regulates both epidermal and intestinal function. In the epidermis it is required for endoreduplication during development and for AMP gene expression (Morita et al. 2002; Zugasti and Ewbank 2009). AMP regulation was shown to depend on the SMA-6 and DAF-4 receptor subunits, as well as on the SMA-3 downstream mediator, but in which tissue these proteins operated could not be resolved (Zugasti and Ewbank 2009). Endoreduplication and body size were shown to depend on epidermal SMA-3, demonstrating a simple "one-step" cell nonautonomous regulation (neurons to epidermis) (Wang et al. 2002). In the intestine, DBL-1 affected expression of immune genes as well as lipid accumulation (Mallo et al. 2002; Clark et al. 2018). In the case of immune genes, earlier work has shown that DBL-1 regulated a gene subset distinct from those regulated by other central immune pathways (p38, DAF-16), which included lectins and lysozymes, and was induced in response to gram-negative pathogens (Alper et al. 2007). The particular cocktail of DBL-1-dependent immune genes may be relevant for *Enterobacter* commensals more than for other bacteria, thus explaining the observed *Enterobacter* bloom in TGFβ mutants. However, the results described in Berg et al. (2019) indicate that the influence of TGFβ signaling on the intestine is indirect and can be exerted from more than one tissue, demonstrating a more complex mode of cell nonautonomous regulation compared to that observed in the epidermis. Interestingly, the same multitissue contributions observed for TGFβ signaling for *Enterobacter* accumulation in the intestine was also observed for TGFβ regulation of lipid accumulation (Clark et al. 2018). In this case, the significance of lipids to worm metabolism may suggest that TGFβ signaling helps integrate information on metabolic states of different tissues to affect the gut, the main fat storage tissue in the worm. In line with such a role, Clark et al., presented data suggesting that TGFβ signaling negatively regulated the insulin signaling pathway (IIS), a central hub of homeostatic control, and alleviated its negative modulation of lipid accumulation. Regulation of IIS by DBL-1 has been reported also in a developmental context, where DBL-1 was shown to promote secretion of insulin-like peptides during L1 arrest, supporting the proposed regulatory pathway (Zheng et al. 2018). Regardless of the involvement of IIS, the similarity in regulatory patterns between TGFβ control of *Enterobacter* abundance and lipid accumulation is intriguing. While it is largely circumstantial, this similarity may be evidence for monitoring of host metabolic state as an integrative proxy for the impact of the gut microbiome, enabling activation of regulatory mechanisms that regulate both host metabolism and its gut microbiome.

An alternative hypothesis, connecting TGFβ signaling with gut commensals, is suggested by a study in mice examining the effects of a TGFβR2 knock-out in dendritic cells. This study found that such disruption caused depletion of mucin-secreting goblet cells through downregulation of Notch signaling, and a subsequent expansion of *Enterobacteriaceae*, promoting colitis (Ihara et al. 2016). Notch signaling is important mostly during *C. elegans* development. However, at least one Notch ligand, OSM-11, was shown to also regulate adult stress resistance and longevity, through activation of the SKN-1 transcription factor (Dresen et al. 2015). Whether similar to vertebrates, the role of TGFβ signaling in *C. elegans* is in maintaining the integrity of intestinal physical barriers is yet to be determined, but this could have an impact on gut commensals, promoting opportunistic pathogenesis.

The discussion above considers the downstream effects DBL-1 secretion. However, what initiates its secretion outside of the intestine is unknown. In experiments addressing behavior and learning in *C. elegans*, DBL-1 was found to enable aversive olfactory learning in response to a pathogen. It was shown to be secreted from AVA interneurons, and for its downstream effects required the expression of its SMA-6 receptor subunit in the epidermis (Zhang and Zhang 2012), presenting an interesting parallel with the role of epidermal TGFβ signaling in controlling *Enterobacter* intestinal abundance. What activates the AVA interneuron is not known, but studies focused on the second *C. elegans* TGFβ ligand, DAF-7, have shown that at least in the case of sensory amphid neurons, secretion could be activated upon recognition of external bacterial metabolites (Meisel et al. 2014). Furthermore, recent results demonstrate that bacteria can be sensed inside the worm, specifically in the pharynx, and assessed for their value as food (Rhoades et al. 2019). This study showed that an acid-sensitive ion channel in serotonergic NSM neurons, which open to the pharynx, responded to ingested bacteria, specifically to the bacterial membranal fraction, and modulated worm locomotion. Whether similar recognition of ingested bacteria could also affect the intestinal niche and its microbiome composition remains to be seen.

C. elegans is particularly suitable for studying the effects of extraintestinal signaling on the gut microbiome, much of this thanks to the ease of generating transgenic animals driving gene expression from tissue-specific promoters. Furthermore, shaping of the worm gut microbiome by TGFβ signaling is but one example of the rapidly increasing repertoire of cell nonautonomous mechanisms shown to take part in maintaining homeostasis in this organism. Cell nonautonomous signaling is important for multicellular host integrity, as it enables integration of metabolic or damage signals to maintain organismal homeostasis. Signals can be initiated in the epidermis or intestine, which are in contact with the environment, as well as by sensory neurons (Pukkila-Worley and Ausubel 2012; Taylor and Dillin 2013; Liu et al. 2018). Neuronal or endocrine circuits subsequently relay the signal to other tissues, including (as feedback loops) to the originating tissue (Cohen et al. 2009; Breton et al. 2016). Signals can be also initiated following inactivation of core cellular functions, such as protein translation, or mitochondrial function, in the context of "cellular surveillance" (Dunbar et al. 2012; Melo and Ruvkun 2012; Berendzen et al. 2016). Such processes are frequent targets of bacterial toxins, but more generally, monitoring enables integration and coordination at the level of the whole organism. It is possible that lack of specialized immune cells in *C. elegans* increases the necessity of such integration, and reliance on neuronal and/or endocrine signal.

8.7 Conclusions and future prospects: Convergence with other systems of host–symbiont interactions

The network of neuronal and endocrine connections between different tissues plays an essential role in maintaining organismal homeostasis. With the growing appreciation of the gut microbiome, as an integral part of the holobiont, it becomes obvious that it too should be connected. Descriptions of the gut–brain axis and the like are common, and often are said to be bidirectional. However, most studies focus on the effect of the microbiome on the host, and rarely on how the microbiome, as an integral part of the holobiont, is regulated by different tissues. *C. elegans* offers the advantage of facile monitoring of gut bacteria *in vivo*, in conjunction with tissue-specific activation of candidate regulatory mechanisms. These capabilities enabled the identification of extraintestinal and indirect contributions of TGFβ signaling to shaping of gut microbiome structure and commensal function. What relays this signal is yet unknown, but both neuronal signaling and endocrine insulin signaling might be involved. Such regulation and the candidate mediators fall well with previous descriptions of cell nonautonomous regulation of stress responses and homeostasis in *C. elegans*. This puts the gut microbiome in a similar category as other worm body tissues, and supports the notion of it being an integral part of the holobiont. However, the molecular mechanisms underlying potential feedback loops between gut microbes and the intestinal and extra-intestinal tissues are still unknown. Deciphering these cellular dialogues is central to understanding the function and significance of host-associated microbiomes. Going forward and addressing hypotheses about these dialogues will benefit from insights about the molecular language used. The model where this language is best characterized is the interactions between the bobtail squid and its bioluminescent symbint *Vibrio fischeri,* described in detail elsewhere in this book. In spite of *de novo* acquisition of symbionts from a complex marine environment each generation, the molecular dialogue between the squid hatchling and *V. fischeri* ensures that the mother's tried and useful commensal is preferred on other available bacteria (Nishiguchi et al. 1998).

Several features of this dialogue are of particular interest: The first is the demonstration of how activation of broad immune responses can direct colonization by a specific symbiont. Reaching the vicinity of the squid host prior to colonization,

V. fischeri can activate host immune responses by secretion of outer membrane vesicles carrying peptidoglycan and lipopolysaccharide. Recognition of these MAMPs induces host secretion of mucus, AMPs, and reactive oxygen species. By inducing this broad response, the symbiont, which is impervious to this toxic cocktail (likely thanks to generations of coevolution), can gain advantage over other environmental bacteria, and move closer to form aggregates on the surface of the ciliary epithelium outside of the light organ (Nyholm et al. 2002; Nyholm and McFall-Ngai 2003; Aschtgen et al. 2016; Schwartzman and Ruby 2016). Presence of small aggregates is sufficient to induce a wave of host gene expression mainly encoding hydrolytic enzymes, including an endochitinase that releases chitobiose from mucosal chitin, serving as food for symbionts and as a signal directing symbiont chemotaxis toward crypts of the light organ (Kremer et al. 2013). Could *C. elegans* DBL-1 induce an AMP cocktail specifically selecting for beneficial bacteria? Could some of the induced *C. elegans* genes identified in Berg et al. (2019) (many of which encoding hydrolytic enzymes) be involved in feeding beneficial bacteria?

Another interesting feature is the coupling of beneficial services to symbiont protection from host immunity. In the case of the squid, the beneficial service is bioluminescence. Oxygen is consumed to produce this bioluminescence. Bacteria incapable of this leave excess of oxygen, which activates phenoloxidases, leading to the immune response of melanization; the incomplete reaction further generates hydrogen peroxide, toxic on its own, but also inducing the generation of additional antibacterial compounds (Schwartzman and Ruby 2016). Thus, while the currency of the squid-*Vibrio* relationship is bioluminescence, the host assesses the quality of its symbiont, and polices against cheaters, by monitoring the presence of chemical byproducts. Whether similar couplings exist between beneficial microbes and their *C. elegans* host remains to be seen. Gene expression studies and genetic screens could help expose such dialogues. The power of *C. elegans* as a genetic model will come in handy for such studies.

Acknowledgments

Work in the Shapira lab is supported by NIH grants R01OD024780 and R01AG061302.

References

Allman, E., D. Johnson, and K. Nehrke. 2009. "Loss of the Apical V-ATPase a-Subunit VHA-6 Prevents Acidification of the Intestinal Lumen during a Rhythmic Behavior in *C. elegans*." *American Journal of Physiology-Cell Physiology* 297 (5): C1071–81.

Alper, S., S. J. McBride, B. Lackford, J. H. Freedman, and D. A. Schwartz. 2007. "Specificity and Complexity of the *Caenorhabditis elegans* Innate Immune Response." *Molecular and Cellular Biology.* doi: 10.1128/mcb.02070-06

Aschtgen, M. S., K. Wetzel, W. Goldman, M. Mcfall-Ngai, and E. Ruby. 2016. "*Vibrio* Fischeri-Derived Outer Membrane Vesicles Trigger Host Development." *Cellular Microbiology* 18 (4): 488–99.

Belzer, C., L. W. Chia, S. Aalvink, B. Chamlagain, V. Piironen, J. Knol et al. 2017. "Microbial Metabolic Networks at the Mucus Layer Lead to Diet-Independent Butyrate and Vitamin B12 Production by Intestinal Symbionts." *mBio* 8 (5). doi: 10.1128/mBio.00770-17

Bender, A., Z. R. Woydziak, L. Fu, M. Branden, Z. Zhou, B. D. Ackley et al. 2013. "Novel Acid-Activated Fluorophores Reveal a Dynamic Wave of Protons in the Intestine of *Caenorhabditis elegans.*" *ACS Chemical Biology* 8 (3): 636–42.

Berendzen, K. M., J. Durieux, L. W. Shao, Y. Tian, H. E. Kim, S. Wolff et al. 2016. "Neuroendocrine Coordination of Mitochondrial Stress Signaling and Proteostasis." *Cell* 166 (6): 1553–63.e10.

Berg, M., D. Monnin, J. Cho, L. Nelson, A. Crits-Christoph, and M. Shapira. 2019. "TGFβ/BMP Immune Signaling Affects Abundance and Function of *C. elegans* Gut Commensals." *Nature Communications* 10 (1): 604.

Berg, M., B. Stenuit, J. Ho, A. Wang, C. Parke, M. Knight et al. 2016a. "Assembly of the *Caenorhabditis elegans* Gut Microbiota from Diverse Soil Microbial Environments." *The ISME Journal* 10 (8): 1998–2009.

Berg, M., X. Y. Zhou, and M. Shapira. 2016b. "Host-Specific Functional Significance of Caenorhabditis Gut Commensals." *Frontiers in Microbiology* 7 (October): 1622.

Block, D. H. S., K. Twumasi-Boateng, H. S. Kang, J. A. Carlisle, A. Hanganu, T. Y.-J. Lai et al. 2015. "The Developmental Intestinal Regulator ELT-2 Controls p38-Dependent Immune Responses in Adult *C. elegans.*" *PLoS Genetics* 11 (5): e1005265.

Botos, I., D. M. Segal, and D. R. Davies. 2011. "The Structural Biology of Toll-like Receptors." *Structure* 19 (4): 447–59.

Breton, J., N. Tennoune, N. Lucas, M. Francois, R. Legrand, J. Jacquemot et al. 2016. "Gut Commensal *E. coli* Proteins Activate Host Satiety Pathways Following Nutrient-Induced Bacterial Growth." *Cell Metabolism* 23 (2): 324–34.

Casals, C., M. A. Campanero-Rhodes, B. García-Fojeda, and D. Solís. 2018. "The Role of Collectins and Galectins in Lung Innate Immune Defense." *Frontiers in Immunology* 9 (September): 1998.

Clark, J. F., M. Meade, G. Ranepura, D. H. Hall, and C. Savage-Dunn. 2018. "*Caenorhabditis elegans* DBL-1/BMP Regulates Lipid Accumulation via Interaction with Insulin Signaling." *G3 (Bethesda)* 8(1): 343–51.

Cohen, M., V. Reale, B. Olofsson, A. Knights, P. Evans, and M. de Bono. 2009. "Coordinated Regulation of Foraging and Metabolism in *C. elegans* by RFamide Neuropeptide Signaling." *Cell Metabolism* 9 (4): 375–85.

Dierking, K., W. Yang, and H. Schulenburg. 2016. "Antimicrobial Effectors in the Nematode *Caenorhabditis elegans*: An Outgroup to the Arthropoda." *Philosophical Transactions of the Royal Society of London. Series B, Biological Sciences* 371 (1695). doi: 10.1098/rstb.2015.0299

Dirksen, P., S. A. Marsh, I. Braker, N. Heitland, S. Wagner, R. Nakad et al. 2016. "The Native Microbiome of the Nematode *Caenorhabditis elegans*: Gateway to a New Host-Microbiome Model." *BMC Biology* 14: 38.

Dresen, A., S. Finkbeiner, M. Dottermusch, J.-S. Beume, Y. Li, G. Walz et al. 2015. "*Caenorhabditis elegans* OSM-11 Signaling Regulates SKN-1/Nrf during Embryonic Development and Adult Longevity and Stress Response." *Developmental Biology* 400 (1): 118–31.

Dunbar, T., Z. Yan, K. Balla, M. Smelkinson, and E. Troemel. 2012. "*C. elegans* Detects Pathogen-Induced Translational Inhibition to Activate Immune Signaling." *Cell Host & Microbe* 11 (April): 375–86.

Etienne-Mesmin, L., B. Chassaing, M. Desvaux, K. De Paepe, R. Gresse, T. Sauvaitre et al. 2019. "Experimental Models to Study Intestinal Microbes-Mucus Interactions in Health and Disease." *FEMS Microbiology Reviews* 43 (5): 457–89.

Félix, M.-A., and C. Braendle. 2010. "The Natural History of *Caenorhabditis elegans.*" *Current Biology: CB* 20 (22): R965–69.

Goldford, J. E., N. Lu, D. Bajić, S. Estrela, M. Tikhonov, A. Sanchez-Gorostiaga et al. 2018. "Emergent Simplicity in Microbial Community Assembly." *Science* 361 (6401): 469–74.

Gottar, M., V. Gobert, T. Michel, M. Belvin, G. Duyk, J. A. Hoffmann et al. 2002. "The Drosophila Immune Response against Gram-Negative Bacteria is Mediated by a Peptidoglycan Recognition Protein." *Nature* 416 (6881): 640–44.

Hedrick, S. M. 2004. "The Acquired Immune System: A Vantage from beneath." *Immunity* 21 (5): 607–15.

Hoving, J. C., G. J. Wilson, and G. D. Brown. 2014. "Signalling C-Type Lectin Receptors, Microbial Recognition and Immunity." *Cellular Microbiology* 16 (2): 185–94.

Hu, S., L. Peng, Y.-T. Kwak, E. M. Tekippe, C. Pasare, J. S. Malter et al. 2015. "The DNA Sensor AIM2 Maintains Intestinal Homeostasis via Regulation of Epithelial Antimicrobial Host Defense." *Cell Reports* 13 (9): 1922.

Ihara, S., Y. Hirata, T. Serizawa, N. Suzuki, K. Sakitani, H. Kinoshita et al. 2016. "TGF-β Signaling in Dendritic Cells Governs Colonic Homeostasis by Controlling Epithelial Differentiation and the Luminal Microbiota." *Journal of Immunology* 196 (11): 4603–13.

Irazoqui, J. E., E. R. Troemel, R. L. Feinbaum, L. G. Luhachack, B. O. Cezairliyan, and F. M. Ausubel. 2010. "Distinct Pathogenesis and Host Responses during Infection of *C. elegans* by *P. Aeruginosa* and *S. Aureus*." *PLoS Pathogens* 6 (7): 1–24.

Juge, N. 2012. "Microbial Adhesins to Gastrointestinal Mucus." *Trends in Microbiology* 20 (1): 30–39.

Kashyap, P. C., A. Marcobal, L. K. Ursell, S. A. Smits, E. D. Sonnenburg, E. K. Costello et al. 2013. "Genetically Dictated Change in Host Mucus Carbohydrate Landscape Exerts a Diet-Dependent Effect on the Gut Microbiota." *Proceedings of the National Academy of Sciences of the United States of America* 110 (42): 17059–64.

Kissoyan, K. A. B., M. Drechsler, E.-L. Stange, J. Zimmermann, C. Kaleta, H. B. Bode et al. 2019. "Natural *C. elegans* Microbiota Protects against Infection via Production of a Cyclic Lipopeptide of the Viscosin Group." *Current Biology: CB* 29 (6): 1030–37.e5.

Kremer, N., E. E. R. Philipp, M.-C. Carpentier, C. A. Brennan, L. Kraemer, M. A. Altura et al. 2013. "Initial Symbiont Contact Orchestrates Host-Organ-Wide Transcriptional Changes That Prime Tissue Colonization." *Cell Host & Microbe* 14 (2): 183–94.

Lee, S.-H., S. Omi, N. Thakur, C. Taffoni, J. Belougne, I. Engelmann et al. 2018. "Modulatory Upregulation of an Insulin Peptide Gene by Different Pathogens in *C. elegans*." *Virulence* 9 (1): 648–58.

Levy, M., C. A. Thaiss, D. Zeevi, L. Dohnalová, G. Zilberman-Schapira, J. A. Mahdi et al. 2015. "Microbiota-Modulated Metabolites Shape the Intestinal Microenvironment by Regulating NLRP6 Inflammasome Signaling." *Cell* 163 (6): 1428–43.

Liu, L., C. Ruediger, and M. Shapira. 2018. "Integration of Stress Signaling in *Caenorhabditis elegans* through Cell-Nonautonomous Contributions of the JNK Homolog KGB-1." *Genetics* 210 (4): 1317–28.

Maduzia, L. L., E. Yu, and Y. Zhang. 2011. "*Caenorhabditis elegans* Galectins LEC-6 and LEC-10 Interact with Similar Glycoconjugates in the Intestine." *The Journal of Biological Chemistry* 286 (6): 4371–81.

Mallo, G. V., C. Léopold Kurz, C. Couillault, N. Pujol, S. Granjeaud, Y. Kohara et al. 2002. "Inducible Antibacterial Defense System in *C. elegans*." *Current Biology: CB* 12 (14): 1209–14.

McFall-Ngai, M. 2007. "Adaptive Immunity: Care for the Community." *Nature* 445 (7124): 153.

McLoughlin, K., J. Schluter, S. Rakoff-Nahoum, A. L. Smith, and K. R. Foster. 2016. "Host Selection of Microbiota via Differential Adhesion." *Cell Host & Microbe* 19 (4): 550–59.

Meisel, J. D., O. Panda, P. Mahanti, F. C. Schroeder, and D. H. Kim. 2014. "Chemosensation of Bacterial Secondary Metabolites Modulates Neuroendocrine Signaling and Behavior of *C. elegans*." *Cell* 159 (2): 267–80.

Melo, J. A., and G. Ruvkun. 2012. "Inactivation of Conserved *C. elegans* Genes Engages Pathogen- and Xenobiotic-Associated Defenses." *Cell* 149 (2): 452–66.

Michel, T., J. M. Reichhart, J. A. Hoffmann, and J. Royet. 2001. "Drosophila Toll Is Activated by Gram-Positive Bacteria through a Circulating Peptidoglycan Recognition Protein." *Nature* 414 (6865): 756–59.

Montalvo-Katz, S., H. Huang, M. D. Appel, M. Berg, and M. Shapira. 2013. "Association with Soil Bacteria Enhances p38-Dependent Infection Resistance in *Caenorhabditis elegans*." *Infection and Immunity* 81 (2): 514–20.

Morita, K., A. J. Flemming, Y. Sugihara, M. Mochii, Y. Suzuki, S. Yoshida et al. 2002. "A *Caenorhabditis elegans* TGF-β, DBL-1, Controls the Expression of LON-1, a PR-Related Protein, That Regulates Polyploidization and Body Length." *The EMBO Journal* 21 (5): 1063–73.

Nishiguchi, M. K., E. G. Ruby, and M. J. McFall-Ngai. 1998. "Competitive Dominance among Strains of Luminous Bacteria Provides an Unusual Form of Evidence for Parallel Evolution in Sepiolid Squid-*Vibrio* Symbioses." *Applied and Environmental Microbiology* 64 (9): 3209–13.

Nyholm, S. V., B. Deplancke, H. Rex Gaskins, M. A. Apicella, and M. J. McFall-Ngai. 2002. "Roles of *Vibrio* Fischeri and Nonsymbiotic Bacteria in the Dynamics of Mucus Secretion during Symbiont Colonization of the Euprymna Scolopes Light Organ." *Applied and Environmental Microbiology* 68 (10): 5113–22.

Nyholm, S. V., and M. J. McFall-Ngai. 2003. "Dominance of *Vibrio* Fischeri in Secreted Mucus Outside the Light Organ of Euprymna Scolopes: The First Site of Symbiont Specificity." *Applied and Environmental Microbiology* 69 (7): 3932–37.

O'Rourke, D., D. Baban, M. Demidova, R. Mott, and J. Hodgkin. 2006. "Genomic Clusters, Putative Pathogen Recognition Molecules, and Antimicrobial Genes Are Induced by Infection of C. elegans with M. Nematophilum." *Genome Research* 16 (8): 1005–16.

Pees, B., W. Yang, A. Zárate-Potes, H. Schulenburg, and K. Dierking. 2016. "High Innate Immune Specificity through Diversified C-Type Lectin-Like Domain Proteins in Invertebrates." *Journal of Innate Immunity* 8 (2): 129–42.

Peschel, A., and H.-G. Sahl. 2006. "The Co-Evolution of Host Cationic Antimicrobial Peptides and Microbial Resistance." *Nature Reviews. Microbiology* 4 (7): 529–36.

Pickard, J. M., C. F. Maurice, M. A. Kinnebrew, M. C. Abt, D. Schenten, T. V. Golovkina et al. 2014. "Rapid Fucosylation of Intestinal Epithelium Sustains Host-Commensal Symbiosis in Sickness." *Nature* 514 (7524): 638–41.

Pujol, N., S. Cypowyj, K. Ziegler, A. Millet, A. Astrain, A. Goncharov et al. 2008. "Distinct Innate Immune Responses to Infection and Wounding in the C. elegans Epidermis." *Current Biology: CB* 18 (7): 481–89.

Pujol, N., E. M. Link, L. X. Liu, C. Léopold Kurz, G. Alloing, M.-W. Tan et al. 2001. "A Reverse Genetic Analysis of Components of the Toll Signaling Pathway in *Caenorhabditis elegans*." *Current Biology: CB* 11 (11): 809–21.

Pukkila-Worley, R., and F. M. Ausubel. 2012. "Immune Defense Mechanisms in the *Caenorhabditis elegans* Intestinal Epithelium." *Current Opinion in Immunology*. doi: 10.1016/j.coi.2011.10.004

Reddy, K. C., T. L. Dunbar, A. M. Nargund, C. M. Haynes, and E. R. Troemel. 2016. "The C. elegans CCAAT-Enhancer-Binding Protein Gamma Is Required for Surveillance Immunity." *Cell Reports* 14 (7): 1581–89.

Rhoades, J. L., J. C. Nelson, I. Nwabudike, S. K. Yu, I. G. McLachlan, G. K. Madan et al. 2019. "ASICs Mediate Food Responses in an Enteric Serotonergic Neuron That Controls Foraging Behaviors." *Cell* 176 (1–2): 85–97.e14.

Roeder, T., M. Stanisak, C. Gelhaus, I. Bruchhaus, J. Grötzinger, and M. Leippe. 2010. "Caenopores Are Antimicrobial Peptides in the Nematode *Caenorhabditis elegans* Instrumental in Nutrition and Immunity." *Developmental and Comparative Immunology* 34 (2): 203–9.

Romoli, O., A. Saviane, A. Bozzato, P. D'Antona, G. Tettamanti, A. Squartini et al. 2017. "Differential Sensitivity to Infections and Antimicrobial Peptide-Mediated Immune

Response in Four Silkworm Strains with Different Geographical Origin." *Scientific Reports* 7 (1): 1048.

Ryan, D. A., R. M. Miller, K. Lee, S. J. Neal, K. A. Fagan, P. Sengupta et al. 2014. "Sex, Age, and Hunger Regulate Behavioral Prioritization through Dynamic Modulation of Chemoreceptor Expression." *Current Biology: CB* 24 (21): 2509–17.

Salger, S. A., K. R. Cassady, B. J. Reading, and E. J. Noga. 2016. "A Diverse Family of Host-Defense Peptides (Piscidins) Exhibit Specialized Anti-Bacterial and Anti-Protozoal Activities in Fishes." *PLOS ONE* 11 (8): e0159423.

Samuel, B. S., H. Rowedder, C. Braendle, M.-A. Félix, and G. Ruvkun. 2016. "*Caenorhabditis elegans* Responses to Bacteria from Its Natural Habitats." *Proceedings of the National Academy of Sciences of the United States of America* 113 (27): E3941–49.

Schmitt, P., R. D. Rosa, and D. Destoumieux-Garzón. 2016. "An Intimate Link between Antimicrobial Peptide Sequence Diversity and Binding to Essential Components of Bacterial Membranes." *Biochimica et Biophysica Acta* 1858 (5): 958–70.

Schwartzman, J. A., and E. G. Ruby. 2016. "A Conserved Chemical Dialog of Mutualism: Lessons from Squid and *Vibrio*." *Microbes and Infection/Institut Pasteur* 18 (1): 1–10.

Shapira, M., B. J. Hamlin, J. Rong, K. Chen, M. Ronen, and M.-W. Tan. 2006. "A Conserved Role for a GATA Transcription Factor in Regulating Epithelial Innate Immune Responses." *Proceedings of the National Academy of Sciences of the United States of America* 103 (38): 14086–91.

Shivers, R. P., D. J. Pagano, T. Kooistra, C. E. Richardson, K. C. Reddy, J. K. Whitney et al. 2010. "Phosphorylation of the Conserved Transcription Factor ATF-7 by PMK-1 p38 MAPK Regulates Innate Immunity in *Caenorhabditis elegans*." *PLoS Genetics* 6 (4): e1000892.

Simonsen, K. T., J. Møller-Jensen, A. R. Kristensen, J. S. Andersen, D. L. Riddle, and B. H. Kallipolitis. 2011. "Quantitative Proteomics Identifies Ferritin in the Innate Immune Response of *C. elegans*." *Virulence* 2 (2): 120–30.

Suzuki, Y., M. D. Yandell, P. J. Roy, S. Krishna, C. Savage-Dunn, R. M. Ross et al. 1999. "A BMP Homolog Acts as a Dose-Dependent Regulator of Body Size and Male Tail Patterning in *Caenorhabditis elegans*." *Development*126 (2): 241–50.

Taylor, R. C., and A. Dillin. 2013. "XXBP-1 Is a Cell-Nonautonomous Regulator of Stress Resistance and Longevity." *Cell* 153 (7): 1435.

Tenor, J. L., and A. Aballay. 2008. "A Conserved Toll-like Receptor Is Required for *Caenorhabditis elegans* Innate Immunity." *EMBO Reports* 9 (1): 103–9.

Thaiss, C. A., N. Zmora, M. Levy, and E. Elinav. 2016. "The Microbiome and Innate Immunity." *Nature* 535 (7610): 65–74.

Tjahjono, E., and N. V. Kirienko. 2017. "A Conserved Mitochondrial Surveillance Pathway Is Required for Defense against *Pseudomonas aeruginosa*." *PLoS Genetics* 13 (6): e1006876.

Twumasi-Boateng, K., and M. Shapira. 2012. "Dissociation of Immune Responses from Pathogen Colonization Supports Pattern Recognition in *C. elegans*." *PLOS ONE* 7 (4): e35400.

Umu, Ö. C. O., K. Rudi, and D. B. Diep. 2017. "Modulation of the Gut Microbiota by Prebiotic Fibres and Bacteriocins." *Microbial Ecology in Health and Disease* 28 (1): 1348886.

Vaishnava, S., M. Yamamoto, K. M. Severson, K. A. Ruhn, X. Yu, O. Koren et al. 2011. "The Antibacterial Lectin RegIII$_\gamma$ Promotes the Spatial Segregation of Microbiota and Host in the Intestine." *Science* 334 (6053): 255–58.

van der Hoeven, R., K. C. McCallum, M. R. Cruz, and D. A. Garsin. 2011. "Ce-Duox1/BLI-3 Generated Reactive Oxygen Species Trigger Protective SKN-1 Activity via p38 MAPK Signaling during Infection in *C. elegans*." *PLoS Pathogens* 7 (12): e1002453.

Van Herreweghen, F., K. De Paepe, H. Roume, F. M. Kerckhof, and T. Van de Wiele. 2018. "Mucin Degradation Niche as a Driver of Microbiome Composition and *Akkermansia muciniphila* Abundance in a Dynamic Gut Model Is Donor Independent." *FEMS Microbiology Ecology* 94 (12): 1–13.

Visvikis, O., N. Ihuegbu, S. A. Labed, L. G. Luhachack, A.-M. F. Alves, A. C. Wollenberg et al. 2014. "Innate Host Defense Requires TFEB-Mediated Transcription of Cytoprotective and Antimicrobial Genes." *Immunity* 40 (6): 896–909.

Wang, J., R. Tokarz, and C. Savage-Dunn. 2002. "The Expression of TGFβ Signal Transducers in the Hypodermis Regulates Body Size in *C. elegans.*" *Development* 129 (21): 4989–98.

Watson, E., L. T. MacNeil, A. D. Ritter, L. S. Yilmaz, A. P. Rosebrock, A. A. Caudy et al. 2014. "Interspecies Systems Biology Uncovers Metabolites Affecting *C. elegans* Gene Expression and Life History Traits." *Cell* 156 (4): 759–70.

Watson, E., V. Olin-Sandoval, M. J. Hoy, C.-H. Li, T. Louisse, V. Yao et al. 2016. "Metabolic Network Rewiring of Propionate Flux Compensates Vitamin B12 Deficiency in *C. elegans.*" *eLife* 5 (July). doi: 10.7554/eLife.17670

Wong, D., D. Bazopoulou, N. Pujol, N. Tavernarakis, and J. J. Ewbank. 2007. "Genome-Wide Investigation Reveals Pathogen-Specific and Shared Signatures in the Response of *Caenorhabditis elegans* to Infection." *Genome Biology* 8 (9): R194.

Yang, W., K. Dierking, and P. C. Rosenstiel. 2016. "GATA Transcription Factor as a Likely Key Regulator of the *Caenorhabditis elegans* Innate Immune Response against Gut Pathogens." *Zoology* 119 (4): 244–53.

Yang, W., C. Petersen, B. Pees, J. Zimmermann, S. Waschina, P. Dirksen et al. 2019. "The Inducible Response of the Nematode *Caenorhabditis elegans* to Members of Its Natural Microbiota across Development and Adult Life." *Frontiers in Microbiology* 10: 1793.

Zhang, J., A. D. Holdorf, and A. J. M. Walhout. 2017. "*C. elegans* and Its Bacterial Diet as a Model for Systems-Level Understanding of Host–microbiota Interactions." *Current Opinion in Biotechnology* 46: 74–80.

Zhang, X., and Y. Zhang. 2012. "DBL-1, a TGF-β, Is Essential for *Caenorhabditis elegans* Aversive Olfactory Learning." *Proceedings of the National Academy of Sciences of the United States of America* 109 (42): 17081–86.

Zhang, Y., W. Li, L. Li, Y. Li, R. Fu, Y. Zhu et al. 2015. "Structural Damage in the *C. elegans* Epidermis Causes Release of STA-2 and Induction of an Innate Immune Response." *Immunity* 42 (2): 309–20.

Zheng, S., H. Chiu, J. Boudreau, T. Papanicolaou, W. Bendena, and I. Chin-Sang. 2018. "A Functional Study of All 40 *Caenorhabditis elegans* Insulin-like Peptides." *The Journal of Biological Chemistry* 293 (43): 16912–22.

Zimmermann, J., N. Obeng, W. Yang, B. Pees, C. Petersen, S. Waschina et al. 2019. "The Functional Repertoire Contained within the Native Microbiota of the Model Nematode *Caenorhabditis elegans.*" *The ISME Journal*, September. doi: 10.1038/s41396-019-0504-y

Zugasti, O., N. Bose, B. Squiban, J. Belougne, C. Léopold Kurz, F. C. Schroeder et al. 2014. "Activation of a G Protein-Coupled Receptor by Its Endogenous Ligand Triggers the Innate Immune Response of *Caenorhabditis elegans.*" *Nature Immunology* 15 (9): 833–38.

Zugasti, O., and J. J. Ewbank. 2009. "Neuroimmune Regulation of Antimicrobial Peptide Expression by a Noncanonical TGF-β Signaling Pathway in *Caenorhabditis elegans* Epidermis." *Nature Immunology* 10 (3): 249–56.

9 Multiple roles of bacterially produced natural products in the bryozoan *Bugula neritina*

Nicole B. Lopanik

Contents

9.1 Introduction

Natural products, or secondary metabolites, can play an important role in the life-history of the producing organism (Puglisi et al. 2019). Sessile organisms that lack a hard outer structure such as plants, sponges, cnidarians, ascidians, and bryozoans, often possess a chemical defense that helps them survive herbivore or predatory activity, or attack from pathogens. As many marine natural products isolated from marine invertebrates have structures similar to compounds produced by terrestrial bacteria, it is hypothesized that microbial symbionts are often the true source of these defensive metabolites (Piel 2009; Lopanik 2014). One of the first documented examples of this type of defensive symbiosis was in two crustaceans, whose eggs were covered with a bacterium producing an antibiotic molecule that protected the eggs from a pathogenic fungus (Gil-Turnes et al. 1989; Gil-Turnes and Fenical 1992). In the terrestrial environment, the eggs of rove beetles are protected from predatory wolf spider by the distasteful compound pederin, which is produced by a bacterial symbiont (Kellner 1999, 2002, 2001). These examples demonstrate that microbial symbionts can participate in complex tritrophic interactions in conjunction with their host and host predators, via the production of defensive compounds.

On the other hand, humans have taken advantage of the complex structure and subsequent bioactivity of natural products for use as antibiotics, anticancer drugs, and anthelmintic drugs (Newman and Cragg 2016; Newman 2019). The ability of microorganisms to produce bioactive compounds is well known. For instance, the anticancer compound mitomycin C is produced by the Actinomycete *Streptomyces lavendulae* (Mao et al. 1999), and the antibiotic erythromycin is biosynthesized by *Saccharopolyspora erythraea* (Cortes et al. 1990). Marine invertebrates such as sponges, tunicates, corals, and bryozoans are another prolific source of novel natural products (recently reviewed in Blunt et al. 2017, 2018). Under closer investigation, it is often discovered that microbial associates are, in fact, responsible for the production of these compounds (Newman and Cragg 2015; Blockley et al. 2017).

While the ecological role of natural products is often characterized by chemical ecologists and their pharmaceutical potential determined by natural product chemists, it is not very common for both attributes to be understood in the environmental context of the source organism, whether metazoan or microorganism. It seems likely, though, that bioactive natural products primarily studied for their pharmaceutical activity, are important in the life history of the producing organism, as these compounds are complex and physiologically expensive to produce. The bioactivity of natural products can affect the producing organism, resulting in, for instance, antibiotic resistance mechanisms in antibiotic-producing microorganisms (Walsh 2003).

Compounds that affect eukaryotic genes or processes produced by a microbial associate in close proximity of a metazoan host have the potential to impact host physiology. The effects of symbiont-produced natural products on eukaryotic cells have been demonstrated for a small number of symbionts that inhabit terrestrial and marine hosts. For instance, swinholide A, a dilactone isolated from unicellular heterotrophic bacteria in the sponge *Theonella swinhoei* (Bewley et al. 1996), interferes with actin filament polymerization (Bubb et al. 1995). The compound, ecteinascidin 743, produced by a microbial symbiont of the marine tunicate *Ecteinascidia turbinata* (Fortman and Sherman 2005; Piel 2009; Schmidt et al. 2012), impedes DNA repair processes through several mechanisms including interfering with DNA transcription factors and binding proteins (van Kesteren et al. 2003). A few studies have investigated how the host responds or tolerates bioactive metabolites produced by the symbiont. *Photorhabdus luminescens*, a symbiont of entomopathogenic nematodes and an insect pathogen, produces an antibiotic stilbene that inhibits the growth of competing bacteria and fungi on the decaying insect (Li et al. 1995). Later studies demonstrated that stilbene suppresses insect host defense by inhibiting phenoloxidase activity (Eleftherianos et al. 2007). Furthermore, juvenile nematodes colonized with *P. luminescens* mutants lacking a stilbene biosynthesis gene were significantly reduced in their ability to develop into the next life stage compared to juveniles with wild-type or complemented *P. luminscens* (Joyce et al. 2008), demonstrating that symbiont-produced natural products may be important to multiple partners. The intracellular β-proteobacterial symbiont of the fungus *Rhizopus* sp. produces the polyketide molecule, rhizoxin, which is the causative agent of rice seedling blight (Partida-Martinez and Hertweck 2005). Rhizoxin binds to tubulin and inhibits polymerization (Hamel 1992), resulting in significant anticancer activity (Hendriks et al. 1992; Kerr et al. 1995; Hanauske et al. 1996; Kaplan et al. 1996; McLeod et al. 1996; Verweij et al. 1996). The fungal *Rhizopus* sp. host is able to resist the β-tubulin-binding rhizoxins by an amino acid substitution in the β-tubulin gene, resulting in a reduced affinity of rhizoxin that is not present in fungi that do not host the symbiont (Schmitt et al. 2008). These studies highlight the dramatic influence of symbiont-produced natural products on their partners, and that these compounds can impact hosts via internal or external mechanisms.

9.2   Bryozoans, *Bugula* spp., and *Bugula neritina*

Bryozoans are filter-feeding colonial invertebrates that primarily inhabit hard substrates in marine environments, and are frequently fouling organisms on the hulls of ships and boats (Figure 9.1A). The colony is composed of zooids, individuals that are typically ~1 mm in length. Bryozoans are distinguished by the presence of a lophophore, a unique feeding structure comprised of a ring of tentacles surrounding the mouth, and are typically protected by a chitinous or calcareous exoskeleton. Because of their small body size, they do not possess any organs for gas exchange or waste removal. Gases and ammonia diffuse across the body wall and other wastes are incorporated into brown bodies, zooids that degenerate after a period of time. One important feature of the colonial body organization of bryozoans are funicular cords that connect the zooids within the colony and allow transport of nutrients and wastes throughout the colony. In some species, funicular cords are also responsible for providing the growing embryo in the ovicell with nutrients (Woollacott and Zimmer 1975). Bryozoans reproduce both sexually and asexually. Most bryozoans are sequential hermaphrodites; colonies may have both female and male zooids, and both self-fertilization and cross-fertilization may occur. Embryos can be brooded either internally or externally. Bryozoans in the genus *Bugula* (Class: Gymnolaemata, Order: Cheilostomata) are arborescent with a worldwide distribution found mostly in temperate habitats, and can be important members in benthic communities. In *Bugula* spp., the embryos are brooded in external calcareous chambers, termed ovicells (Figure 9.1) (Woollacott and Zimmer 1972). Only some of the zooids are reproductive at a given time, with ovicell-free zooids typically at the base and growing tips, and ovicell-bearing zooids in the middle of the branches. The embryo develops for 2–3 weeks before it is released into the water column by the adult zooid. *Bugula neritina* is found in temperate marine habitats throughout the world (Figure 9.1B) (Banta 1980) and is well-studied, in part, because of its accessibility, and because it is the source of bioactive natural products, the bryostatins (see below). *B. neritina* larvae are lecithotrophic and contain energy reserves that allow them to swim, settle, and metamorphose without feeding until they develop into a juvenile (Figure 9.1D) (Woollacott and Zimmer 1971). The larvae reside in the water column for only a short time (0.5–2 hrs) before they settle, and, as such, larval dispersal is fairly limited (Keough 1989; Keough and Chernoff 1987). However, *B. neritina* is thought to also

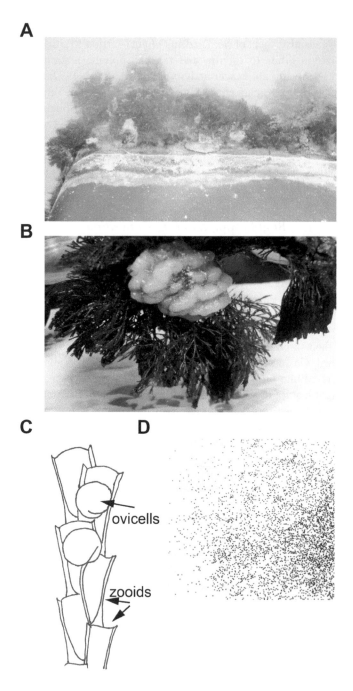

FIGURE 9.1 *Bugula neritina.* (A) Colonies of *B. neritina* attached to a floating dock in Beaufort, NC. (B) Full-sized colonies growing with two tunicates. (C) Drawing of branch with zooids and reproductive ovicells. (D) *B. neritina* larvae.

be dispersed anthropogenically as it is often found on the hulls of ships (Mackie et al. 2006), and can be invasive in some areas (Ryland et al. 2011). Once a larva settles, it rapidly proceeds through metamorphosis ($\sim$48 hrs) and the juvenile asexually divides and bifurcates to form a full-sized colony.

9.3 Bryostatins

Bugula neritina is the source of the bioactive polyketide metabolites, bryostatins. Interest in *B. neritina* began in the late 1960s when researchers were prospecting for pharmaceutical leads from marine animals, and discovered that extracts had anticancer activity (Pettit et al. 1970). Approximately 12 years later, the same research group published the structure of bryostatin 1 (Figure 9.2A), the first compound isolated from *B. neritina* with anticancer activity (Pettit et al. 1982). Over the subsequent 30 years, 20 different structures of bryostatins have been elucidated from different populations of *B. neritina* (Figure 9.2) (Pettit 1991; Pettit et al. 1996; Lin et al. 1998; Lopanik et al. 2004a; Yu et al. 2015). Besides anticancer activity (Newman 2012), bryostatin 1 has been tested for activity against Alzheimer's disease (Nelson et al. 2009, 2017; Farlow et al. 2019), HIV (Mehla et al. 2010; Martinez-Bonet et al. 2015; Gutierrez et al. 2016), and multiple sclerosis (Kornberg et al. 2018).

Bryostatin 1 binds with high affinity to the C1b region of the diacyl glycerol (DAG) binding regulatory region of protein kinase C (PKC) (Kraft et al. 1986; De Vries et al. 1988), a serine-threonine kinase that is essential in signaling cascades and implicated in a variety of processes including protein secretion, cell pH alteration, cell growth, and modification to the cytoskeleton (Newton 1995; Akita 2008). Of the 10 different isoenzymes of PKC, bryostatin binds to the conventional (cPKC, α, β_I, β_{II}, γ) and the novel (nPKC, δ, ε, η, θ) isoforms, but not to the atypical forms (aPKC, ξ, λ/ι), as they lack the C1b binding domain (Mutter and Wills 2000). Inactive PKC is found in the cytosol where the pseudosubstrate blocks the active site (Newton 2001). For c- and nPKCs, DAG released by phospholipases binds to PKC, which is allosterically activated. Its affinity for phosphatidyl serine is increased and the affinity for Ca^{2+} is shifted to the physiological range, resulting in release of the pseudosubstrate by a conformational change and migration to the cell membrane, the location of its substrates and regulators. Differences in the regulatory domains of the three isoforms dictate the essential cofactors necessary for activation. After binding to bryostatin, PKC is activated briefly, autophosphorylated, translocated to the cell membrane, and is then downregulated by ubiquitination and subsequently degraded by proteasomes (Clamp and Jayson 2002). Bryostatins with slight structural variations display a dramatic difference in activating PKC isoenzymes (Wender et al. 2011). For instance, bryostatin 1 induces rapid translocation of PKCβ conjugated to green fluorescent protein (GFP) to the membrane of CHO cells, whereas bryostatin 2, which only differs from bryostatin 1 by an acetate group on C7, has no effect on PKCβ. Further, bryostatin 1 induces greater PKCα degradation after prolonged activation, compared to PKCδ and PKCε. The selectivity of the bryostatins for different PKC isoforms suggests that these compounds could be used as regulators of PKC.

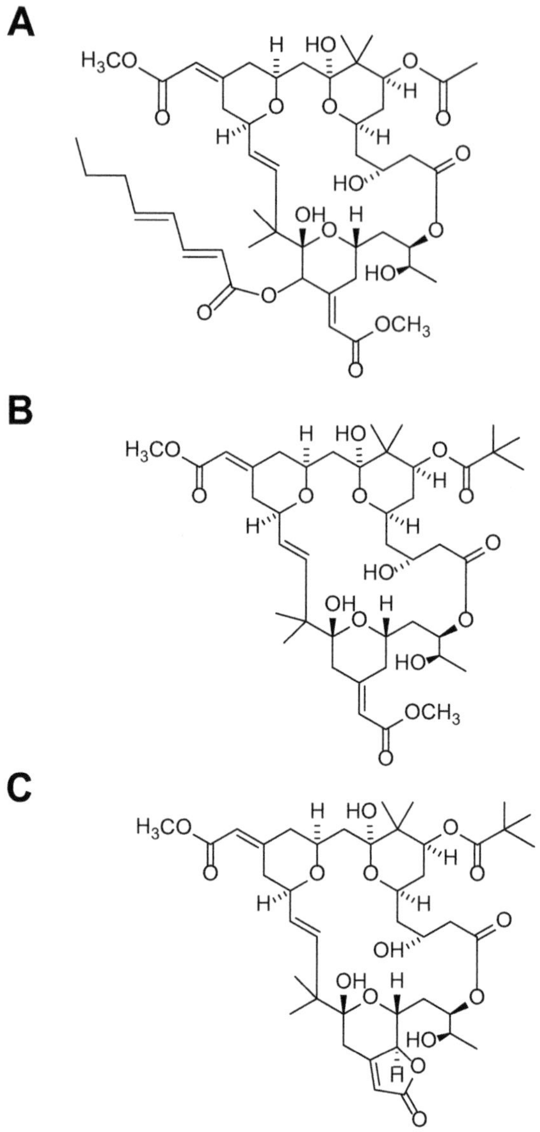

FIGURE 9.2 Some of the bryostatins found in different populations of *B. neritina*. (A) Bryostatin 1. (B) Bryostatin 10. (C) Bryostatin 20.

9.4 Bryostatin production by the bacterial symbiont of *B. neritina*

Extracellular rod-shaped bacteria were observed in the funicular cords of adult *B. neritina* colonies (Woollacott and Zimmer 1975) and in the pallial sinus (a groove on the aboral pole) of larvae (Woollacott 1981). The presence of only one bacterial phylotype, a γ-Proteobacterium, was confirmed in larvae via *in situ* hybridization

(Haygood and Davidson 1997). As the symbiont could not be cultured, it was named "*Candidatus* Endobugula sertula." *In situ* hybridizations in adult zooids and rhizoids have confirmed that *E. sertula* resides in the funicular cords (Sharp et al. 2007). Adult colonies antibiotic-cured of symbionts contained 50% less bryostatins than control animals, suggesting that the symbiont produces the bryostatins (Davidson et al. 2001). Next-generation larvae of cured adult colonies did not possess *E. sertula* or bryostatins (including the deterrent bryostatins 10 and 20), and were not deterrent to predators (Lopanik et al. 2004b). Size was similar in symbiotic and aposymbiotic colonies, suggesting that *E. sertula* does not contribute nutritionally to *B. neritina* (Lopanik et al. 2004b). The polyketide synthase gene cluster that putatively prescribes biosynthesis of the bryostatins was sequenced in two different populations of *B. neritina* (Type S from NC, and Type D from California [CA], see below) with minor variations in bryostatin structure (Sudek et al. 2007). Portions of this gene cluster were also shown to be absent in aposymbiotic larvae (Davidson et al. 2001; Lopanik et al. 2006a). Although analyses of two independent antibiotic curing experiments with populations from the western Atlantic and eastern Pacific support the hypothesis that *E. sertula* produces bryostatins (Davidson et al. 2001; Lopanik et al. 2004b, 2006a; Sudek et al. 2007), it has not been unequivocally demonstrated as cultivation of the symbiont has been unsuccessful to date.

9.5   Defensive role of bryostatins

While the pharmacological activity of bryostatins has been studied since the 1980s, chemical ecology researchers demonstrated that bryostatins act as a chemical defense for *B. neritina*, as larvae and their crude extracts from shallow populations in North Carolina (NC) are unpalatable to both vertebrate and invertebrate predators, while adult extracts are not deterrent (Lindquist 1996; Lindquist and Hay 1996; Tamburri and Zimmer-Faust 1996). Bryostatins are much more concentrated in larvae ($\sim$20$\times$) than in adult ovicell-free colonies (Lopanik et al. 2006b), explaining patterns of larval palatability (Lindquist and Hay 1996), and appear to be localized to the outside of the larva (Lopanik et al. 2004b; Sharp et al. 2007), where they would be most effective as a chemical defense. More than 90% of *B. neritina* larvae that were rejected by vertebrate and invertebrate predators successfully settled and metamorphosed, indicating that their chemical defense is extremely potent (Lindquist 1996; Lindquist and Hay 1996). Using bioassay-guided fractionation, three deterrent and three non-deterrent bryostatins were isolated from NC *B. neritina* (Lopanik et al. 2004a,b). Chemically defended larvae use strong photic cues to choose appropriate microenvironments that reduce juvenile exposure to environmental and biological stresses (Olson 1985; Young 1986; Young and Bingham 1987; Young and Chia 1987; Walters 1992; Lindquist and Hay 1996). For colonial adults able to regenerate after bouts of predation (Jackson 1985), a chemical defense may be too costly. This may be the case in adult *B. neritina*.

   Despite the utility of possessing defensive bryostatins, *B. neritina* colonies without *E. sertula* have been documented in different locations. The first "aposymbiotic" *B. neritina* colonies were collected in Delaware and Connecticut, whereas all colonies analyzed from North Carolina, Florida, and Louisiana possessed the symbiont

(McGovern and Hellberg 2003). Interestingly, the colonies collected from the northern locations varied genetically from their more southern counterparts. Previous work by Haygood and collaborators showed a similar genetic divergence in *B. neritina* in California populations (Davidson and Haygood 1999). These two studies established a complex of three *B. neritina* sibling species in the western Atlantic and eastern Pacific (Type D, Type N, and Type S), which was further confirmed by other studies (Mackie et al. 2006; Fehlauer-Ale et al. 2014). Moreover, the two sibling species in California possessed symbionts with minor genetic variation in their 16S rDNA sequences, in addition to differences in bryostatin composition (Davidson and Haygood 1999).

The variation in symbiotic status in *B. neritina* in the western Atlantic, with symbiotic colonies predominately at low latitudes and "aposymbiotic" colonies at higher latitudes, coupled to the role of bryostatins as a defensive compound, led to the hypothesis that biogeographical variation in predation pressure (i.e., lower predation pressure at higher latitudes [Vermeij 1978]) impacts the distribution of the symbiont. In this scenario, in the presence of predators, the symbiont imposes a physiological cost to the host that is smaller than the benefit of defensive bryostatins, so the symbiont is retained. In areas of low predation pressure, the physiological cost of hosting the symbiont is too high, and the symbiont is removed. In previous studies of *B. neritina* populations along the East Coast, the host colonies that possessed the symbiont were all a specific genotype, Type S, while Type N colonies lacked the symbiont. A more recent and extensive investigation of these populations revealed that a small number of colonies from low latitudes did not possess the symbiont, and that some collected at higher latitudes did possess the symbiont (Linneman et al. 2014). The host genotypes were also more widespread than previously thought, with Type S colonies at higher latitudes, and Type N at lower. Moreover, we found Type S colonies without the symbiont and Type N colonies that had appeared to obtain the symbiont. In California, Haygood and coworkers showed that the sibling species of *B. neritina* (Type D and Type S) had variants of the symbiont, *E. sertula* (Davidson and Haygood 1999). Interestingly, they found that bryostatin composition varied as well, suggesting that the different strains of symbiont produce different bryostatins. These variant host/symbiont/ bryostatin colonies were found at the same location, but distributed with depth: Type D colonies were found deeper, while the Type S colonies were in shallow locations, suggesting that a biotic or abiotic environmental factor may affect their distribution. No Type N colonies were found in California in that study, so their discovery in the Western Atlantic was unexpected.

Phylogenetic analysis by Lim-Fong and Haygood (Lim-Fong et al. 2008) had suggested that the host sibling species and symbiont strains had coevolved, which fit with the hypothesized vertical transmission of the symbiont from parent to offspring via the larval stage (Woollacott 1981; Haygood and Davidson 1997). Interestingly, the symbiont in the Type N colonies from the Western Atlantic were found to be identical to those from the Type S colonies based on 16S rDNA and ITS sequencing (Linneman et al. 2014), which has important implications regarding symbiont transmission and environmental reservoirs. Further, the Type N larvae appeared to possess bryostatins, suggesting that these larvae are defended as well. No *E. sertula* sequences were found in high-throughput sequencing microbial community analysis of seawater collected adjacent to *B. neritina* colonies (Patin et al. 2019), so it is unclear how Type S strain *E. sertula* has been incorporated into Type N colonies, and suggests that the relationship between *B. neritina* and *E. sertula* may be more flexible than previously thought.

More recently, we have performed a more systematic approach to sampling populations along the Western Atlantic to assess the frequency of the symbiont and host type. We found that symbiont frequency in colonies is positively correlated with latitude, while host type is not (Lopanik et al., in prep), further supporting the hypothesis that symbiont distribution is impacted by biogeographical patterns in predation. There were, however, significant differences among host/symbiont populations in close geographic proximity suggesting that other factors can impact the distribution of host and symbiont types. While symbiont-produced bryostatins may play a role in *B. neritina* survival, other environmental factors obviously impact this as well.

9.6   Impacts of symbiont and symbiont-produced metabolites on host physiology

The bryostatins bind with high affinity to protein kinase C (PKC), a eukaryotic protein involved in many different cellular processes. In adult colonies, symbiont cells are located in funicular cords that connect the individual zooids for the transport of nutrients and waste through the colony. Moreover, *B. neritina* larvae possess high concentrations of bryostatins (Lopanik et al. 2004b, 2006b), although the bryostatins appear to be localized to the outside of the larva (Lopanik et al. 2004a, b; Sharp et al. 2007). It is possible that symbiont-produced bryostatins may affect PKC in the proximal host cells, which could, in turn, impact colony traits. To test this, *B. neritina* larvae from individual colonies collected in NC were treated with an antibiotic for four days to reduce symbiont titers, and then grown in unfiltered seawater for five months. While colonies with reduced symbiont titers did not vary significantly in size compared to untreated control colonies, they had significantly fewer ovicells, the reproductive brood structures for embryo growth and development (Mathew et al. 2016). This was further confirmed in colonies with naturally reduced symbiont titers. In addition, PKC expression differed in crude protein extracts of the symbiont-reduced and control colonies, suggesting that the absence of the symbiont and/or symbiont-produced bryostatins can affect host physiology. Molecular and morphological studies showed that host reproduction was not affected at the individual zooid level, but that bryostatins may influence early differentiation of female germinal cells (Mathew et al. 2018).

One interesting aspect of this system is that the symbiont and/or symbiont-produced bryostatins appears to be important for fecundity and reproduction only in some colonies. The majority of colonies collected at higher latitudes in VA and DE do not possess the symbiont, and they are significantly more fecund than their symbiont-reduced counterparts from NC (Lopanik, unpub. data). These observations suggest that the symbiont-dependent reproduction may only be important under certain conditions. For *B. neritina* and its defensive symbiont, this may occur in regions where predation levels are higher, presumably at lower latitudes, and where water temperatures are warmer. The potential mechanism that drives this is unknown, but it seems that these host–symbiont interactions vary under differing environmental contexts.

9.7 Bryostatins and symbionts in closely related genera

Since the discovery of bryostatins in *Bugula neritina*, the symbiotic status and natural products of other closely related bryozoan species have been investigated. Many of these were originally identified as species in the genus *Bugula*, but have since been reclassified (Fehlauer-Ale et al. 2015). In 1981, Woollacott showed that like *B. neritina*, larvae of *Bugula pacifica* (now *Crisularia pacifica*) and *Bugula simplex* (now *Bugulina simplex*) possessed bacterial cells in their pallial sinus (Woollacott 1981). In contrast, larvae of *Bugula stolonifera* (now *Bugulina stolonifera*) and *Bugula turrita* (now *Crisularia turrita*) do not possess bacterial symbionts. Of the species that possess potential symbionts, Lim and Haygood (2004) found that *Bugulina simplex* in Massachusetts, USA, possesses a closely related symbiont, "*Candidatus* Endobugula glebosa," as well as bryostatin-like compounds. Later, Lim-Fong and Haygood performed a systematic investigation of closely related bugulid bryozoans to determine the evolutionary relationships between host and symbionts, and bryostatin production. They showed that *C. pacifica*, which possesses a symbiont, does not have bryostatin-like compounds (Lim-Fong et al. 2008), suggesting that bryostatin production is not intrinsic in the Endobugula symbionts.

Another area of interest is how the bugulid host PKCs have adapted to the presence of symbiont-produced bryostatins. The cellular targets of the bryostatins, classical and novel forms of protein kinase C, are signaling proteins involved in multiple processes. It seems likely that the presence of bryostatins in host tissue would reduce PKC signaling efficacy. One possible mechanism to ameliorate the effects of bryostatin on host PKC is to modify the bonding site, which is the C1b region of the diacylglycerol binding domain (Lorenzo et al. 1999). *In vitro* studies of C1b domains from differing PKC isoforms showed that amino acid variation affects bryostatin binding affinity (Irie et al. 2002). To investigate this, we collected different bugulid species colonies from locations in the Western Atlantic and Eastern Pacific, and determined their identity using COI sequences (Lim-Fong et al. 2008). Primer design was based on *B. neritina* PKC protein sequences (Mathew et al. 2016), and the C1b domain region of *Crisularia pacifica* PKCδ was sequenced from colony cDNA. Differences in PKCδ C1b domain amino acid sequences from *B. neritina*, *C. pacifica*, and the mammal *Rattus norvegicus* suggest that bryostatin binding affinity could vary (Figure 9.3A). Interestingly, the C1b domain from *C. pacifica*, which has a symbiont, but does not possess bryostatins, is more similar to the C1b domain *in R. norvegicus* (bryostatin-naïve) than the *B. neritina* C1b domain sequence is to *R. norvegicus* (Table 9.1). Molecular modeling (The PyMOL Molecular Graphics System, Version 2.0 Schrödinger, LLC) of the PKCδ C1b domain binding to phorbol ester reveals two regions with close proximity to the activator molecule (Figure 9.3B). While there is little variation in the amino acid sequences of one region (aa 22–26), another region is more variable (aa 8–12) (Figure 9.3). Using molecular modeling

A

```
                          10          20          30          40
  B. neritina  HRFKPHNF IGPHFCDHCGSLLVGI IRQGLKCEACGTNCHKRCEKLMPNLC
  C. pacifica  HRLKVHNYLTPTFCDHCGMLLVGI IRQGLKCEICGINCHKKCEKLLPNLC
R. norvegicus  HRFKVYNYMSPTFCDHCGTLLWGLVKQGLKCEDCGMNVHHKCREKVANLC
```

B

FIGURE 9.3 (A) Alignment of *B. neritina*, *C. pacifica*, and *R. norvegicus* PKCδ C1b domains. The colored lines correspond with the amino acid residues noted in (B). (B) Model of *R. norvegicus* PKCδ C1b domain structure with phorbol ester in binding site. Amino acid residues in close contact with phorbol ester highlighted in magenta and orange.

to compare the amino acid sequences, again the *B. neritina* C1b domain structure is more divergent from that of *R. norvegicus* than *C. pacifica* (Table 9.2). It seems likely that differences in amino acid residue in the PKCd C1b domains could affect bryostatin binding affinity (Irie et al. 2002), but this hypothesis should be tested experimentally.

TABLE 9.1

Percent Identity of PKCδ C1b Domain Amino Acid Sequences

% Identity	C. pacifica	R. norvegicus
B. neritina	78	58
C. pacifica	–	64

TABLE 9.2

Root Mean-Square Deviation in PKCδ C1b Domain Model Structures

RMSD	C. pacifica	R. norvegicus
B. neritina	0.220	0.396
C. pacifica	–	0.259

9.8   Future directions

While many aspects of the *B. neritina*/"*Ca*. Endobugula sertula" symbiosis system have been fairly well-studied, there are many questions that remain to be answered regarding the impacts of the symbiont and symbiont-produced bryostatins on the host ecology and physiology. For instance, are the bryostatin-like compounds found in *Bugulina simplex* similarly deterrent? Ecological studies have not been performed on other bugulid species, but the different bryostatins may vary in deterrence to specific predators. While some bryostatins extracted from *B. neritina* were deterrent to the fish predator *Lagodon rhomboides*, others were not (Lopanik et al. 2004a,b), including the bioactive bryostatin 1 (Lopanik, unpub. data). However, bryostatin 1 is not regularly found in the East coast populations of *B. neritina* used in this study. It is possible that bryostatin 1 is deterrent to predators that cooccur with producing populations. The suite of bryostatins within a host could have evolved to protect against sympatric predators. Evolutionary variation in the presence of the symbiont and bryostatins within bugulid hosts leads to many questions about host–symbiont coevolution. For instance, is reproduction in *Bugulina simplex* affected by bryostatin-like compounds as it is in *B. neritina*, and/or does the symbiont provide a chemical defense for the host? Typically *B. simplex* occurs at higher latitudes than *B. neritina*, so it is possible that these compounds may provide a chemical defense to a different suite of predators. Why does *Bugulina stolonifera*, which co-occurs with *B. neritina,* and could potentially benefit from a chemical defense, not possess a symbiont, when the closely related *B. simplex* does?

Still unanswered is how the symbiont/symbiont-produced bryostatins affect the host under different contexts, as colonies from lower latitudes with lower symbiont titers have reduced fecundity, while those from higher latitudes do not appear to be affected by symbiont absence. Is the presence of bryostatins in colonies at lower latitudes needed to trigger reproductive processes? Techniques such as transcriptomics on symbiotic and symbiont-reduced colonies from various locations may reveal how the symbiont impacts the host under different contexts. Genome editing (although this is not yet available for bryozoans) could be used to modify the host PKC C1b domains to determine how host physiology is affected by the presence of the symbiont/bryostatins under different environmental situations. In a recent study, we found that symbiont titer varies significantly (from 0% to 63% of the host microbiome [Patin et al. 2019]) within individual colonies. Transcriptomics may provide insights into how this variation impacts host physiology. Clearly, there is a complex relationship between the bugulid bryozoans and their Endobugula symbionts and symbiont-produced bryostatins. Much more needs to be done to understand all of the nuances of these host/symbiont interactions.

Acknowledgments

Much of the work described in here was conducted by Kayla Bean, Jonathan Linneman, and Meril Mathew. As usual, Niels Lindquist from the UNC-Chapel Hill Institute of Marine Sciences was very generous with use of wet and dry laboratory space. Funding was provided by NSF Biological Oceanography (1608709) and Advances in Biological Informatics (1564559).

References

Akita, Y. 2008. "Protein kinase Cε: Multiple roles in the function of, and signaling mediated by, the cytoskeleton." *FEBS Journal* 275 (16):3995–4004. doi: 10.1111/j.1742-4658.2008.06557.x.

Banta, W. C. 1980. "Bryozoa." In *Common Intertidal Invertebrates of the Gulf of California*, edited by R. C. Brusca, 359–90. Tuscon, AZ: The University of Arizona Press.

Bewley, C. A., N. D. Holland, and D. J. Faulkner. 1996. "Two classes of metabolites from *Theonella swinhoei* are localized in distinct populations of bacterial symbionts." *Experientia* 52 (7):716–22.

Blockley, A., D. R. Elliott, A. P. Roberts, and M. Sweet. 2017. "Symbiotic microbes from marine invertebrates: Driving a new era of natural product drug discovery." *Diversity-Basel* 9 (4). doi: 10.3390/d9040049.

Blunt, J. W., A. R. Carroll, B. R. Copp, R. A. Davis, R. A. Keyzers, and M. R. Prinsep. 2018. "Marine natural products." *Natural Product Reports* 35 (1):8–53. doi: 10.1039/c7np00052a.

Blunt, J. W., B. R. Copp, R. A. Keyzers, M. H. G. Munro, and M. R. Prinsep. 2017. "Marine natural products." *Natural Product Reports* 34 (3):235–94. doi: 10.1039/c6np00124f.

Bubb, M. R., I. Spector, A. D. Bershadsky, and E. D. Korn. 1995. "Swinholide A is a microfilament-disrupting marine toxin that stabilizes actin dimers and severs actin filaments." *Journal of Biological Chemistry* 270 (8):3463–6.

Clamp, A., and G. C. Jayson. 2002. "The clinical development of the bryostatins." *Anti-Cancer Drugs* 13 (7):673–83.

Cortes, J., S. F. Haydock, G. A. Roberts, D. J. Bevitt, and P. F. Leadlay. 1990. "An unusually large multifunctional polypeptide in the erythromycin-producing polyketide synthase of *Saccharopolyspora erythraea*." *Nature* 348 (6297):176–8.

Davidson, S. K., S. W. Allen, G. E. Lim, C. M. Anderson, and M. G. Haygood. 2001. "Evidence for the biosynthesis of bryostatins by the bacterial symbiont 'Candidatus Endobugula sertula' of the bryozoan *Bugula neritina*." *Applied and Environmental Microbiology* 67 (10):4531–7.

Davidson, S. K., and M. G. Haygood. 1999. "Identification of sibling species of the bryozoan *Bugula neritina* that produce different anticancer bryostatins and harbor distinct strains of the bacterial symbiont 'Candidatus Endobugula sertula'." *Biological Bulletin* 196 (3):273–80.

De Vries, D. J., C. L. Herald, G. R. Pettit, and P. M. Blumberg. 1988. "Demonstration of sub-nanomolar affinity of bryostatin 1 for the phorbol ester receptor in rat brain." *Biochemical Pharmacology* 37 (21):4069–73.

Eleftherianos, I., S. Boundy, S. A. Joyce, S. Aslam, J. W. Marshall, R. J. Cox et al. 2007. "An antibiotic produced by an insect-pathogenic bacterium suppresses host defenses through phenoloxidase inhibition." *Proceedings of the National Academy of Sciences, USA* 104 (7):2419–24.

Farlow, M. R., R. E. Thompson, L. J. Wei, A. J. Tuchman, E. Grenier, D. Crockford et al. 2019. "A randomized, double-blind, placebo-controlled, Phase II Study Assessing Safety, tolerability, and efficacy of bryostatin in the treatment of moderately severe to severe Alzheimer's disease." *Journal of Alzheimers Disease* 67 (2):555–70. doi: 10.3233/jad-180759.

Fehlauer-Ale, K. H., J. A. Mackie, G. E. Lim-Fong, E. Ale, M. R. Pie, and A. Waeschenbach. 2014. "Cryptic species in the cosmopolitan *Bugula neritina* complex (Bryozoa, Cheilostomata)." *Zoologica Scripta* 43 (2):193–205. doi: 10.1111/zsc.12042.

Fehlauer-Ale, K. H., J. E. Winston, K. J. Tilbrook, K. B. Nascimento, and L. M. Vieira. 2015. "Identifying monophyletic groups within *Bugula* sensu lato (Bryozoa, Buguloidea)." *Zoologica Scripta* 44 (3):334–47. doi: 10.1111/zsc.12103.

Fortman, J. L., and D. H. Sherman. 2005. "Utilizing the power of microbial genetics to bridge the gap between the promise and the application of marine natural products." *ChemBioChem* 6 (6):960–78.

Gil-Turnes, M. S., and W. Fenical. 1992. "Embryos of *Homarus americanus* are protected by epibiotic bacteria." *Biological Bulletin* 182 (1):105–8.

Gil-Turnes, M. S., M. E. Hay, and W. Fenical. 1989. "Symbiotic marine bacteria chemically defend crustacean embryos from a pathogenic fungus." *Science* 246 (4926):116–8.

Gutierrez, C., S. Serrano-Villar, N. Madrid-Elena, M. J. Perez-Elias, M. E. Martin, C. Barbas et al. 2016. "Bryostatin-1 for latent virus reactivation in HIV-infected patients on antiretroviral therapy." *AIDS* 30 (9):1385–92. doi: 10.1097/qad.0000000000001064.

Hamel, E. 1992. "Natural products which interact with tubulin in the vinca domain— Maytansine, rhizoxin, Phomopsin A, Dolastatin 10 and Dolastatin 15, and Halichondrin B." *Pharmacology & Therapeutics* 55 (1):31–51. doi: 10.1016/0163-7258(92)90028-x.

Hanauske, A. R., G. Catimel, S. Aamdal, W. T. Huinink, R. Paridaens, N. Pavlidis et al. 1996. "Phase II clinical trials with rhizoxin in breast cancer and melanoma." *British Journal of Cancer* 73 (3):397–9. doi: 10.1038/bjc.1996.68.

Haygood, M. G., and S. K. Davidson. 1997. "Small-subunit rRNA genes and *in situ* hybridization with oligonucleotides specific for the bacterial symbionts in the larvae of the bryozoan *Bugula neritina* and proposal of 'Candidatus Endobugula sertula'." *Applied and Environmental Microbiology* 63 (11):4612–6.

Hendriks, H. R., J. Plowman, D. P. Berger, K. D. Paull, H. H. Fiebig, O. Fodstad et al. 1992. "Preclinical antitumor activity and animal toxicology studies of rhizoxin, a novel tubulin-binding agent." *Annals of Oncology* 3 (9):755–63.

Irie, K., A. Nakahara, Y. Nakagawa, H. Ohigashi, M. Shindo, H. Fukuda et al. 2002. "Establishment of a binding assay for protein kinase C isozymes using synthetic C1 peptides and development of new medicinal leads with protein kinase C isozyme and C1 domain selectivity." *Pharmacology & Therapeutics* 93 (2–3):271–81. doi: 10.1016/s0163-7258(02)00196-1.

Jackson, J. B. C. 1985. "Distribution and ecology of clonal and aclonal benthic invertebrates." In *Population Biology and Evolution of Clonal Organisms*, edited by J. B. C. Jackson, L. W. Buss and R. E. Cook, 297–355. New Haven, CT: Yale University Press.

Joyce, S. A., A. O. Brachmann, I. Glazer, L. Lango, G. Schwar, D. J. Clarke et al. 2008. "Bacterial biosynthesis of a multipotent stilbene." *Angewandte Chemie-International Edition* 47 (10):1942–5. doi: 10.1002/anie.200705148.

Kaplan, S., A. R. Hanauske, N. Pavlidis, U. Bruntsch, A. teVelde, J. Wanders et al. 1996. "Single agent activity of rhizoxin in non-small-cell lung cancer: A phase II trial of the EORTC Early Clinical Trials Group." *British Journal of Cancer* 73 (3):403–5. doi: 10.1038/bjc.1996.70.

Kellner, R. L. L. 1999. "What is the basis of pederin polymorphism in *Paederus riparius* rove beetles? The endosymbiotic hypothesis." *Entomologia Experimentalis et Applicata* 93 (1):41–9.

Kellner, R. L. L. 2001. "Suppression of pederin biosynthesis through antibiotic elimination of endosymbionts in *Paederus sabaeus*." *Journal of Insect Physiology* 47 (4–5):475–83.

Kellner, R. L. L. 2002. "Molecular identification of an endosymbiotic bacterium associated with pederin biosynthesis in *Paederus sabaeus* (Coleoptera: Staphylinidae)." *Insect Biochemistry and Molecular Biology* 32 (4):389–95.

Keough, M. J. 1989. "Dispersal of the bryozoan *Bugula neritina* and effects of adults on newly metamorphosed juveniles." *Marine Ecology Progress Series* 57 (2):163–71.

Keough, M. J., and H. Chernoff. 1987. "Dispersal and population variation in the bryozoan *Bugula neritina*." *Ecology* 68 (1):199–210.

Kerr, D. J., G. J. Rustin, S. B. Kaye, P. Selby, N. M. Bleehen, P. Harper et al. 1995. "Phase II trial of rhizoxin in advanced ovarian, colorectal, and renal cancer." *British Journal of Cancer* 72 (5):1267–9. doi: 10.1038/bjc.1995.498.

Kornberg, M. D., M. D. Smith, H. A. Shirazi, P. A. Calabresi, S. H. Snyder, and P. M. Kim. 2018. "Bryostatin-1 alleviates experimental multiple sclerosis." *Proceedings of the National Academy of Sciences of the United States of America* 115 (9):2186–91. doi: 10.1073/pnas.1719902115.

Kraft, A. S., J. B. Smith, and R. L. Berkow. 1986. "Bryostatin, an activator of the calcium phospholipid-dependent protein kinase, blocks phorbol ester-induced differentiation of human promyelocytic leukemia cells HL-60." *Proceedings of the National Academy of Sciences, USA* 83:1334–8.

Li, J. X., G. H. Chen, H. M. Wu, and J. M. Webster. 1995. "Identification of 2 pigments and a hydroxystilbene antibiotic from *Photorhabdus luminescens*."*Applied and Environmental Microbiology* 61 (12):4329–33.

Lim, G. E., and M. G. Haygood. 2004. " '*Candidatus* Endobugula glebosa', a specific bacterial symbiont of the marine bryozoan *Bugula simplex*." *Applied and Environmental Microbiology* 70 (8):4921–9.

Lim-Fong, G. E., L. A. Regali, and M. G. Haygood. 2008. "Evolutionary relationships of '*Candidatus* Endobugula' bacterial symbionts and their *Bugula* bryozoan hosts." *Applied and Environmental Microbiology* 74 (11):3605–9. doi: 10.1128/aem.02798-07.

Lin, H., Y. Yi, W. Li, X. Yao, and H. Wu. 1998. "Bryostatin 19: A new antineoplastic compound from *Bugula neritina* in south China sea." *Zhongguo Haiyang Yaowu* 17 (1):1–3.

Lindquist, N. 1996. "Palatability of invertebrate larvae to corals and sea anemones." *Marine Biology* 126 (4):745–55.

Lindquist, N., and M. E. Hay. 1996. "Palatability and chemical defense of marine invertebrate larvae." *Ecological Monographs* 66 (4):431–50.

Linneman, J., D. Paulus, G. Lim-Fong, and N. B. Lopanik. 2014. "Latitudinal variation of a defensive symbiosis in the *Bugula neritina* (Bryozoa) sibling species complex." *PLOS ONE* 9 (10):e108783. doi: 10.1371/journal.pone.0108783.

Lopanik, N., K. R. Gustafson, and N. Lindquist. 2004a. "Structure of bryostatin 20: A symbiont-produced chemical defense for larvae of the host bryozoan, *Bugula neritina*." *Journal of Natural Products* 67 (8):1412–4.

Lopanik, N., N. Lindquist, and N. Targett. 2004b. "Potent cytotoxins produced by a microbial symbiont protect host larvae from predation." *Oecologia* 139 (1):131–9.

Lopanik, N. B. 2014. "Chemical defensive symbioses in the marine environment." *Functional Ecology* 28 (2):328–40. doi: 10.1111/1365-2435.12160.

Lopanik, N. B., N. M. Targett, and N. Lindquist. 2006a. "Isolation of two polyketide synthase gene fragments from the uncultured microbial symbiont of the marine bryozoan *Bugula neritina*." *Applied and Environmental Microbiology* 72 (12):7941–4.

Lopanik, N. B., N. M. Targett, and N. Lindquist. 2006b. "Ontogeny of a symbiont-produced chemical defense in *Bugula neritina* (Bryozoa)." *Marine Ecology Progress Series* 327:183–91.

Lorenzo, P. S., K. Bogi, K. M. Hughes, M. Beheshti, D. Bhattacharyya, S. H. Garfield et al. 1999. "Differential roles of the tandem C1 domains of protein kinase C delta in the biphasic down-regulation induced by bryostatin 1." *Cancer Research* 59 (24):6137–44.

Mackie, J. A., M. J. Keough, and L. Christidis. 2006. "Invasion patterns inferred from cytochrome oxidase I sequences in three bryozoans, *Bugula neritina*, *Watersipora subtorquata*, and *Watersipora arcuata*." *Marine Biology* 149 (2):285–95.

Mao, Y. Q., M. Varoglu, and D. H. Sherman. 1999. "Molecular characterization and analysis of the biosynthetic gene cluster for the antitumor antibiotic mitomycin C from *Streptomyces lavendulae* NRRL 2564." *Chemistry & Biology* 6 (4):251–63.

Martinez-Bonet, M., M. I. Clemente, M. J. Serramia, E. Munoz, S. Moreno, and M. A. Munoz-Fernandez. 2015. "Synergistic activation of latent HIV-1 expression by novel histone deacetylase inhibitors and bryostatin-1." *Scientific Reports* 5. doi: 10.1038/srep16445.

Mathew, M., K. I. Bean, Y. Temate-Tiagueu, A. Caciula, I. Mandoiu, A. Zelikovsky et al. 2016. "Influence of symbiont-produced bioactive natural products on holobiont fitness in the marine bryozoan, *Bugula neritina* via protein kinase C (PKC)." *Marine Biology* 163:1–17.

Mathew, M., T. Schwaha, A. N. Ostrovsky, and N. B. Lopanik. 2018. "Symbiont-dependent sexual reproduction in marine colonial invertebrate: Morphological and molecular evidence." *Marine Biology* 165 (1):14. doi: 10.1007/s00227-017-3266-y.

McGovern, T. M., and M. E. Hellberg. 2003. "Cryptic species, cryptic endosymbionts, and geographical variation in chemical defences in the bryozoan *Bugula neritina*." *Molecular Ecology* 12 (5):1207–15.

McLeod, H. L., L. S. Murray, J. Wanders, A. Setanoians, M. A. Graham, N. Pavlidis et al. 1996. "Multicentre phase II pharmacological evaluation of rhizoxin." *British Journal of Cancer* 74 (12):1944–8. doi: 10.1038/bjc.1996.657.

Mehla, R., S. Bivalkar-Mehla, R. N. Zhang, I. Handy, H. Albrecht, S. Giri et al. 2010. "Bryostatin modulates latent HIV-1 infection via PKC and AMPK signaling but inhibits acute infection in a receptor independent manner." *PLOS ONE* 5 (6): e11160. doi: 10.1371/journal.pone.0011160.

Mutter, R., and M. Wills. 2000. "Chemistry and clinical biology of the bryostatins." *Bioorganic & Medicinal Chemistry* 8 (8):1841–60.

Nelson, T. J., C. H. Cui, Y. Luo, and D. L. Alkon. 2009. "Reduction of β-amyloid levels by novel protein kinase C ε activators." *Journal of Biological Chemistry* 284 (50):34514–21. doi: 10.1074/jbc.M109.016683.

Nelson, T. J., M. K. Sun, C. Lim, A. Sen, T. Khan, F. V. Chirila et al. 2017. "Bryostatin effects on cognitive function and PKCε in Alzheimer's disease phase IIa and expanded access trials." *Journal of Alzheimers Disease* 58 (2):521–34. doi: 10.3233/jad-170161.

Newman, D. J. 2012. The Bryostatins. In *Anticancer Agents from Natural Products*, edited by G. M. Cragg, D. G. I. Kingston and D. J. Newman, 2nd edition.

Newman, D. J. 2019. "From natural products to drugs." *Physical Sciences Reviews* 4 (4). doi: 10.1515/psr-2018-0111.

Newman, D. J., and G. M. Cragg. 2015. "Endophytic and epiphytic microbes as 'sources' of bioactive agents." *Frontiers in Chemistry* 3. doi: 10.3389/fchem.2015.00034.

Newman, D. J., and G. M. Cragg. 2016. "Natural products as sources of new drugs from 1981 to 2014." *Journal of Natural Products* 79 (3):629–61. doi: 10.1021/acs.jnatprod.5b01055.

Newton, A. C. 1995. "Protein kinase C: Structure, function, and regulation." *Journal of Biological Chemistry* 270 (48):28495–8.

Newton, A. C. 2001. "Protein kinase C: Structural and spatial regulation by phosphorylation, cofactors, and macromolecular interactions." *Chemical Reviews* 101 (8):2353–64.

Olson, R. R. 1985. "The consequences of short-distance larval dispersal in a sessile marine invertebrate." *Ecology* 66 (1):30–9.

Partida-Martinez, L. P., and C. Hertweck. 2005. "Pathogenic fungus harbours endosymbiotic bacteria for toxin production." *Nature* 437 (7060):884–8.

Patin, N. V., S. Locklear, F. J. Stewart, and N. B. Lopanik. 2019. "Symbiont frequency predicts microbiome composition in a model bryozoan-bacterial symbiosis." *Aquatic Microbial Ecology* 83 (1):1–13. doi: 10.3354/ame01901.

Pettit, G. R. 1991. "The Bryostatins." *Fortschritte Der Chemie* 57:153–95.

Pettit, G. R., J. F. Day, J. L. Hartwell, and H. B. Wood. 1970. "Antineoplastic components of marine animals." *Nature* 227 (5261):962–3.

Pettit, G. R., F. Gao, P. M. Blumberg, C. L. Herald, J. C. Coll, Y. Kamano et al. 1996. "Antineoplastic agents. 340. isolation and structural elucidation of bryostatins 16–18." *Journal of Natural Products* 59 (3):286–9.

Pettit, G. R., C. L. Herald, D. L. Doubek, D. L. Herald, E. Arnold, and J. Clardy. 1982. "Isolation and structure of bryostatin 1." *Journal of the American Chemical Society* 104 (24):6846–8.

Piel, J. 2009. "Metabolites from symbiotic bacteria." *Natural Product Reports* 26 (3):338–62. doi: 10.1039/b703499g.

Puglisi, M. P., J. M. Sneed, R. Ritson-Williams, and R. Young. 2019. "Marine chemical ecology in benthic environments." *Natural Product Reports* 36 (3):410–29. doi: 10.1039/c8np00061a.

Ryland, J. S., J. D. D. Bishop, H. De Blauwe, A. El Nagar, D. Minchin, C. A. Wood et al. 2011. "Alien species of *Bugula* (Bryozoa) along the Atlantic coasts of Europe." *Aquatic Invasions* 6 (1):17–31. doi: 10.3391/ai.2011.6.1.03.

Schmidt, E. W., M. S. Donia, J. A. McIntosh, W. F. Fricke, and J. Ravel. 2012. "Origin and variation of tunicate secondary metabolites." *Journal of Natural Products* 75 (2):295–304. doi: 10.1021/np200665k.

Schmitt, I., L. P. Partida-Martinez, R. Winkler, K. Voigt, E. Einax, F. Dolz et al. 2008. "Evolution of host resistance in a toxin-producing bacterial-fungal alliance." *ISME Journal* 2 (6):632–41. doi: 10.1038/ismej.2008.19.

Sharp, K. H., S. K. Davidson, and M. G. Haygood. 2007. "Localization of 'Candidatus Endobugula sertula' and the bryostatins throughout the life cycle of the bryozoan *Bugula neritina*." *ISME Journal* 1 (8):693–702. doi: 10.1038/ismej.2007.78.

Sudek, S., N. B. Lopanik, L. E. Waggoner, M. Hildebrand, C. Anderson, H. B. Liu et al. 2007. "Identification of the putative bryostatin polyketide synthase gene cluster from 'Candidatus Endobugula sertula', the uncultivated microbial symbiont of the marine bryozoan *Bugula neritina*." *Journal of Natural Products* 70 (1):67–74.

Tamburri, M. N., and R. K. Zimmer-Faust. 1996. "Suspension feeding: Basic mechanisms controlling recognition and ingestion of larvae." *Limnology and Oceanography* 41 (6):1188–97.

van Kesteren, C., M. M. M. de Vooght, L. Lopez-Lazaro, R. A. A. Mathot, J. H. M. Schellens, J. M. Jimeno et al. 2003. "Yondelis (R) (trabectedin, ET-743): The development of an anticancer agent of marine origin." *Anti-Cancer Drugs* 14 (7):487–502. doi: 10.1097/00001813-200308000-00001.

Vermeij, G. J. 1978. *Biogeography and Adaptation: Patterns of Marine Life*. Cambridge, MA: Harvard University Press.

Verweij, J., J. Wanders, T. Gil, P. Schoffski, G. Catimel, A. teVelde et al. 1996. "Phase II study of rhizoxin in squamous cell head and neck cancer." *British Journal of Cancer* 73 (3):400–2. doi: 10.1038/bjc.1996.69.

Walsh, C. 2003. "Natural and producer immunity versus acquired resistance." In *Antibiotics: Actions, Origins, Resistance*, 91–105. Washington, DC: ASM Press.

Walters, L. J. 1992. "Post-settlement success of the arborescent bryozoan *Bugula neritina* (L.): The importance of structural complexity." *Journal of Experimental Marine Biology and Ecology* 164 (1):55–71.

Wender, P. A., J. L. Baryza, S. E. Brenner, B. A. DeChristopher, B. A. Loy, A. J. Schrier et al. 2011. "Design, synthesis, and evaluation of potent bryostatin analogs that modulate PKC translocation selectivity." *Proceedings of the National Academy of Sciences of the United States of America* 108 (17):6721–6. doi: 10.1073/pnas.1015270108.

Woollacott, R. M. 1981. "Association of bacteria with bryozoan larvae." *Marine Biology* 65 (2):155–8.

Woollacott, R. M., and R. L. Zimmer. 1971. "Attachment and metamorphosis of cheilo-ctenostome byozoan *Bugula neritina* (Linne)." *Journal of Morphology* 134 (3):351–82.

Woollacott, R. M., and R. L. Zimmer. 1972. "Origin and structure of the brood chamber in *Bugula neritina* (Bryozoa)." *Marine Biology* 16 (2):165–70.

Woollacott, R. M., and R. L. Zimmer. 1975. "A simplified placenta-like system for the transport of extraembryonic nutrients during embryogenesis of *Bugula neritina* (Bryozoa)." *Journal of Morphology* 147 (3):355–78.

Young, C. M. 1986. "Direct observations of field swimming behavior in larvae of the colonial ascidian *Ecteinascidia turbinata*." *Bulletin of Marine Science* 39 (2):279–89.

Young, C. M., and B. L. Bingham. 1987. "Chemical defense and aposematic coloration in larvae of the ascidian *Ecteinascidia turbinata*." *Marine Biology* 96 (4):539–44.

Young, C. M., and F. S. Chia. 1987. "Abundance and distribution of pelagic larvae as influenced by predation, behavior and hydrographic factors." In *Reproduction of Marine Invertebrates*, edited by A. C. Giese, J. S. Pearse and V. B. Pearse, 385–463. Palo Alto, CA: Blackwell Scientific.

Yu, H. B., F. Yang, Y. Y. Li, J. H. Gan, W. H. Jiao, and H. W. Lin. 2015. "Cytotoxic bryostatin derivatives from the South China Sea Bryozoan *Bugula neritina*." *Journal of Natural Products* 78 (5):1169–73. doi: 10.1021/acs.jnatprod.5b00081.

10 The molecular dialogue through ontogeny between a squid host and its luminous symbiont

Margaret J. McFall-Ngai

Contents

10.1 Introduction

Bacterial light organs are present in several species of the squid family Sepiolidae (the bobtail squids), where they occur in association with the ink sac, often as a conspicuous bilobed feature. Nishiguchi et al. (2004), using both morphological and molecular characters, resolved the phylogeny of 21 sepiolid species, 17 in nine genera with light organs and 4 in two genera without; the species without light organs nest among those that have them. In the species with light organs, some have broad and some have narrow geographic and depth ranges, but as a group they have been captured in most of the world's major marine regions from 65°N to 50°S, from less than 1 m to 1700+ m depth (Reid and Jereb 2005). The females of this family have another symbiotic organ, the accessory nidamental gland (ANG), which has a consortium of bacteria (Collins et al. 2012; Kerwin and Nyholm 2017); a recent study of this system has provided strong evidence that the ANG symbionts protect the egg clutches from fouling by marine fungi (Kerwin et al. 2019). Because of their close proximity in the host's mantle cavity, early studies had suggested that the light organ could have evolved from the ANG (Naef 1923; Nishiguchi et al. 2004). However, recent genomic analyses have strongly suggested that these two organs have divergent evolutionary histories (Belcaid et al. 2019), and that the light organ is a prime example of "evolutionary tinkering," using the existing gene repertoire of the eye to evolve another light modulating organ somewhere else in the body (Tong et al. 2009).

The symbionts of most sepiolid squid light organs have not been characterized. The host species studied thus far harbor the symbiont(s) *Vibrio fischeri* and/or *Vibrio logei*, which also occur as free-living constituents of the bacterioplankton. (It should be noted that these bacteria are also called *Aliivibrio fischeri or logei*; the genus has changed several times in the last few decades, and most practitioners in the field do not agree with the separation of this *Vibrio* species away from the other species of this genus.) *V. fischeri* is a "mesophile," that is, a temperate species, with growth optima above 15°C, and *V. logei* is a "psychrophile," that is, a cold-water species, with growth optima below 15°C (Fidopiastis et al. 1998). The symbiont species that have been characterized in the *Euprymna* spp. are solely *V. fischeri*, but the *Sepiola* spp. of the Mediterranean have a mixture of *V. fischeri* and *V. logei*; the ratios of these species in the organ correlates with depth of the species and/or seasonal changes in environmental temperatures (Nishiguchi 2002). The symbionts in the light organs of the most northern and southern latitudes are predicted to be psychrophiles, but their identity remains to be determined. In addition to being in symbioses with squid and members of the bacterioplankton, certain strains of *V. fischeri* also occur in fishes of the West Pacific fish family, the Monocentridae, or pinecone fishes, which harbor light organs in the lower jaw.

10.2   Features of the *Euprymna scolopes-Vibrio fischeri* association as a model symbiosis

The symbiosis between *V. fischeri* and the host *Euprymna scolopes* Berry 1913, a sepiolid endemic to the Hawaiian archipelago, has been studied for over 30 years. In the first examinations of the symbiosis in the 1980s, the symbiont was identified (Hastings and Nealson 1981) and the horizontal transmission of the association was described (Wei and Young 1989). The field has grown in the intervening decades, such that currently over a dozen laboratories in a dozen states in the United States have research programs focused on the *E. scolopes-V. fischeri* symbiosis. Thus, an advantage of the squid–vibrio model is its advanced state of development. As of mid-2019, over 340 papers have been published on the system.

The symbiosis has several features that render it a good experimental model. The host females lay egg clutches that yield from a few dozen to hundreds of individual hatchlings. Thus, the "n" for experiments can be large, and intra- and interclutch variation in symbiotic phenotypes can be characterized. Size of the juvenile animal, ~2 mm in total length, the transparency of the nascent symbiotic tissue, and the timing of colonization, ~12-h following inoculation to full population of the organ, make the system a good subject for microscopic analysis of symbiosis onset. The study of this association has often pushed the boundaries of microscopic analysis of a symbiosis. For example, the colonized juvenile was the first symbiotic association to use hybridization chain reaction fluorescent *in-situ* hybridization (HCR-FISH) to simultaneously localize rare messages of the host and symbiont within tissues (Nikolakakis et al. 2015). The system can be chemically and pharmacologically manipulated by adding reagents (e.g., antibiotics, immune modulators, lectins, and derivatives of bacterial cells) to the surrounding seawater. Because the partnership provides the animal host with luminescence, it is ecologically, but not physiologically obligate, both partners can be cultured independently in the laboratory with no detectable effect on their fitness. The "currency" of luminescence also allows the noninvasive evaluation of the presence/absence and extent of symbiosis by quantifying light output by the host in a photometer.

The adult hosts are 2–3 cm in mantle length and are easily cultured in the laboratory. They bury in the sand during the day and emerge at night to forage in the water column, where they use the bacterial luminescence to counterilluminate, that is, as a camouflaging behavior against down-welling moonlight and starlight. In the adult light organ, the epithelium containing the symbionts (the "central core") can be cleanly dissected from the accessory tissues (i.e., the lens, reflector and ink sac; see Figure 10.3A), which allows for molecular and biochemical analysis of the host–symbiont dialogue in those tissue layers where the partners directly interact. Further, similar to vertebrates, the squids have a closed circulatory system; >100 μL of hemolymph (blood) can be extracted from the host for analysis of the macrophage-like cells and the serum. For example, recent metabolomics studies of the adult host hemolymph revealed differences in the metabolite profiles dependent on symbiosis, sex of the host, and time of day (Koch et al. in prep).

A variety of molecular resources are available for both the host and symbiont. For the host, these resources include an annotated genome sequence and 31 transcriptomes from 10 host organs at different colonization states of the light organ (symbiotic vs. nonsymbiotic, or colonized by different strains or mutants of the symbiont), stages of maturation, and time of day (Belcaid et al. 2019). NanoString analyses have also been developed for the host, which has provided the opportunity to query a subset of candidate genes for change in gene expression under a variety of conditions (Moriano-Gutierrez et al. 2019). While genetic manipulation of *Euprymna scolopes* is not currently routine, CRISPR-Cas has been applied in this animal at the Marine Biological Laboratory, Woods Hole, MA, (Reardon 2019). *Vibrio fischeri* was among the first nonpathogenic aquatic bacteria sequenced (Ruby et al. 2005), and currently two closed, and >50 draft, genomes are available from strains collected around the world and from various ecological niches. In addition, Tn-mutant libraries have been constructed (e.g., Brooks et al. 2014), as well as dozens of clean-deletion mutations that impact specific aspects of symbiont–symbiont and symbiont–host interaction.

10.3 Host activities before symbiont colonization: Embryogenesis and early posthatching

The *E. scolopes* embryonic period ranges from 18–26 d, proceeding more slowly or more quickly at cooler or warmer temperatures, respectively. During this period, the nascent light organ forms, poising the tissues for *V. fischeri* colonization immediately upon hatching (Montgomery and McFall-Ngai 1993). About half way through embryogenesis, two invaginations begin to form, one on either side of the region of the hind-gut ink-sac complex, that develop into the tissues to be colonized at hatching. Then, in sequence over the next days before hatching, two more invaginations occur on each side of the organ; this staggered development results in three regions on either side that are of differing maturation at hatching. Concomitantly with these surface invaginations, complex, juvenile-specific ciliated fields form on the lateral surfaces of the nascent organ that will potentiate colonization by the symbiont; each field comprises an area surrounding the three pores that result from the surface invaginations and two appendages, one anterior and one posterior (Figure 10.1A).

In addition to these anatomical features, because light organ colonization occurs within hours of hatching and is specific to *V. fischeri*, biochemical determinants that

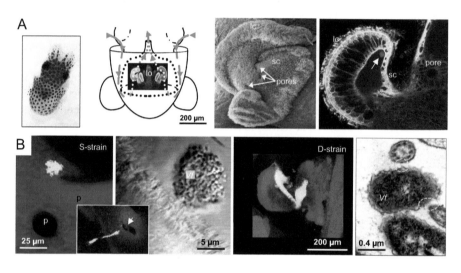

FIGURE 10.1   The initial engagement of the symbiotic partners. (A) The hatchling light organ. *Far left*, a newly hatched juvenile; the light organ can be seen through the mantle cavity as a dark area; *middle left*, a confocal image of the hatchling light organ, which resides in the center of the host's body cavity surrounded by the funnel (black dotted lines); during ventilation, environmental seawater is drawn into the body cavity and over the organ (blue and yellow arrows/lines); *middle right*, scanning electron micrograph of the light organ surface, showing the field of long cilia (lc) on the outside of the anterior and posterior appendages and along the medial edge; these cilia circumscribe the region of short cilia (sc) around the pores, where the symbionts will enter host tissues; *far right*, confocal micrograph showing the anatomy of the ciliated surface, which occurs as a single layer of ciliated epithelia overlying a blood sinus into which hemocytes (white arrow) migrate during symbiont colonization. (B) Symbiont aggregation. *Left* and *left middle*, GFP-labeled *V. fischeri* of two strain types (S and D) that behave differently during aggregation; blue, mucus; red, host tissues; p, pore; *inset, lower middle*, bacteria chemotaxing from an aggregate into the pores (white arrow). *Middle right*, differential interference contrast (DIC) image showing an aggregate of *V. fischeri* (Vf) cells entering the pores. Right, a transmission electron micrograph of *V. fischeri* (Vf) cells associating with cilia. Green circles indicate areas of symbiont–cilium contact.

promote symbiont selection are also produced during the embryonic period, notably stores of antimicrobials (Nyholm and McFall-Ngai 2004; Troll et al. 2010; Heath-Heckman et al. 2014; Chen et al. 2017). For example, nitric oxide (NO), which has antimicrobial activity, can be detected in high concentrations in the embryonic and hatchling light organ (Davidson et al. 2004). At hatching, the organ sheds copious mucus from its surface in response to a conserved molecule of the environmental bacteria, peptidoglycan (PGN) (Nyholm et al. 2002), the molecule that is the backbone of the bacterial cell wall. The shed mucus contains vesicles with the embryonically synthesized nitric oxide synthase and nitric oxide. Studies with *V. fischeri* mutants defective in resistance to NO do not colonize normally (Wang et al. 2010a,b). Also in this shed mucus is a peptidoglycan-recognition protein, EsPGRP2, which has an amidase activity that denatures PGN (Troll et al. 2010). Since PGN comprises the surface of Gram-positive bacteria, this amidase activity is likely responsible for sanctioning association with the organ surface of this subset of the environmental bacteria; thus, the data suggest that this activity drives the first step of the specific selection of *V. fischeri,* that is, by restricting association to Gram-negative bacteria.

10.4 Early posthatching activity that mediates species and strain specificity of the association

The successful onset of symbiosis requires synergism between biophysical and biochemical features of the organ. Within seconds of hatching, the juvenile begins to ventilate its mantle cavity, and the cilia on the surface of the nascent light organ begin to beat (Figure 10.2A). In addition, the pathway through which *V. fischeri* cells will migrate to reach the crypts, their site of residence, opens and expands (Essock-Burns et al. in press). Two kinds of cilia on each lateral field participate in symbiont selection: (1) the metachronal beating of long cilia, which cover the outer surfaces of the anterior and posterior appendages, as well as the medial edge of the ciliated field, draw water across the light organ and focus Gram-negative bacteria into a stagnant zone above the pores; and, (2) the random beating of short cilia, which line the inner surfaces of the appendages and the regions immediately surrounding the pores, appears to mix the chemicals provisioned into those regions (Nawroth et al. 2017).

The first ~25 years of study of the squid–vibrio symbiosis focused on colonization by *V. fischeri* ES114, the parent strain used for the development of quality molecular genetics in the bacterial symbiont. In the first 3 h following inoculation into the host's environment, 5–10 cells of *V. fischeri* ES114 attach to the host cilia (Altura et al. 2013; Figure 10.1B). This attachment causes a specific, robust change in host gene expression (Kremer et al. 2013), which is remarkable in light of the fact that this transcriptional change occurs in response to a few *V. fischeri* cells in a background of about one million nonspecific environmental bacteria per milliter of seawater. In addition, the *V. fischeri* cells are interacting with only a few host cells, but the transcriptome was generated from whole light organs, each of which is estimate to have ~10,000 cells. Because of the robustness and magnitude of the transcriptomic changes, these data suggest that the triggers inducing these changes in transcription are powerful enough to have an effect that radiates across the light organ tissues.

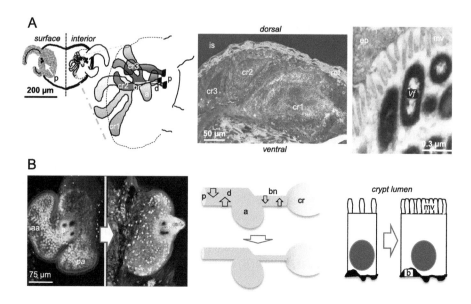

FIGURE 10.2   Colonization and development. *Left,* each side of the light organ has three pores (white arrow). After moving into the pores, the *V. fischeri* cells travel down ducts into an antechamber. Then, on average, a single cell goes through a bottleneck into each crypt. Crypt 1 (cr1, yellow) is the most superficial, being ~10–20 μm below the surface, and most mature; crypts 2 and 3 (cr2, blue; cr3, red) occur more deeply and are more immature. *Middle,* a histological section showing the three crypts full of *V. fischeri* cells. *Right,* a transmission electron micrograph showing the contact of *V. fischeri* cells with the microvilli of the epithelium of cr1. a, antechamber; aa, anterior appendage; b, basement membrane; bn bottleneck, cr, crypt; d, duct; ep, epithelia; is, ink sac; mv, microvilli; p, pore; pa, posterior appendage; ref, reflector; Vf, V. fischeri.

This change in host gene expression has two major consequences: (1) an upregulation of transcription of antimicrobials; and (2) the upregulation of a chitinase, which breaks polymeric chitin present in the mucus into chitobiose. The *V. fischeri* cells then aggregate on the surface of the organ, a behavior that requires the presence of the *V. fischeri* capsule (Visick 2009). These cells then pause for 1–2 hours before migrating into host tissues, which requires bacteria motility (for review see, Aschtgen and Ruby 2019). Whereas early in aggregation other Gram-negative bacteria are often present and, in the absence of *V. fischeri*, other Gram-negative bacteria can aggregate, when *V. fischeri* cells are present, other bacteria are eventually excluded from the aggregations (Nyholm et al. 2000). Studies with mutants of *V. fischeri* suggest that the host presents a cocktail of antimicrobials against which the symbiont is uniquely capable of withstanding. In addition, *V. fischeri* chemotaxes into the host organ in response to the sensing of chitobiose, but chemotaxing to this molecule requires that the *V. fischeri* cells be primed, which would occur through exposure of the dimeric chitin in the mucus, which is produced by the activity of the upregulated and secreted host chitinase (Kremer et al. 2013).

Recent studies of strain variation in *V. fischeri* have revealed alternative strategies for host colonization (Bongrand and Ruby 2019a). Some strains have the behavior of strain ES114, which, when inoculated with the levels of *V. fischeri* that occur in the field (~500–5000 cells/mL of seawater), make the typical aggregate of 5–10 cells that pauses before colonizing. In addition, when strains of this strain type are competed against one another in colonization experiments, a significant proportion of them will "share" the light organ ("S" strains), although generally occupying different crypts within one organ. In contrast, other strains dominate in competition ("D" strains), that is., they are the sole colonizer of the organ. Such competition analyses revealed a dominance hierarchy among the D-type strains; those at top of the hierarchy would outcompete those below them for exclusive colonization of the organ. The mechanisms underlying these differences are determined by the aggregation behavior and the speed of colonization. D strains tend to make very large aggregates in comparison to S strains (Figure 10.1B) (Koehler et al. 2018), and they colonize the light organ more quickly (Bongrand et al. 2016; Bongrand and Ruby 2019b). Current studies are being conducted to determine whether D-strains occur in the plankton as "preprimed" for colonization, that is, not in need of a pause to change gene expression, or if selection for large aggregates is a mechanism that functions to swamp out the effects of host antimicrobials and obviate the need for chemotaxis priming.

10.5   Colonization and early development

In the colonization process (for review, see McFall-Ngai 2014), the *V. fischeri* cells enter into host tissues through the six pores on the surface, three on each side of the organ; no other environmental bacteria migrate into the pores. Each pore leads to an independent pathway through which the symbionts migrate; each pathway includes a narrow duct that opens into a broader antechamber, which has an opening to a narrow bottleneck on the medial side (Figure 10.2A). It is through this narrow feature that *V. fischeri* cells enter the crypts (Essock-Burns et al. in press). Once the symbionts access the crypts, their populations are restricted to these regions, only going back through the migration pathways each morning when ~90% of the symbiont cells are expelled into the environment (see Section 10.2.1 below). The areas of the light organ where symbionts migrate and then populate are also ultrastructurally different, that is, whereas the epithelia of the duct, antechamber, and bottleneck have densely ciliated apical surfaces, the population of *V. fischeri* interfaces with microvillous crypt epithelia (Figure 10.2A; Essock-Burns et al. in press).

Although many *V. fischeri* cells may migrate through a given pathway, experiments with isogenic symbiont cells labeled with different fluorochromes have demonstrated that, on average, a single *V. fischeri* cell makes it through the bottleneck and grows out in the crypt space (Figure 10.2A) (Wollenberg and Ruby 2009). These data suggest that negotiation through the bottleneck imposes a physical challenge to the colonizing cells. Nonisogenic strains are less likely to cocolonize, but the array of mechanisms for this sanctioning are not known. However, a recent study determined that some strains of *V. fischeri* carry a type VI secretion system, which behaves somewhat like a microscopic harpoon that can be directed against other cells, giving the cells carrying this system a competitive edge on colonization of the crypts (Speare et al. 2018).

As a whole, the light organ responds to colonization at the molecular, biochemical, physiological, ultrastructural, anatomical, and morphological levels. In addition to the symbionts influencing host light-organ gene expression with their first interactions with the host cell surface, early colonization events are also reflected in changes in the host transcriptome across tissues comprising the light organ. Specifically, studies of host gene expression have been conducted on whole light organs at 18 h post colonization (Chun et al. 2008), when the wild-type symbionts are in log-phase growth in the crypt spaces and dimly luminous, and at 24 h (Moriano-Gutierrez et al. 2019), when they achieve their initial full colonization and are brightly luminous. The transcriptomes of these wild-type colonized animals were compared to those of aposymbiotic animals (exposed to environmental bacteria in seawater without introduced *V. fischeri* cells) and those of animals colonized with a mutant strain defective in light production. At 18 h the largest driver of host gene expression is the presence of colonizing *V. fischeri*. In contrast, at 24 h, the largest driver in differential gene expression is the luminescence of the symbionts. In both cases, many of the changes are associated with the regulation of immune genes (McFall-Ngai et al. 2010), such as those encoding antimicrobial peptides/proteins, and transcripts shared by the eye and light organ; as mentioned above, the eye and light organ, as two organs that modulate light, are highly convergent in form and function (Tong et al. 2009; McFall-Ngai et al. 2012; Belcaid et al. 2019).

A recent study has shown that colonization of the light-organ crypts also influences the expression of dozens of genes in other host organs (Moriano-Gutierrez et al. 2019). These analyses examined the gene expression of the gills, which are an immune organ in cephalopods, and the eye, which is non-only convergent with the light organ, but also is critical in the coordination between the light organ and the eye for effective counterillumination behavior. The transcriptomes of these tissues were examined at 24 h in animals uncolonized, colonized with wild type *V. fischeri*, or colonized by *V. fischeri* mutants defective in light production. Each organ had a unique response. Further, the colonization-induced change in the gill transcriptome is independent of light production, only responding to the presence or absence of the bacteria. In contrast, the change in the eye transcriptome with light-organ colonization is entirely reliant on light production by the bacteria. How the presence of the symbionts in the light organ is communicated to these remote tissues remains to be determined.

Whereas the transcriptomic studies thus far have focused on the whole light organ, numerous studies have been done on symbiosis-induced development of the constituents of the organ (for review see McFall-Ngai 2014). The most conspicuous such change is the regression of the ciliated fields that potentiate symbiont recruitment (Figure 10.2B, left). At ~12 h following colonization, mucus shedding ceases from these fields and the bacterial symbionts deliver an irreversible signal from the crypt spaces that drives a gradual morphogenesis of these features over the next ~4 d. Two cell types are involved in this process: (i) those of the ciliated fields themselves, which undergo apoptosis and are sloughed, a process that results from an increase in the activity of a matrix metalloproteinase (Koropatnick et al. 2014) and a cathepsin (Peyer et al. 2018) along the basement membrane of these epithelial cells; and, (ii) hemocytes (Figure 10.1A, right), whose migration into the blood sinuses of the superficial ciliated appendages completely fills these spaces. Manipulation of this migration behavior has demonstrated that changes in gene expression of the hemocytes, in response to the ~12 h irreversible signal, is essential for morphogenesis. It should be noted here that while nearly all of these developmental studies were done with colonization by the S strain ES114, the timing of host developmental phenotypes is accelerated by the rapid colonization of D strains (Bongrand and Ruby 2019b).

How the "message" gets from the crypt space to the superficial epithelium remains unknown. However, the nature of the bacterial cues has been well characterized. The lipid A of the symbiont's lipopolysaccharide (LPS), an abundant molecule in the outer membrane of Gram-negative bacteria, and the cell wall, or peptidoglycan (PGN), monomer work in synergy with symbiont light production to trigger morphogenesis (for review see McFall-Ngai 2014). The LPS and PGN are microbe-associated molecular patterns, or MAMPs, that have also been implicated in symbiosis-induced development of mammals (Bouskra et al. 2008). These morphogenetic cues are delivered to the host cells as either freely soluble molecules (Koropatnick et al. 2004) or constituents of outer membrane vesicles (OMVs; Aschtgen et al. 2016a,b). Further, although this development is irreversibly triggered at 12 h, if the light organ is cured of its symbionts within the first few days of colonization, it will begin to shed mucus from the remaining ciliated fields and recruit symbionts again. Further, this field of cells slowly regresses in aposymbiotic animals, likely due to the sensing of low levels of MAMPs of environmental bacteria. Together these data provide evidence that the ciliated fields are not required for colonization, but rather increase the effectiveness of recruitment. However, aposymbiotic animals remain susceptible to colonization for several weeks.

The regions of the migration pathways and the crypts themselves are also influenced by symbiont colonization. However, in contrast with the irreversible signal for morphogenesis of the light organ's superficial ciliated fields, most of the symbiont-induced developmental changes in these internal portions of the juvenile light organ are reversible by antibiotic curing. As mentioned above, because of their sequential embryonic development, each of the three migration paths and three crypts are of different maturity (Essock-Burns et al. in press). The impact of these differences is currently under investigation. Thus, the discussion here will be restricted to developmentally induced changes in the pathway leading to crypt 1 and in crypt 1 itself, the most mature and the most deeply studied landscape. Along the symbiont migration path, the most conspicuous changes are the constriction of the duct (Kimbell and McFall-Ngai 2004), and bottleneck regions (Essock-Burns et al. in press) (Figure 10.2B, middle). In the duct, a two to threefold loss in circumference correlated with an increase in actin abundance and a concomitant decrease in the number of cells interfacing the duct lumina. The $\sim$25-μm long, narrow bottleneck ($\sim$7 μm wide at hatching, narrowing to $\sim$2 μm upon colonization) appears to have two functions: (i) restricting the number of colonizing V. fischeri, which is, on average a single cell/crypt, which grows out to populate the crypt space (Wollenberg and Ruby 2009); and, (ii) restricting the population of V. fischeri cells to the crypts. The mechanisms by which the bacteria trigger duct and bottleneck constriction remain to the determined.

Once the bacteria are in the crypt spaces, they induce cell swelling (a fourfold increase in volume) of the epithelial cells (Visick et al. 2000), and an increase in microvillar density (Lamarcq and McFall-Ngai 1998) (Figure 10.2B, right) that results in each symbiont cell being nearly surrounded by host membranes. Unlike the symbiont-induced duct and bottleneck phenotypes, bacterial triggers for these developmental changes have been determined. Mutants in *V. fischeri* defective in light production fail to induce cell swelling (Visick et al. 2000) and the MAMP LPS induces the increase in microvillar density (Heath-Heckman et al. 2016)

Several host features that are responses to symbiont cues have also been defined. Notably mutants of *V. fischeri* defective in light production are also defective in normal triggering of nearly all symbiosis-induced developmental phenotypes (McFall-Ngai et al. 2012; McFall-Ngai 2014). The light organ has two light-responsive systems, both of which have sensitivities that are congruent with the $\sim$490-nm light emission of *V. fischeri*. Specifically, all elements of both the ciliary and rhabdomeric phototransduction systems are present in the light organ, including the visual pigment rhodopsin; furthermore, the tissues are responsive to light, as demonstrated by electroretinogram (Tong et al. 2009). Secondly, the light organ tissues have the light-receptive protein cryptochrome (Heath-Heckman et al. 2013). Other genes or proteins involved in developmental regulation of the eye have also been implicated in symbiosis-induced development of the light organ, notably, *pax6* (Peyer et al. 2014) and *crumbs* (Peyer et al. 2017).

Several receptors of bacterial MAMPs are also present in the light organ, most notably proteins that sense LPS or PGN derivatives, since these MAMPs induce several host developmental phenotypes either alone or in synergy. These proteins include members of two protein families: (i) LBP/BPI, lipopolysaccharide-binding/ bacteriocidal-permeability-increasing, EsLBP/BPI 1-4 (Krasity et al. 2011); and, (ii) PGRP, peptidoglycan-recognition, EsPGRP1-5 (Goodson et al. 2005; Collins et al. 2012). Some of these MAMPs receptors have been studied in depth, specifically EsLBP1, 2, and 4 and EsPGRP1 and 2. On symbiont colonization, the transcript expression and protein production of EsLBP1 increase in the crypt epithelium, where *V. fischeri* cells induce LPS-mediated morphogenesis (Krasity et al. 2015). EsLBP2 and 4 behave as a BPIs, that is, they are antimicrobial, and also have high expression in the juvenile light animal, but their roles appear to be in tissue defense rather than development (Chen et al. 2017). EsPGRP1 is directly involved in the events of development (Troll et al. 2009). This protein occurs in the nucleus of the cells of the juvenile-specific superficial ciliated epithelium. In response to colonization by *V. fischeri*, or exposure to PGN, EsPGRP1 leaves the nucleus and the cells go into the typical apoptosis that underlies morphogenesis of this tissue. Further, mutants in *V. fischeri* that are defective in presentation of PGN derivatives were defective in this EsPGRP1-driven phenotype.

Exactly how the bacterial cues get to their target tissues remains poorly understood. The fact that, in response to only $\sim$5 *V. fischeri* cells associating with 1–2 host cells on the organ's ciliated surface, the animal responds with a specific and robust change in gene expression across the entire organ (Kremer et al. 2013) suggests that many of the responses occur as a result of amplification and transduction of bacterial cues. In a recent study, colonizing *V. fischeri* cells were labeled with stable isotopes, and Nano Secondary-Ion Mass Spectrometry (NanoSIMS) analysis was used to follow labeled materials exported from these cells (Cohen et al. 2020). These studies revealed trafficking of exported biomolecules of the symbiont into host epithelia, which concentrated in the nucleolus and euchromatin of the host cells. This phenomenon could be recapitulated by exposure to OMVs alone. Interestingly, three proteins in the protein pool of the OMVs have eukaryotic nuclear localization signals, two of which have been implicated in nuclear localization of materials exported by bacterial pathogens. The activity of these proteins and that of other biomolecules imported into host nuclei offers an exciting avenue for future study.

10.6 The basis of a stable symbiosis: Daily rhythms and maturation of the symbiotic organ

Over the first 20+ years of developing the squid–vibrio system, all experimental work focused on the juvenile animals. Studies of symbiosis in animals >1 week was entirely descriptive of the maturation state in wild-caught animals, from the anatomy, which showed that the structure of the mature light organ retains the intimate contact with host cells (Figure 10.3B) (McFall-Ngai and Montgomery 1990) to the transcriptome (Figure 10.4, left) (Wier et al. 2010). The effort developed over the last decade to raise animals through their life cycle (Koch et al. 2014) has allowed for the study of the maturation of the symbiosis.

An understanding of one particular aspect of the symbiosis has been greatly enriched by this capability: daily rhythms as a mechanism of symbiosis stability. These daily rhythms actually begin on the very first day following colonization. The data suggest a diel rhythm superimposed on a circadian rhythm. Each day at dawn, in response to the dawn light cue, the animal vents 70%–90% of the symbionts from the crypts (Graf and Ruby 1998). Studies of this phenomenon in juvenile animals have revealed that the bottleneck transiently opens at this time to allow for the

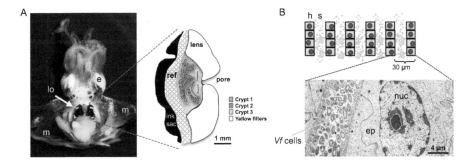

FIGURE 10.3 The mature symbiosis. (A) *Left*, a ventral dissection revealing light organ (lo) in the center of the mantle cavity; m, mantle tissue that has been reflected back; e, eye; dashed lines, the portion of light organ in diagram the right. *Right*, a sagittal section through the organ showing the relationship of the tissues. The three crypts are surrounded by accessory tissues, the ink sac, reflector (ref), and lens, that modulate light emission from the symbiont into the environment; the crypts open into the environment through a pore. McFall-Ngai, M. 2014. *Annual Reviews of Microbiology*, 68:177–94.) *Right*, position of *V. fischeri* cells relative to host cells. Upper, sheets of host cells (s) alternate with thin extracellular spaces where the symbionts (s) reside in close proximity to host cell surfaces. A transmission electron micrograph showing the intimate relationship of V. fischeri (*Vf*) cells (green arrow) to host cells; ep, host epithelial cell cytoplasm; nuc, host epithelial cell nucleus.

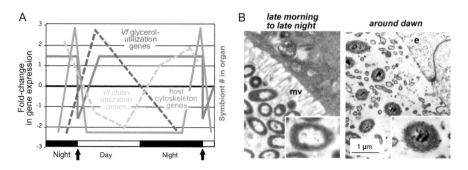

FIGURE 10.4   The daily rhythm of the mature squid–vibrio symbiosis. (A) In the adult squid, the pattern of the change in gene expression by the symbionts indicates that host-derived nutrients switch from day (glycerol, red) to night (chitin, green). Preceding the early morning change in symbiont gene expression, in the hours before dawn, the host upregulates nearly all genes encoding constituents and regulators of the cytoskeleton. (B) The transcriptomic data suggested ultrastructural changes in the host epithelial cells. *Left*, throughout much of the day-night cycle, the epithelial cells are strongly polarized with abundant microvilli interacting with the symbionts. *Right*, with the light cue of dawn, the cells are effaced, shedding the microvilli into the crypt space. Insets, the electron density of the symbiont cells at dawn suggests differences in biochemistry at the different times of day.

passage of the bacteria out of the crypts and through the pores on the organ surface (Essock-Burns et al. in press). During these cycles, not only are symbionts vented, but also expelled are host proteins that have been exported into the crypt spaces. Two notable examples are EsPGRP2, which has an amidase activity that detoxifies symbiont peptidoglycan monomer, or TCT (tracheal cytotoxin) (Troll et al. 2010), and an alkaline phosphatase (EsAP), which cleaves phosphate groups off of the sugars of the lipid A moiety of LPS, thereby detoxifying the lipid A (Rader et al. 2012). These detoxifying molecules function to prevent the MAMPs from perturbing the host epithelium; after being diminished at venting, they build up again each day in the crypt space as the bacteria regrow.

A circadian rhythm is suggested by the day-night patterns of bioluminescence in the host (Boettcher et al. 1996). Per cell luminescence of light-organ *V. fischeri* is low during daylight hours, when the animal is buried in the sand, but begins to increase just before night in anticipation of the animal emerging from the sand to forage in the water column and use its luminescence for camouflaging; then, just before dawn, in anticipation of a quiescent period, the per cell luminescence of the symbiont decreases. This circadian-type behavior is reflected in the adult transcriptome, where changes in host gene expression are most pronounced just before the dawn light cue (Figure 10.4, left) (Wier et al. 2010). Also suggestive of a circadian component to these daily rhythms is the light-organ cycling in the expression of the clock gene *cry* (Heath-Heckman et al. 2013), which begins immediately upon initial colonization. This cycling requires the presence of the symbionts and also requires light production by those symbionts; mutants of *V. fischeri* defective in light production do not induce *cry* cycling.

The host and symbiont cells have a profound rhythm in the adult symbiosis. Just before venting, nearly all of the genes that are recognized as encoding proteins involved in host cytoskeleton are upregulated (Figure 10.4, left) (Wier et al. 2010). Then, in response to the dawn light cue, the microvilli of the host crypt epithelia efface (Figure 10.4, right), presenting to the remaining *V. fischeri* cells membranes on which to grow to repopulate the light organ. In response, *V. fischeri* upregulates genes associated with using these membranes as a nutrient source through anaerobic respiration (Figure 10.4, left). The host cells then repolarize their membranes, restoring the complex microvilli to their apical surfaces. With the membrane nutrient source depleted, the bacteria turn on genes associated with the anaerobic fermentation of chitin as a nutrient source (Figure 10.4, left), suggesting that the host is presenting chitin to the symbionts.

This cycling of nutrients only begins after 3–4 weeks following colonization of the organ (Schwartzman et al. 2015), which is concomitant with the maturation of the host's behavior of burying in the sand during the day and foraging at night. In support of this finding is the observation that mutants in the utilization of chitin as a nutrient source are not defective in colonization in the first weeks of symbiosis, but are defective at 3–4 weeks. In addition, these analyses demonstrated that the chitin, which is high in host hemocytes (Heath-Heckman and McFall-Ngai 2011), is brought to the light organ through the trafficking of hemocytes to the crypt spaces, where they release their chitin and die; this daily rhythm of trafficking of hemocytes does not occur in juveniles. The anaerobic fermentation of chitin changes the biochemistry of the crypt spaces. Most critically, it reduces the pH of the environment. Also secreted into the crypt space is host hemocyanin, which, under acidic conditions, releases more of its oxygen in a classic Bohr effect; the increase in oxygen enhances bacterial luminescence during the evening hours (Kremer et al. 2014). Interestingly, this acidification is likely to have another effect. The EsPGRP2 and EsAP, which protect the host from the perturbing effects of MAMPs, are inactive at acidic pH. As such, the environment of the adult host crypts is predicted to be harsher to the host tissues than that of the juvenile organ.

How is hemocyte trafficking to the crypt space regulated on a day-night cycle? A recent study (Koch et al. in press) has demonstrated that a host cytokine, macrophage-migration inhibitory factor, or MIF, is expressed at high levels during the day and low levels at night in the adult light organ crypts. Immunocytochemistry with an antibody to MIF showed that the protein follows the same pattern. Studies of the behavior of the hemocytes confirmed that their migration is inhibited by the squid's MIF protein. Taken together, the data provide evidence that MIF upregulation is responsible for inhibiting the trafficking of the chitin-bearing hemocytes during the day, and the downregulation of this cytokine at night allows these cells to migrate toward the bacteria-rich environments of the crypts.

In addition to maturation of the light organ itself, the animal displays systemic maturation in response to symbiosis. Studies of gene expression in tissues remote from the light organ show that response to symbiosis is different in juvenile and adult animals (Moriano-Gutierrez et al. 2019). In addition, the circulating blood cells or hemoctyes also mature in response to the presence of the symbiont (Nyholm et al. 2009; Rader et al. 2019). Maturation of the hemocytes involves a change in the specific recognition of *V. fischeri*. Whereas naïve hemocytes recognize the symbiont, this recognition is abrogated by persistent exposure to the symbionts; naïve and "educated" symbionts see other bacterial cells similarly. Interestingly, this maturation of the host hemocytes requires the presence of the *V. fischeri* outer membrane protein OmpU, which is a protein implicated in pathogenesis of *Vibrio* spp. (see e.g., Baliga et al. 2018; Yang et al. 2018).

10.7 Conclusions

The 30+ years of study of the squid–vibrio system has taught us how intricate a symbiosis can be, even a binary one. Data to date provide evidence that the processes of specificity determination, host–symbiont recognition, and partner development and maturation require complex host–symbiont dialogue. However, for none of these processes in the light organ do we have anywhere near a complete picture. Much work remains to be done to unravel the mysteries that drive this association. In addition, the mechanisms by which information on the state of symbiosis is conveyed, not only throughout the organ itself, but also across the remote tissues of the body remain to be determined, that is, are they neural, humoral, both, or other? Most exciting will be the introduction of host genetics to the system, which is currently underway and promises to open whole new vistas for the study of this system (Reardon 2019).

Acknowledgments

I thank the many principal investigators, students, and postdocs who have contributed to our understanding of the squid–vibrio symbiosis. I thank Edward Ruby for helpful suggestions on the manuscript. Research performed in the laboratories of the author (MMN) and Edward Ruby (EGR) that were described in this chapter were funded by the National Institutes of Health (NIH) and the National Science Foundation (NSF), currently NIH R37-R01AI50661 (to MMN and EGR) and NIH OD11024 and GM135254 (EGR and MMN) and NSF DBI 1828262. Acquisition of the Leica TCS SP8 X confocal, used for recent imaging, was supported by NSF DBI 1828262.

References

Altura, M. A., E. A. C. Heath-Heckman, A. Gillette, N. Kremer, A.-M. Krachler, C. A. Brennan, E. G. Ruby, K. Orth, and M. J. McFall-Ngai. 2013. The first engagement of partners in the *Euprymna scolopes-Vibrio fischeri* symbiosis is a two-step process initiated by a few environmental symbiont cells. *Environmental Microbiology* 15:2937–50. doi: 10.1111/1462-2920.12179. PMID: 23819708.

Aschtgen, M. S., J. B. Lynch, E. Koch, J. Schwartzman, M. McFall-Ngai, and E. G. Ruby. 2016a. Rotation of *Vibrio fischeri* flagella produces outer membrane vesicles that induce host development. *Journal of Bacteriology* 198:2156–65. doi: 10.1128/JB.00101-16. PMID: 27246572.

Aschtgen, M. S., and E. G. Ruby. 2019. Insights into flagellar function and mechanism from the squid-vibrio symbiosis. *N. J. P. Biofilms Microbiomes* 5:32. doi: 10.1038/s41522-019-0106-5. PMID: 31666982.

Aschtgen, M. S., K. Wetzel, W. Goldman, M. McFall-Ngai, and E. Ruby. 2016b. *Vibrio fischeri*-derived outer membrane vesicles trigger host development. *Cellular Microbiology*. 18:488–99. doi: 10.1111/cmi.12525. PMID: 26399913.

Baliga, P., M. Shekar, and M. N. Venugopal. 2018. Potential outer membrane protein candidates for vaccine development against the pathogen *Vibrio anguillarum*: A reverse vacinology based identification. *Current Microbiology* 75:368–77. doi: 10.1007/s00284-017-1390-z. PMID: 29119233.

Belcaid, M., G. Casaburi, S. J. McAnulty, H. Schmidbaur, A. M. Suria, S. Moriano-Gutierrez, M. S. Pankey et al. 2019. Symbiotic organs by shaped by distinct modes of genome evolution in cephalopods. *Proceedings of the National Academy of Sciences USA* 116:3030–35. doi: 10.1073/pnas.1817322116. PMID: 30635418.

Boettcher, K. J., E. G. Ruby, and M. J. McFall-Ngai. 1996. Bioluminescence in the symbiotic squid *Euprymna scolopes* is controlled by a daily biological rhythm. *Journal of Comparative Physiology* 179:65–73. doi: 10.1007/BF00193435.

Bongrand C., E. J. Koch, S. Moriano-Gutierrez, O. X. Cordero, M. McFall-Ngai, M. F. Polz, and E. G. Ruby. 2016. A genomic comparison of 13 symbiotic Vibrio fischeri isolates from the perspective of their host source and colonization behavior. *ISME Journal* 10:2907–17. doi: 10.1038/ismej.2016.69. Epub 2016 Apr 29. PMID: 27128997.

Bongrand, C., and E. G. Ruby. 2019a. The impact of *Vibrio fischeri* strain variation on host colonization. *Current Opinion in Microbiology* 50:15–9. doi: 10.1016/j.mib.2019.09.002. PMID: 31593868.

Bongrand, C., and E. G. Ruby. 2019b. Achieving a multi-strain symbiosis: Strain behavior and infection dynamics. *ISME Journal* 13:798–806. doi: 10.1038/s41396-018-0305-8. PMID: 30353039.

Bouskra, D., C. Brézillon, M. Bérard, C. Werts, R. Varona, I. G. Boneca, and G. Eberl. 2008. Lymphoid tissue genesis induced by commensals through NOD1 regulates intestinal homeostasis. *Nature* 456:507–10. doi: 10.1038/nature07450. PMID: 18987631.

Brooks Ii, J. F., M. C. Gyllborg, D. C. Cronin, S. J. Quillin, C. A. Mallama, R. Foxall, C. Whistler, A. L. Goodman, and M. J. Mandel. 2014. Global discovery of colonization determinants in the squid symbiont *Vibrio fischeri*. *Proceedings of the National Academy of Sciences USA* 111:17284–9. doi: 10.1073/pnas.1415957111. PMID: 25404340.

Chen, F., B. C. Krasity, S. M. Peyer, S. Koehler, E. G. Ruby, X. Zhang, and M. J. McFall-Ngai. 2017. Bactericidal permeability-increasing proteins shape host-microbe interactions. *mBio* 8:e00040–17. doi: 10.1128/mBio.00040-17. PMID: 28377525.

Chun C. K., J. V. Troll, I. Koroleva, B. Brown, L. Manzella, E. Snir, H. Almabrazi et al. 2008. Effects of colonization, luminescence, and autoinducer on host transcription during development of the squid-vibrio association. *Proceedings of the National Academy of Sciences USA* 105:11323–8. doi: 10.1073/pnas.0802369105. PMID: 18682555.

Cohen, S. K., M.-S. Aschtgen, J. B. Lynch, S. Koehler, F. Chen, J. Escrig, J. Daraspe, E. G. Ruby, A. Meibom, and M. McFall-Ngai. 2020. Tracking the cargo of extracellular symbionts into the host tissues with correlated electron microscopy and NanoSIMS imaging. *Cellular Microbiology* 22:e13177. doi: 10.1111/cmi.13177. PMID: 32185893.

Collins, A. J., B. A. LaBarre, B. S. Won, M. V. Shah, S. Heng, M. H. Choudhury, S. A. Haydar, J. Santiago, and S. V. Nyholm. 2012. Diversity and partitioning of bacterial populations within the accessory nidamental gland of the squid Euprymna scolopes. *Applied and Environmental Microbiology* 78:4200–8. doi: 10.1128/AEM.07437-11. PMID: 22504817.

Davidson, S. K., T. A. Koropatnick, R. Kossmehl, L. Sycuro, and M. J. McFall-Ngai. 2004. NO means 'yes' in the squid-vibrio symbiosis: The role of nitric oxide in the initiation of a beneficial association. *Cellular Microbiology* 6:1139–51. doi: 10.1111/j1462-5822.2004.00429.x. PMID: 15527494.

Essock-Burns, T., W. E. Goldman, E. G. Ruby, and M. J. McFall-Ngai. 2020. Interactions of the host with both the environment and the symbiont drives the development of a complex biogeography that restricts symbiont localization. *mBio*. (in press).

Fidopiastis, P. M., S. V. Boletzky, and E. G. Ruby. 1998. A new niche for *Vibrio logei*, the predominant light organ symbiont of squids in the genus *Sepiola*. *Journal of Bacteriology* 180:59–64. PMID: 9422593.

Goodson, M. S., M. Kojadinovic, J. V. Troll, T. E. Scheetz, T. L. Casavant, M. B. Soares, and M. J. McFall-Ngai. 2005. Identifying components of the NF-kappaB pathway in the beneficial *Eurpymna scolopes-Vibrio fischeri* light organ symbiosis. *Applied and Environmental Microbiology* 71:6934–46. doi: 10.1128/AEM.71.11.6934-6946.2005. PMID: 16269728.

Graf, J., and E. G. Ruby. 1998. Host-derived amino acids support the proliferation of symbiotic bacteria. *Proceedings of the National Academy of Sciences USA* 95:1818–22. doi: 10.1073/pnas.95.4.1818. PMID: 9466100.

Hastings, J. W., and K. H. Nealson. 1981. The symbiotic luminous bacteria. In: Starr. M. P., H. Stolp, H. G. Truper, A. Balows, and H. G. Schlegel (eds.) *The Prokaryotes*. Springer-Verlag, New York, Heidelberg, Berlin, pp. 1332–45.

Heath-Heckman, E. A., J. Foster, M. A. Apicella, W. E. Goldman, and M. McFall-Ngai. 2016. Environmental cues and symbiont microbe-associated molecular patterns functions in concert to drive the daily remodeling of the crypt-cell brush border of the *Euprymna scolopes* light organ. *Cellular Microbiology* 18:1642–52. doi: 10.1111/cmi.12602. PMID: 27062511.

Heath-Heckman, E. A., A. A. Gillette, R. Augustin, M. X. Gillette, W. E. Goldman, and M. J. McFall-Ngai. 2014. Shaping the microenvironment: Evidence for the influence of a host galaxin on symbiont acquisition and maintenance in the squid-vibrio symbiosis. *Environmental Microbiology* 16:3669–82. doi: 10.1111/1462-2920.12496. PMID: 24802887.

Heath-Heckman, E. A., and M. J. McFall-Ngai. 2011. The occurrence of chitin in the hemocytes of invertebrates. *Zoology (Jena)* 114:191–8. doi: 10.1016/j.zool.2011.02.002. PMID: 21723107.

Heath-Heckman, E. A., S. M. Peyer, C. A. Whistler, M. A. Apicella, W. E. Goldman, and M. J. McFall-Ngai. 2013. Bacterial bioluminescence regulates expression of a host cryptochrome gene in the squid-vibrio symbiosis. *mBio* e00167–13. doi: 10.1128/mBio.00167-13. PMID: 23549919.

Kerwin A. H., S. M. Gromek, A. M. Suria, R. M. Samples, D. J. Deoss, K. O'Donnell, S. Frasca, Jr. et al. 2019. Shielding the next generation: Symbiotic bacteria from a reproductive organ protect bobtail squid eggs from fungal fouling. *MBio* 2019 Oct 29;10(5): pii: e02376-19. doi: 10.1128/mBio.02376-19. PMID: 31662458.

Kerwin, A. H., and S. V. Nyholm. 2017. Reproductive system symbiotic bacteria are conserved between two distinct populations of *Euprymna scolopes* from Oahu, Hawaii. *mSphere* 2018 Mar 28;3(2): pii: e00531-17. doi: 10.1128/mSphere.00531-17. PMID: 29600280.

Kimbell, J. R., and M. J. McFall-Ngai. 2004. Symbiont-induced changes in host actin during the onset of a beneficial animal-bacterial association. *Applied and Environmental Microbiology* 70:1434–41. doi: 10.1128/aem.70.3.1434-1441.200. PMID: 15006763.

Koch, E. J., T. I. Miyashiro, M. J. McFall-Ngai, and E. G. Ruby. 2014. Features governing long-term persistence in the squid-vibrio symbiosis. *Molecular Ecology* 23:1624–34. doi: 10.1111/mec.12474. PMID: 24118200.

Koch, E. J., E. G. Ruby, M. McFall-Ngai, and M. Liebeke. In press. The impact of persistent colonization by *Vibrio fischeri* on the metabolome of the host squid *Euprymna scolopes*. *Journal of Experimental Biology*.

Koehler S., R. Gaedeke, C. Thompson, C. Bongrand, K. L. Visick, E. Ruby, and M. McFall-Ngai. 2018. The model squid-vibrio symbiosis provides a window into the impact of strain-and species-species-level differences during the initial stages of symbiont engagement. *Environmental Microbiology* 21:3269–83. doi: 10.1111.1462-2920.14392. PMID: 30136358.

Koropatnick, T. A., J. T. Engle, M. A. Apicella, E. V. Stabb, W. E. Goldman, and M. J. McFall-Ngai. 2004. Microbial factor-mediated development of a host-bacterial mutualism. *Science* 306:1186–8. doi: 10.1126/science.1102218. PMID 15539604.

Koropatnick, T., M. S. Goodson, E. A. Heath-Heckman, and M. McFall-Ngai. 2014. Identifying the cellular mechanisms of symbiont-induced epithelial morphogenesis in the squid-vibrio association. *Biological Bulletin* 26:56–68. doi: 10.1086/BBLv226n1p56. PMID: 24648207.

Krasity, B. C., J. V. Troll, E. M. Lehnert, K. T. Hackett, J. P. Dillard, M. A. Apicella, W. E. Goldman, J. P. Weiss, and M. J. McFall-Ngai. 2015. Structural and functional features of a developmentally regulated lipopolysaccharide-binding protein. *mBio* 6:e00193–15. doi: 10.1128/mBio.01193-15. PMID: 26463160.

Krasity, B. C., J. V. Troll, J. P. Weiss, and M. J. McFall-Ngai. 2011. LBP/BPI proteins and their relatives: Conservation over evolution and roles in mutualism. *Biochemical Society Transactions* 39:1039–44. doi: 10.1042/BST0391039. PMID: 21787344.

Kremer, N., E. E. R. Philipp, M.-C. Carpentier, C. A. Brennan, L. Kraemer, M. A. Altura, R. Augustin et al. 2013. Initial symbiont contact orchestrates host organ-wide transcriptional changes that prime tissue colonization. *Cell Host & Microbe* 14:183–94. doi: 10.1016/j.chom.2013.07.006. PMID: 23954157.

Kremer, N., J. Schwartzman, R. Augustin, L. Zhou, E. G. Ruby, S. Hourdez, and M. J. McFall-Ngai. 2014. The dual nature of haemocyanin in the establishment and persistence of the squid-vibrio symbiosis. *Proceedings of the Royal Society Biological Sciences.* 281:20140504. doi: 10.1098/rspb.2014.0504. PMID: 24807261.

Lamarcq, L. H., and M. J. McFall-Ngai. 1998. Induction of a gradual, reversible morphogenesis of its host's epithelial brush border by *Vibrio fischeri. Infection and Immunity* 66:777–85. PMID: 9453641.

McFall-Ngai, M. J. 2014. The importance of microbes in animal development: Lessons from the squid-vibrio symbiosis. *Annual Reviews of Microbiology* 68:177–94. doi: 10.1146/annurev-micro-091313-103654. PMID: 24995875.

McFall-Ngai, M., E. A. Heath-Heckman, A. A. Gillette, S. M. Peyer, and E. A. Harvie. 2012. The secret languages of coevolved symbioses: Insights from the *Euprymna scolopes-Vibrio fischeri* symbiosis. *Seminars in Immunology* 24:3–8. doi: 10.1016/j.smim.2011.11.006. PMID: 22154556.

McFall-Ngai, M. J., and M. K. Montgomery. 1990. The anatomy and morphology of the adult bacterial light organ of *Euprymna scolopes* Berry (Cephalopoda:Sepiolidae). *The Biological Bulletin* 179:332–9. doi: 10.2307/1542325. PMID: 29314961.

McFall-Ngai, M., S. V. Nyholm, and M. G. Castillo. 2010. The role of the immune system in the initiation and persistence of the *Euprymna scolopes-Vibrio fischeri* symbiosis. *Seminars in Immunology* 22:48–53. doi: 10.1016/j.smim.2009.11.003. PMID: 20036144.

Montgomery, M. K., and M. J. McFall-Ngai. 1993. Embryonic development of the light organ of the sepiolid squid *Euprymna scolopes* Berry. *Biological Bulletin* 184:296–308. doi: 10.2307/1542448. PMID: 29300543.

Moriano-Gutierrez, S., E. J. Koch, H. Bussan, K. Romano, M. Belcaid, F. E. Rey, E. G. Ruby, and M. J. McFall-Ngai. 2019. Critical symbiont signals drive both local and systemic changes in diel and developmental host gene expression. *Proceedings of the National Academy of Sciences USA* 116:7990–9. doi: 10.1073/pnas.1819897116. PMID: 30833394.

Naef, A. 1923. Die Cephalopoden (Systemik). *Fauna Flora Golf Napoli (Monograph)* 35:1–863.

Nawroth, J., H. Guo, E. Koch, E. A. Heath-Heckman, J. C. Hermanson, E. Ruby, J. Dabiri, E. Kanso, and M. McFall-Ngai. 2017. Motile cilia create fluid-mechanical microhabitats for the active recruitment of the host microbiome. *Proceedings of the National Academy of Sciences USA* 114:9510–6. doi: 10.1073/pnas.1706926114. PMID: 28835539.

Nikolakakis, K., E. Lehnert, M. J. McFall-Ngai, and E. G. Ruby. 2015. Use of hybridization chain reaction-fluorescent *in situ* hybridization to track gene expression by both partners during initiation of symbiosis. *Applied and Environmental Microbiology* 81:4728–35. doi: 10. 1128/AEM.00890-15. PMID: 25956763.

Nishiguchi, M. K. 2002. The use of physiological data to corroborate cospeciation events in symbiosis. In: DeSalle, R., G. Giribet, and W. Wheeler (eds.) *Molecular Systematics and Evolution: Theory and Practice.* Birkhauser Verlag, Switzerland, pp. 237–44.

Nishiguchi, M. K., J. E. Lopez, and S. V. Boletzky. 2004. Enlightenment of old ideas from new investigations: More questions regarding the evolution of bacteriogenic light organs in squids. *Evolution and Development* 6:41–9. doi: 10.1111/j.1525-142x.2004.04009.x. PMID: 15108817.

Nyholm, S. V., B. DePlanke, H. R. Gaskins, M. Apicella, and M. J. McFall-Ngai. 2002. The roles of Vibrio fischeri and nonsymbiotic bacteria in the dynamics of mucus secretion during symbiont colonization of the Euprymna scolopes light organ. *Applied Environmental Microbiology* 68:5113–22. doi: 10.1128/aem.68.10.5113-5122.2002. PMID: 12324362.

Nyholm, S. V., and M. J. McFall-Ngai. 2004. The winnowing: Establishing the squid-vibrio symbiosis. *Nature Reviews Microbiology* 2:632–42. doi: 10.1038/nrmicro957. PMID: 15263898.

Nyholm, S. V., E. V. Stabb, E. G. Ruby, and M. J. McFall-Ngai. 2000. Establishment of an animal-bacterial association: Recruiting symbiotic vibrios from the environment. *Proceedings of the National Academy of Sciences USA* 97:10231–5. doi: 10.1073/pnas.97.18.10231. PMID: 10963683.

Nyholm, S. V., J. J. Stewart, E. G. Ruby, and M. J. McFall-Ngai. 2009. Recognition between symbiotic *Vibrio fischeri* and the haemocytes of *Euprymna scolopes. Environmental Microbiology* 11:483–93. doi: 10.1111/j.1462-2920.2008.01788.x. PMID: 19196278.

Peyer, S. M., E. A. C. Heath-Heckman, and M. J. McFall-Ngai. 2017. Characterization of the cell polarity gene *crumbs* during early development and maintenance of the squid-vibrio light organ symbiosis. *Development, Genes, and Evolution* 227:375–87. doi: 10.1007/s00427-017-0576-5. PMID: 28105525.

Peyer, S. M., N. Kremer, and M. J. McFall-Ngai. 2018. Involvement of a host Cathepsin L in symbiont-induced cell death. *Microbiologyopen* 7:e00632. doi: 10.1002/mbo3.632. PMID: 29692003.

Peyer, S. M., M. S. Pankey, T. H. Oakley, and M. J. McFall-Ngai. 2014. Eye-specification genes in the bacterial light organ of the bobtail squid *Euprymna scolopes*, and their expression in response to symbiont cues. *Mechanisms of Development* 131:111–26. doi: 10.1016/j.mod.2013.09.004. PMID: 24157521.

Rader, B. A., N. Kremer, M. A. Apicella, W. E. Goldman, and M. J. McFall-Ngai. 2012. Modulation of symbiont lipid A signaling by host alkaline phosphatase in the squid-vibrio symbiosis. *mBio* 3:e00093–12. doi: 10.1128/mBio.00093-12. PMID: 22550038.

Rader, B., S. J. McAnulty, and S. V. Nyholm. 2019. Persistent symbiont colonization leads to a maturation of hemocyte response in the *Euprymna scolopes/Vibrio fischeri* symbiosis. *Microbiologyopen* 8:e858. doi: 10.1002/mbo3.858. PMID: 31197972.

Reardon, S. 2019. CRISPR gene-editing creates wave of exotic model organisms. *Nature* 568: 441–42.

Reid, A., and P. Jereb. 2005. Family Sepiolidae. In P. Jereb and C. F. E. Roper (eds.) *Cephalopods of the World. An Annotated and Illustrated Catalogue of Species Known to Date.* Vol.1. Chambered nautiluses and sepioids (Nautilidae,Sepiidae, Sepiolidae, Sepiadariidae, Idiosepiidae and Spirulidae). FAO Species Catalogue for Fishery Purposes. No. 4, Vol. 1. FAO, Rome, pp. 153–203.

Ruby, E.G., M. Urbanowski, J. Campbell, A. Dunn, M. Faini, R. Gunsalus, P. Lostroh et al. 2005. Complete genome sequence of *Vibrio fischeri*: A symbiotic bacterium with pathogenic congeners. *Proceedings of the National Academy of Sciences USA* 102:3004–9. doi: 10.1073/pnas.0409900102. PMID: 15703294.

Schwartzman, J. A., E. Koch, E. A. Heath-Heckman, L. Zhou, N. Kremer, M. J. McFall-Ngai and E. G. Ruby. 2015. The chemistry of negotiation: Rhythmic, glycan-driven acidification in a symbiotic conversation. *Proceedings of the National Academy of Sciences USA* 112:566–71. doi: 10.1073/pnas.1418580112. PMID: 25550509.

Speare L., A. G. Cecere, K. R. Guckes, S. Smith, M. S. Wollenberg, M. J. Mandel, T. Miyashiro, and A. N. Septer. 2018. Bacterial symbionts us a type VI secretion system to eliminate competitors in their natural host. *Proceedings of the National Academy of Sciences USA* 115:E8528–37. doi: 10.1073/pnas.1808302115. PMID: 30127013.

Tong, D., N. S. Rozas, T.H. Oakley, J. Mitchell, N.J. Colley, and M. J. McFall-Ngai. 2009. Evidence for light perception in a bioluminescent organ. *Proceedings of the National Academy of Sciences USA* 106:9836–41. doi: 10.1073/pnas.0904571106. PMID: 19509343.

Troll, J. V., D. M. Adin, A. M. Wier, N. Paquette, N. Silverman, W. E. Goldman, F. J. Stadermann, E. V. Stabb, and M. J. McFall-Ngai. 2009. Peptidoglycan induces loss of a nuclear peptidoglycan recognition protein during host tissue development in a beneficial animal-bacterial symbiosis. *Cellular Microbiology* 11:1114–27. doi: 10.1111/j.1462-5822.2009.01315.x. PMID: 19416268.

Troll, J. V., E. H. Bent, N. Pacquette, A. M. Wier, W. E. Goldman, N. Silverman, and M. J. McFall-Ngai. 2010. Taming the symbiont for coexistence: A host PGRP neutralizes a bacterial symbiont toxin. *Environmental Microbiology* 12:2190–203. doi: 10.1111/j.1462-2920.2009.02121.x. PMID: 21969913.

Visick, K. L. 2009. An intricate network of regulators controls biofilm formation and colonization by *Vibrio fischeri*. *Molecular Microbiology* 74:782–9. doi: 10.1111/j.1365-2958.2009.06899.x. PMID: 19818022.

Visick, K. L., J. Foster, J. Doino, M. McFall-Ngai, and E. G. Ruby. 2000. *Vibrio fischeri lux* genes play an important role in colonization and development of the host light organ. *Journal of Bacteriology* 182:4578–86. doi: 10.1128/jb.182.16.4578-4586.2000. PMID: 10913092.

Wang, Y., Y. S. Dufour, H. K. Carlson, T. J. Donohue, M. A. Marletta, and E. G. Ruby. 2010a. H-NOX-mediated nitric oxide sensing modulates symbiotic colonization by *Vibrio fischeri*. *Proceedings of the National Academy of Sciences USA* 107:8375–80. doi: 10.1073/pnas.1003571107. PMID: 20404170.

Wang, Y., A. K. Dunn, J. Wilneff, M. J. McFall-Ngai, S. Spiro, and E. G. Ruby. 2010b. *Vibrio fischeri* flavohaemoglobin protects against nitric oxide during initiation of the squid-vibrio symbiosis. *Molecular Microbiology* 78:903–15. doi: 10.1111/j.1365-2958.2010.07376.x. PMID: 20815823.

Wei, S. L., and R. E. Young. 1989. Development of symbiotic bacterial bioluminescence in a nearshore cephalopod, *Euprymna scolopes*. *Marine Biology* 103:541–6.

Wier, A. M., S. V. Nyholm, M. Mandel, P. Massengo-Tiassé, A. L. Schaefer, I. Koroleva, S. Splinter-BonDurant et al. 2010. Transcriptional patterns in both host and bacterium underlie a daily rhythm of ultrastructural and metabolic change in a beneficial symbiosis. *Proceedings of the National Academy of Sciences USA* 107:2259–64. doi: 10.1073/pnas.0909712107. PMID: 20133870.

Wollenberg, M. S., and E. G. Ruby. 2009. Population structure of *Vibrio fischeri* within the light organs of *Euprymna scolopes* squid from two Oahu (Hawaii) populations. *Applied and Environmental Microbiology* 75:193–202. doi: 10.1128/AEM.01792-08. PMID: 18997024.

Yang, J. S., J. H. Jeon, M. S. Jang, S. S. Kang, K. B. Ahn, M. Song, C. H. Yun, and S. H. Han. 2018. *Vibrio cholerae* OmpU induces IL-8 expression in human intestinal epithelial cells. *Molecular Microbiology* 93:47–54. doi: 10.1016/j.molimm.2017.11.005. PMID: 29145158.

11 Evolving integrated multipartite symbioses between plant-sap feeding insects (Hemiptera) and their endosymbionts

Gordon Bennett

Contents

> ...my study on the symbiotic adaptations of the cicadas [Hemiptera] opened up a veritable fairyland of insect symbiosis...
>
> **Buchner (1965)**

11.1 Introduction

Many animals establish obligate interactions with heritable microbes that provide essential metabolic services (McFall-Ngai et al. 2013; Hacquard et al. 2015). These benefits often provide key evolutionary traits that permit hosts to exploit ecological resources otherwise unsuitable for animal life (Moran 2007). A prominent example is in the insects, which have leveraged microbial symbionts to exploit a broad range of niches to become one of the most abundant and diverse animal groups on earth (Douglas 2009; Feldhaar 2011; Bennett and Moran 2015; Sudakaran, Kost, and Kaltenpoth 2017). In particular, most species in the plant-sap feeding insect order, the Hemiptera (>80,000 described species), obligately rely on beneficial microbes for essential amino acids and vitamins that are depauperate in phloem and xylem saps (Figure 11.1; Moran and Baumann 2000; Baumann 2005; Douglas 2006). Species in the Hemiptera have wide-ranging interactions with symbionts that generally live within their bodies and are either (i) intracellularly restricted to specialized host tissues and transovarially transmitted, or (ii) extracellularly located to the midgut and are transmitted through various mechanisms outside of the body (Buchner 1965; Salem et al. 2015; Sudakaran et al. 2015). Many lineages within the Hemiptera further obligately rely on multiple complimentary symbionts, whose combined metabolisms furnish all essential nutrients required by their hosts (McCutcheon and Moran 2010; Salem et al. 2013; Douglas 2016). Obligate symbioses in the Hemiptera are generally maintained on the order of tens to hundreds of millions of years and are associated with the diversification of major subgroups within the order (e.g., Cicadas, Aphids, Leafhoppers, etc., Figure 11.1).

Improving genomic and molecular techniques have now made it possible to investigate a wide range of hemipteran symbioses, including those recalcitrant to direct experimentation (McFall-Ngai 2015). We are now beginning to understand some generalities about the evolutionary origins, establishment, and maintenance of these symbioses (Bennett and Moran 2015; Skidmore and Hansen 2017; McCutcheon, Boyd, and Dale 2019). Fungal and bacterial symbionts are likely derived from diverse origins, including insect pathogens, vectored microbes (e.g., plant pathogens), or other facultative microbes that can provide insect hosts with occasional benefits. Over time, antagonistic or commensalistic traits give way to those that provide hosts with consistent beneficial services, leading to eventual codependence (McCutcheon, Boyd, and Dale 2019). As these relationships evolve towards obligate reliance, hosts and symbionts establish highly integrated molecular and cellular mechanisms that permit a relatively stable symbiosis to persist over many host generations. This chapter reviews important aspects of how both microbial symbionts and their insect hosts coevolve to integrate and sustain their interactions.

11.2 Roles of Hemipteran symbionts: Nutrition and beyond

The primary reason hemipterans engage in obligate symbiosis with bacterial and fungal partners is for nutritional benefits. The proliferation of genome sequences for these microbial symbionts has revealed that indeed, even though their genomes become extremely reduced (discussed below), they maintain genes and metabolic pathways that complement the nutritional limitations of their hosts' specialized diets (Moran and Bennett 2014; Bennett and Moran 2015). For example, the vast majority of species in the Hemiptera, and particularly in the Auchenorrhyncha and Sternorrhyncha (Figure 11.1), specialize on xylem and phloem plant-sap diets (Shigenobu et al. 2000; Nakabachi et al. 2006; McCutcheon and Moran 2010). Both diet types are depauperate in essential amino acids (EAA) and B vitamins, although

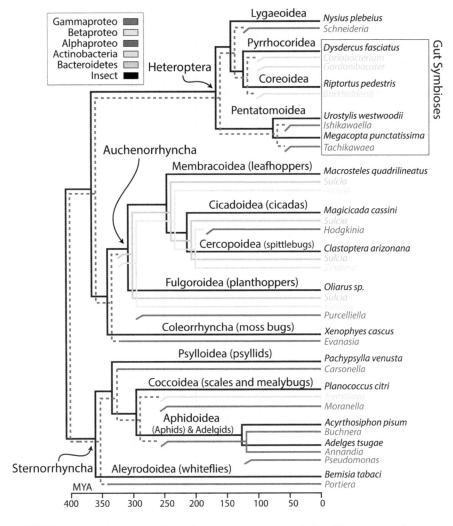

FIGURE 11.1 Phylogenetic schematic summarizing the relationships between the major lineages of the Hemiptera order and their heritable microbial symbionts. Select host species and their particular microbial partners for which we have genomic evidence of host–symbiont coevolution and nutritional integration are shown (see text for discussions and citations of these particular systems). Symbiont lineages and names are color coded according to their higher-taxonomic classification (see inset box; proteobacteria class names are truncated). Dashed lines represent predicted relationships and origins, and gray dashed lines show where ancestral symbionts are unknown. Symbionts that infect the host midgut are labeled with a gray box; all others are intracellular and endocytosed in bacteriocytes. The scale bar illustrates divergence times in millions of years (MYA=millions of years ago) for the major host lineages and the origins of their symbionts. (Host phylogeny and divergence times are adapted from Johnson, K.P. et al. 2018. *Proceedings of the National Academy of Sciences USA* 115:12775–80.)

to varying degrees; and, xylem is further limited in sugar and other essential nutrients (Douglas 2006; Ankrah, Chouaia, and Douglas 2018). To obtain these resources, hosts have allied with a diversity of microbial symbionts (Hansen and Moran 2014; Bennett and Moran 2015; Sudakaran, Kost, and Kaltenpoth 2017).

Some Hemipteran insect lineages have switched to feed on more nitrogen-rich diets (i.e., EAAs), such as plant tissues and animal blood. However, even in these cases, hosts still maintain obligate symbioses with microbes whose genomes reveal metabolic streamlining tailored to supplement other missing, but essential nutrients. An illustrative example occurs in the bacterial symbiont of the moss bugs (Coleorrhyncha; Figure 11.1) that encode enriched gene sets to synthesize additional nonessential amino acids and organic sulfur, which are rarely retained by other hemipteran symbionts (Santos-Garcia et al. 2014). The obligate food source of moss bugs, bryophytes, is limited in both of these resources compared with higher vascular plants. In other cases, symbionts are dedicated to vitamin provisioning. For example, heteropteran species that feed on nutrient rich plant tissues, as occurs in seed feeding species in the family Pyrrhocoridae (Heteroptera; Figure 11.1), generally rely on their microbes for B vitamins (Salem et al. 2014). Similarly, species that specialize on blood diets (e.g., bed bugs and kissing bugs) also require their microbes for B vitamins (Beard, Cordon-Rosales, and Durvasula 2002; Nikoh et al. 2014). In more extreme cases, some Hemiptera have transitioned to feed on nitrogen-rich parenchyma cells (e.g., Typhlocybinae leafhoppers) and appear to have lost their symbionts entirely (Buchner 1965; Moran, Tran, and Gerardo 2005).

Hemipteran symbionts may also have broad roles in other systemic host functions. Yeast-like symbionts in the brown planthopper (Fulgoroidea; *Nilaparvata lugens*) maintain the capability to synthesize sterols, which animals also require from exogenous sources (Gibson and Hunter 2010; Fan et al. 2015). In insects, sterols are important in cell membrane formation, cell signal transduction, and developmental and molting hormone production (Noda and Koizumi 2003). Symbionts also occasionally retain non-EAAs, including tyrosine-related synthesis genes essential in neuromodulation, and exoskeleton formation and hardening (True 2003; Simonet et al. 2016b). While the functional role of hemipteran symbionts on host neurobiology is not fully known, curing insects of their obligate microbes often leads to weakened and discolored cuticles that suggests their intrinsic role in its formation (Kikuchi et al. 2016; Anbutsu et al. 2017; Hirota et al. 2017). Recent work in pea aphids (Aphidoidea; *Acyrthosiphon pisum*) showed that the tyrosine pathway is shared between the host and their bacterial symbiont, *Buchnera* (Figure 11.1; Wilson et al. 2010). During embryonic development, aphid-encoded enzymes involved in tyrosine synthesis are more highly expressed in the bacteriome, with a commensurate increase in expression of other cuticle formation genes (Rabatel et al. 2013). Knockdown of this pathway with RNAi leads to severe deformities in nymphs, confirming its central role in host development (Simonet et al. 2016b). The symbionts of several other hemipteran host species also retain genes that can synthesize tyrosine (e.g., some heteropterans [pyrrhocorids] and Sternorrhyncha [adelgids]) (Figure 11.1; Nikoh et al. 2011; Santos-Garcia et al. 2017; Weglarz et al. 2018), highlighting a potentially important role of some symbionts in host cuticle development.

Extracellular symbionts of the Heteroptera that establish in the gut are also known to have broad systemic effects on their hosts. In the bean bug (Coreoidea; *Riptortus pedestris*), *Burkholderia* bacterial symbionts increase production of proteins important in nutrient stocks in adult insect hemolymph that in turn influence development time and the number of eggs laid (Lee et al. 2017). Establishment of *Burkholderia* also prime the host humoral immune system by increasing expression of antimicrobial peptides, which improve host survival when challenged with potentially pathogenic bacteria (Kim et al. 2015). Moreover, the failure of beneficial gut microbes to establish in their hosts can lead to systemic stress responses, highlighting their broad roles in homeostatic balance of their hosts. For example, in the African cotton stainer (Pyrrhocoridae: *Dysdercus fasciatus*), individuals deprived of their symbionts exhibit environmental and nutritional stress responses by overexpressing heat shock proteins, glucose and B vitamin transporters, and B vitamin processing genes (Salem et al. 2014). Thus, although extracellular symbionts are located to the gut, their activities have wide-ranging health effects on hosts vis-à-vis balancing systemic resources and nutrition.

Finally, obligate symbionts can also provide their hemipteran hosts with essential services other than nutrition. In several cases, microbes encode pathways that provide environmental protection from abiotic or biotic factors. For example, the whitefly bacterial symbiont (Figure 11.1; Aleyroididea: *Bemisia tabaci*), *Portiera*, encodes carotenoid biosynthetic pathways that can aid in light absorption, prevent oxidative damage to cells, and may also help to stabilize symbiont DNA and genomic structures (Sloan and Moran 2012a). In other cases, beneficial microbes provide defensive secondary compounds that may help to protect hosts from parasitism and predation. The co-obligate bacterial symbiont, *Proftella*, found in the Asian citrus psyllid (Figure 11.1; Psylloidea; *Diaphorina citri*) encodes a polyketide synthesis pathway that has cytotoxic effects (Nakabachi et al. 2013). In insects, polyketides are known defensive compounds that accumulate in the hemolymph and can functionally deter predators (Kellner and Dettner 1996). Similarly, yeast-like symbionts in the Cerataphidinae aphids also encode polyketide synthesis pathways (Vogel and Moran 2013), suggesting that these kinds of obligate defensive symbioses may be relatively widespread.

11.3   Genome evolution in Hemipteran symbionts

Obligate symbionts of the Hemiptera are characterized by extreme genome reduction. The genomic encoded capabilities of symbionts have significant implications for how they interact and communicate with their hosts. The eldest symbionts generally have the trimmest genomes, as evolutionary forces have rent them down to the minimum requirements of a sustainable symbiosis (Figure 11.2; McCutcheon and Moran 2011). In the most extreme cases, ancient intracellular symbionts have lost >90% of their genetic content, with genomes ranging from 0.1 to 1 megabases (Moran and Bennett 2014). In contrast, the genomes of younger symbionts, and those with extracellular and open-environment life stages, are typically larger and encode more cellular capabilities; however, if maintained, they will also experience gene losses over evolutionary time (Figure 11.2; Koga and Moran 2014; Bennett et al. 2016; Weglarz et al. 2018). Regardless of age, obligate symbionts consistently exhibit some level of genome reduction and reduced cellular capabilities, including cell envelop synthesis, nonessential amino acid and vitamin synthesis pathways, nucleotide synthesis, generation of cellular energy, and DNA mutation repair and recombination,

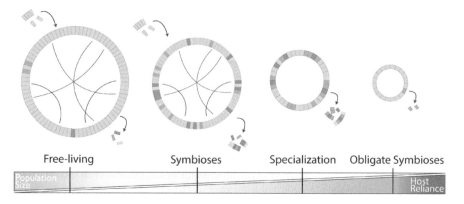

Free-living   Symbioses   Specialization   Obligate Symbioses

FIGURE 11.2   Model of genome degeneration in the heritable beneficial symbionts of the Hemiptera. Genomic rings and labeled scales illustrate the evolutionary changes as bacteria go from a free-living state with a large population size to an obligate symbiont with small population sizes. Initially genomes are able to acquire novel genes and recombine (inner black lines) in order to retain adaptive flexibility. However, after bacteria become fixed symbionts, they can only lose genes and become increasingly dependent on hosts for cellular and metabolic support. (Figure and model adapted from McCutcheon, J.P., and N.A. Moran. 2011. *Nature Reviews Microbiology* 10:13–26.)

among others (McCutcheon and Moran 2011). The genome-encoded capabilities of symbionts have critical implications for how hosts must adapt to integrate and maintain a functional symbiosis with their partners.

In recent years, the number of hemipteran symbionts with sequenced genomes has expanded greatly, providing a window into the process of genome degeneration (Figure 11.2; Bennett and Moran 2015; Wernegreen 2015; McCutcheon, Boyd, and Dale 2019). Initially, host-associated microbes experience rapid and widespread gene losses that removes traits required for free-living and that are redundant with the host environment. After microbes become obligately established and rely on the host for persistent vertical propagation, they are subject to intense genetic drift due to small population sizes that are frequently bottlenecked. As this ratchet turns, randomly accrued deleterious mutations become fixed, driving the impairment and random loss of genes—even for those genes seemingly essential to the symbioses (Moran 1996; Wernegreen 2002; Moran, McLaughlin, and Sorek 2009). The effects of this process are exacerbated by the fact that symbionts generally lose the ability acquire novel genes through recombination.

Although the genomes of ancient symbionts tend to become structurally conserved, gene loss is ongoing and asymmetrical between symbiont species that are shared by related host lineages (Moran, McLaughlin, and Sorek 2009; McCutcheon and Moran 2010; Sloan and Moran 2012b; Bennett et al. 2016; Bennett and Mao 2018; Łukasik et al. 2018; Otero-Bravo, Goffredi, and Sabree 2018; Chong, Park, and Moran 2019). Several mechanisms can contribute to these differential gene losses, including (i) adaptive response to host shifts to new niches that render certain bacterial metabolisms unnecessary, (ii) acquisition of additional symbionts (discussed below), and (iii) ongoing genetic drift. Indeed, unique loss of nutrient-provisioning genes are observed in *Buchnera* of aphid species that have transitioned to feeding in more nutritive plant galls (van Ham et al. 2003). Similarly, the very smallest symbiont genomes occur in symbiotic bacteria that are partnered with co-obligate microbes, as is commonly observed in the Auchenorrhyncha species (McCutcheon, McDonald, and Moran 2009b; McCutcheon and Moran 2010; Bennett and Moran 2013; Bennett and Mao 2018). In these systems, nutrition synthesis responsibilities are nonredundantly partitioned between partners. Nevertheless, drift is considered to be the main driving forces differentially shaping strain variation among all symbiont lineages, sometimes with very bizarre outcomes that include genome fragmentation and expansion (e.g., Van Leuven et al. 2014; Campbell et al. 2015).

Finally, although genome reduction is best known for intracellular symbionts, this process is also observed in Heteropteran systems, despite the extracellular localization of microbes to a more complex gut environment, and their transmission through mechanisms outside of the host body (Nikoh et al. 2011; Kenyon, Meulia, and Sabree 2015; Otero-Bravo, Goffredi, and Sabree 2018). Presumably genome reduction in these symbionts occurs because they persist in a nutrient rich and protected environment, inside and outside of the host. The excretion factors that symbionts are usually transmitted with likely create a stable, resource rich environment reducing the need for bacteria to maintain certain capabilities (Hosokawa et al. 2005; Nikoh et al. 2011).

11.4   Symbiont bearing organs: Transmission and development

An essential aspect of establishing a successful symbiosis is the evolution of an interface where hosts and their microbes can engage in a dialogue, and exchange resources throughout their lifecycles. These organs must also be paired with a reliable transmission strategy, which faithfully propagates microbial symbionts between generations (Moran, McCutcheon, and Nakabachi 2008; Kaiwa et al. 2014). Depending on the modality of symbiont acquisition, the organs and mechanisms that underlie this process differ significantly. Generally, intracellular symbionts are transovarially transmitted and established in specialized organs (bacteriomes) and cell-types (bacteriocytes). In contrast, extracellular symbionts are usually transmitted through mechanisms that involve a life stage outside of the host body (Figures 11.2 and 11.3; Buchner 1965; Engel and Moran 2013; Douglas 2014; Salem et al. 2015). The transmission mechanisms hosts use have further been implicated in the degree to which hosts and symbionts coevolve and codiversify: for example, hosts that invest more resources and control over symbiont transmission are expected to be more tightly coevolved with their microbes (reviewed by Salem et al. 2015).

11.4.1   Intracellular symbioses: Transovarial transmission and bacteriome development

In systems where symbionts are generally intracellular, hosts undertake the rather dramatic process of evolving entirely new organs in their bodies to house their microbes (i.e., bacteriomes or mycetomes; Figure 11.3b; Buchner 1965). Bacteriome formation is fully integrated into the developmental program of the host (Braendle

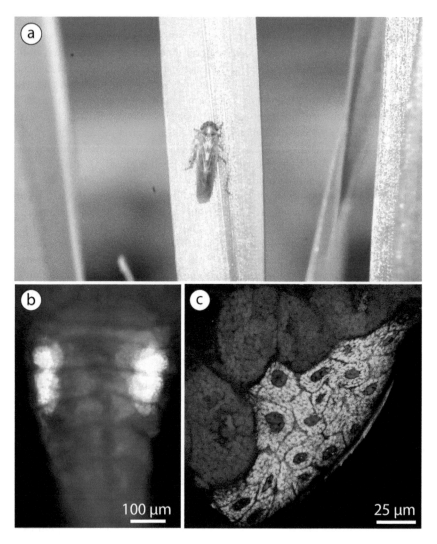

FIGURE 11.3 Beneficial symbiosis in the leafhopper, Macrosteles quadrilineatus (a) and related Deltocephalinae species (see also Figure 11.1). (b) Fluorescence in situ hybridization (FISH) of the bacteriome symbiont organs at the lateral edges of the host abdomen. (Host DNA is counter-stained blue, Sulcia shown in red, and Nasuia shown in green). (c) FISH of one host bacteriome showing the distinct bacterial species within their dedicated bacteriocytes. Panel (b) was provided Ryuchi Koga.

et al. 2003; Skidmore and Hansen 2017; Simonet et al. 2018). Within these tissues, microbes are usually intracellularly endocytosed within bacteriocytes that serve as the direct interface for communicating and exchanging cellular resources (Figures 11.3c and 11.4). With few exceptions, distinct symbiont species are further segregated into specific bacteriocyte types that evolve *de novo* with the acquisition of their resident symbionts (Buchner 1965; Koga et al. 2013). This segregation appears to be a host-level adaptation to specifically regulate symbiotic interactions with particular microbial species. Bacteriocytes have distinctive gene expression profiles that comprises hundreds of genes involved in essential metabolism, metabolite, transport, and other basic cellular functions required to maintain and support particular microbial species (discussed below; e.g., Husnik et al. 2013; Sloan et al. 2014; Ankrah, Luan, and Douglas 2017; Mao et al. 2017).

The mechanisms hemipterans use to transovarially transmit their intracellular symbionts between generations vary considerably. In many cases, they are released from bacteriocytes into the hemolymph and traverse the body to infect oocytes or embryonic tissues (Buchner 1965; Koga et al. 2012). However, in some host species (e.g., whitefly and *Putoidae* mealybugs) bacteria are inherited along with somatic bacteriocytes (Normark 2003; Luan et al. 2018). The inherited bacteriocytes appear to maintain genotypes that are distinct from other host germline and somatic tissues. Variation in transmission mechanisms likely reflects the independent origins of symbiotic associations, the discrete evolutionary development of bacteriome organs and cell-types, and the distinct coevolutionary trajectories of host and symbiont lineages. Given the general age of symbioses in the Hemiptera, however, it is not entirely clear how transovarial transmission mechanisms evolved. On possible scenario is that emerging symbionts may initially have independent capabilities to widely infect host tissues and the host germline, ensuring their own persistence across generations. The host may then co-opt this strategy, evolving tissues and mechanisms to sequester symbionts to particular locations in the body and to stabilize their inheritance. In other cases, as a hypothetical example, transovarially transmitted symbionts may emerge from persistently acquired gut microbes (see below). As gut tissues evolve into fixed bacteriome-like organs, they may then establish reliable transovarial transmission mechanisms. The organs and cells that are ultimately dedicated to housing and transmitting symbionts may depend on early infection patterns and also presymbiotic resource availability that can initially support the metabolisms of nascent symbionts (Wilson and Duncan 2015). Regardless, as hosts come to depend on these symbionts, it is within their interest to wrest control of tissue localization and the transmission schedule in order to guarantee their inheritance by offspring.

Although transovarial transmission of symbionts has been observed in a wide diversity of hemipteran hosts using microscopy techniques (e.g., Buchner 1965; Koga et al. 2012; Dan et al. 2017; Kobiałka et al. 2018), the genetic mechanisms, molecular signaling, or other systematic cues that initiate these processes are not well understood. Nevertheless, general patterns emerge that include distinct bacteriocyte involvement, transformational changes in bacterial cell morphology, and regeneration of bacteriocytes, which collectively point to a strictly regulated process. Emerging evidence from several systems provides some insight into the tightly evolved coordination underlying symbiont transmission across hemipteran hosts. For example, in some cicada species (Figure 11.1; Cicadoidea), it was recently shown that different host species dynamically evolve to transmit distinct bacterial titers to ensure offspring receive the requisite amount of bacteria (Campbell et al. 2018). Most cicadas rely on two bacteria, *Sulcia* and *Hodgkinia*, but in some species *Hodgkinia*'s genome has fragmented with bacterial cells carrying distinct gene sets (Van Leuven et al. 2014; Campbell et al. 2015). Species that require more *Hodgkinia* cells have evolved to transmit more of them, in order to ensure that developing embryos acquire the entire essential gene repertoire. Remarkably, despite evolutionary changes in *Hodgkinia* transmission, cicada embryos appear to obtain roughly the same titer of *Sulcia*. The transmission mechanisms for *Sulcia* are likely conserved across host species since this symbiont retains a single conserved genome (Campbell et al. 2018). The evolution of controlled transmission of particular symbiont numbers also appears to be widespread in other hemipteran hosts, underlying an important mechanism of control during early host development (Vogel and Moran 2011; Simonet et al. 2016a,b).

The process for transovarial transmission of an intracellular symbiont is best detailed in the pea aphid-*Buchnera* symbiosis. Aphid species have evolved a highly coordinated system that transfers *Buchnera* cells to embryos at a specific time during their early development (Mira and Moran 2002; Braendle et al. 2003). Particular bacteriocytes in proximity to the ovarioles exocytose bacterial cells into the hemolymph that are directly taken up into embryonic syncytium (Koga et al. 2012). Other facultative symbionts (e.g., *Serratia symbiotica*) are able to take advantage of this system and simultaneously infect developing embryos along with *Buchnera*. This feature may lead to the eventual establishment of these bacteria as co-obligate symbionts that participate in essential nutrition provisioning, as has been observed in the conifer aphid (*Cinara cedri*; Pérez-Brocal et al. 2006; Lamelas et al. 2011).

In other hemipteran species with intracellular symbionts, particular portions of the bacteriome are responsible for their selective transmission. For example, in some leafhoppers (e.g., Deltocephalinae species; Figure 11.3), which generally rely on two bacterial species, one bacterium (*Sulcia*) accumulates in a budding mass of host cells that is then releases bacteria directly into the hemolymph (Buchner 1965; Kaiser 1980; Kobiałka et al. 2016, 2018). The other bacterium (*Nasuia*) is expelled within whole bacteriocytes that then breakdown and release symbionts into the hemolymph. Lost bacteriocytes and bacteria are quickly replaced by cell division (Buchner 1965). In either case, vacuole-like structures are observed to initiate distinctive transformations in bacterial cell morphology, which are maintained only during the transmission process (Buchner 1965). Bacterial cell alterations may help to protect them from degradation while traversing the host hemolymph. After expulsion from their respective bacteriocytes, both bacterial symbionts eventually accumulate at follicular cells of the ovariole, and are then simultaneously inducted into oocytes and embryonic bacteriocytes (Szklarzewicz et al. 2015; Kobiałka et al. 2018). The two divergent transmission modes likely reflect the independent origins and distinct coevolutionary processes underlying the integration of *Sulcia* and *Nasuia*.

Developmental work in the pea aphid-*Buchnera* symbioses have shed some light on the genetic programs that ready embryos for symbiont infection. Cells destined to become bacteriocytes are prespecified and undergo a consistent schedule of homeobox transcription factor expression (e.g., *distal-less*, *ultrabithorax*, *engrailed*) that initiate bacteriocyte formation (Braendle et al. 2003). The expression of these genes occurs even if *Buchnera* has been removed, and also in other aphid species that have recently replaced *Buchnera* with fungal symbionts. It was further observed that hosts begin to more highly express relatively large sets of aphid-specific secreted proteins in bacteriocytes that likely target *Buchnera* following induction into the embryo (Shigenobu and Stern 2012). The expression of these genes is maintained throughout the bacteriocyte lifecycle and may play important roles in regulating *Buchnera*. Intriguingly, some of these pea aphid genes are convergently used in bacteriome formation in other Hemiptera species (e.g., developmental transcription factors). For example, the homeobox genes, *Ultrabithorax* and *abdominal-A*, are similarly essential for bacteriocyte development, positioning, and structure in the Lygaeoidea species, *Nysius plebius* (Figure 11.1; Matsuura et al. 2015). Thus, the evolutionary development of symbiotic tissues may follow shared and predictable processes.

11.4.2 Extracellular symbioses: External transmission and the midgut

In many species of the Heteroptera, beneficial symbionts are acquired through nymphal feeding and maintained extracellularly in the gut. To ensure their acquisition and integration of the right microbes, host lineages have evolved diverse transmission strategies that include feces smearing, bacterial encapsulation on egg surfaces, and embedding in nutritional jelly (Hosokawa et al. 2005; Kaltenpoth, Winter, and Kleinhammer 2009; Kaiwa et al. 2014; Otero-Bravo and Sabree 2015; Salem et al. 2015). However, in other species such as the bean bug, symbiotic bacteria strains are acquired by filtering from free-living bacterial populations in the open environment (e.g., soil and plants; Kikuchi, Hosokawa, and Fukatsu 2007; Kikuchi, Hosokawa, and Fukatsu 2011b; Kikuchi and Yumoto 2013). In both acquisition modes, bacteria establish in specialized portions of the midgut that appear to have little function in food digestion and nutritional breakdown (Ohbayashi et al. 2015). Bacteria are integrated into the midgut during particular developmental stages when gut tissues are able to sequester and regulate the symbioses (Kikuchi, Hosokawa, and Fukatsu 2011b; Bauer et al. 2014; Park et al. 2018). Colonized tissues then go through metabolic and physical transformations into loci of host–bacterial interactions akin to bacteriomes (Kaltenpoth, Winter, and Kleinhammer 2009; Kikuchi, Hosokawa, and Fukatsu 2011a; Futahashi et al. 2013; Kim et al. 2013a; Ohbayashi et al. 2015; Lee et al. 2017). These patterns suggest that the symbionts themselves play an intrinsic role in priming the host to developmentally alter tissues for the purpose of maintaining a stable symbiosis.

Since many heteropteran species that harbor their symbionts in the gut acquire them from the environment, they require selective mechanisms to exclude other opportunistic and pathogenic microbes. To this end, heteropteran species have evolved a highly specific filtration organ in their gut tract—a constricted passageway lined with microvilli and a mucosal membrane—that selectively permits bacterial symbionts to enter the midgut, while excluding others and even food (Ohbayashi et al. 2015). While it remains unknown exactly how this organ identifies and admits permissible bacteria, they must be motile (i.e., retaining flagella) in order to reach the midgut (Ohbayashi et al. 2015). Bacteria may also be required to encode specific digestive enzymes that can degrade host mucosal layers (Ohbayashi et al. 2015), and maintain adapted resistance to host-derived antimicrobials (Futahashi et al. 2013; Kim et al. 2013a). Thus, while the filtration organ modulates inoculation of the midgut with beneficial symbiont strains, it also appears to provide an immune system structure that protects heteropteran hosts from other exploitative environmental microbes.

In order for symbiotic bacteria to stably persist in the midgut, they require certain cellular capabilities. For example, in the bean bug, *Burkholderia* symbionts are unable to establish in the gut despite the ability to reach it if they have weakened cell wall components (disruption of the *uppP* gene), lack certain metabolisms (e.g., purine biosynthesis), or are unable to synthesize endocellular storage vesicles important in bacterial stress response (i.e., polyhydroxyalkanoate synthesis genes) (Kim et al. 2013b; Kim et al. 2014). Several recent studies have shed light on host-level mechanisms that select for these bacterial traits. Hosts have evolved a number of regulatory functions that appear to manage the symbiosis and the midgut environment, including regulating expression of lysosomal genes, proteases, and specific microbial recognition factors and antimicrobial peptides (Futahashi et al. 2013; Kim et al. 2015; Park et al. 2018). For example, hosts in the Pyrrochoridae, when infected with their beneficial bacteria, highly express certain pattern recognition factors that may help to distinguish beneficial symbionts from other gut bacteria (Bauer et al. 2014). These molecules may further direct host immune response to selectively regulate symbiont cell numbers and to protect against opportunistic infections. Taken together, the midgut environment may be seen as a highly regulated and challenging environment which selects for adaptively suited microbes.

11.5   Maintaining and regulating microbial symbionts

11.5.1   Evolution of mechanisms to maintain and regulate symbionts

Although seemingly beneficial, bacterial symbionts with tiny genomes impose challenges both to themselves and to their hosts (Bennett and Moran 2015). In order to persist, bacteria require extensive assistance derived from their own genome-encoded cellular capabilities, and also from those of their prokaryotic and eukaryotic partners (see Figure 11.4). Symbionts generally do retain core sets of genes and pathways which provide them with some autonomy (e.g., DNA replication, transcription, translation); however, these often become impaired and incomplete (McCutcheon and Moran 2011; Moran and Bennett 2014). While bacteria themselves have evolved some independent strategies to maintain these core functions (discussed below), their gene losses are so extensive that it is not feasible for them to persist without substantial genetic, cellular, and metabolic inputs from their insect hosts. The mechanisms and evolutionary processes that underlie the maintenance and regulation of beneficial symbionts is best understood for intracellular symbioses (e.g., Sternorrhyncha and Auchenorrhyncha; Figure 11.1) and are largely reviewed below. However, several studies provide an emerging picture of how these relationships may also evolve among the extracellular symbioses common in heteropterans.

The role of hemipteran hosts in regulating their symbionts, and the mechanisms they use, are coming into focus for a handful of host species. Broadly, host genotype has been linked to the maintenance of bacterial population size, as has been observed in both pea aphids and pyrrhocorid cotton stainers (Vogel and Moran 2011; Salem et al. 2013; Chong and Moran 2016). Experimental mismatch between coevolved host lines and their particular bacterial strain leads to apparent dysregulation of symbiont numbers and resultant declines in host fitness. A breakdown in host ability

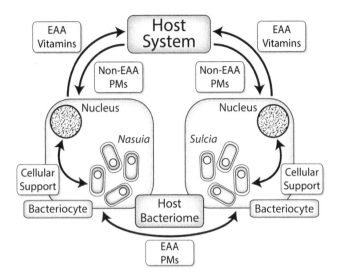

FIGURE 11.4 Model of microbiome integration between intracellular bacteria and insects. The predicted nutritional interactions of the *Macrosteles quadrilineatus* leafhopper, and its symbionts, *Sulcia* and *Nasuia*, are shown. Gray boxes illustrate different cellular and bodily compartments involved in nutritional exchange. Arrows and white boxes show integration of metabolite exchange and cellular activities (EAA, essential amino acid; PM, precursor metabolite; Non-EAA, nonessential amino acids).

to regulate their symbioses is likely due to genetic incompatibilities between partners, and an inability of hosts to successfully manage bacterial activities. A recent series of studies have shed light on the mechanisms underlying the successful integration of these systems. In intracellular symbioses, the bacteriomes differentially express hundreds to thousands of host-encoded genes that can complement genes and entire metabolisms missing from symbiont genomes (Figure 11.4; Hansen and Moran 2011; Husnik et al. 2013; Sloan et al. 2014; Luan et al. 2015; Mao, Yang, and Bennett 2018). Similarly, hosts of extracellular symbionts also differentially express a wide range of genes in the midgut that are involved in nutrition transport and processing, and host immune functions that may further regulate microbial populations (Futahashi et al. 2013; Bauer et al. 2014; Ioannidis et al. 2014; Salem et al. 2014).

The evolution of host support mechanisms is a complex process, often involving the evolution of novel functional traits in hemipteran hosts. Support systems are specifically delivered to symbionts through the bacteriocytes, or the dedicated gut tissues, that harbor them. The genetic underpinnings of these systems are generally derived from four evolutionary processes: (i) selective and enriched expression of native host genes capable of participating in essential symbiotic functions and metabolism, (ii) acquisition of novel genes through insect gene duplications, (iii) reassignment of mitochondrial support genes to the symbiont interface, and (iv) gain of novel host traits through the horizontal transfer of genes to its genome from other infecting bacteria (reviewed by Mao, Yang, and Bennett 2018). These mechanisms closely parallel evolutionary features used to define mitochondrial and plastid organelles, leading to debates as to whether or not hemipteran symbionts should be classified as such (McCutcheon and Keeling 2014; McCutcheon 2016). Nevertheless, while these mechanisms appear to be broadly and repeatedly used across the Hemiptera, the question of which genes are recruited when and through what mechanisms is specific to individual host-symbiont lineages (Husnik et al. 2013; Sloan et al. 2014; Luan et al. 2015). Host organs that sustain particular symbiont lineages have distinct origins and evolve specifically matched support systems to prop-up their residents, even when hosts obligately depend on multiple microbes (Ankrah, Chouaia, and Douglas 2018; Mao, Yang, and Bennett 2018).

11.5.2   Symbiont self-help and self-regulation

Despite massive gene losses, hemipteran symbionts generally maintain capabilities for certain core cellular functions, including DNA replication, transcription and translation, and chaperonins, among others (Moran and Bennett 2014). Even though these functional categories also lose some genes among different symbiont lineages (Bennett and Moran 2015), their core capabilities appear to be maintained and evolutionarily conserved (Sabater-Muñoz et al. 2017). The maintenance of these functions is something of a departure from the organelle systems that insect symbioses are often compared with (McCutcheon 2016). It suggests that hemipteran symbionts maintain some level of cellular autonomy possibly because (i) hosts cannot (yet) supplant these functions, (ii) it is more efficient for symbionts to maintain them, or (iii) symbionts selfishly retain them to avoid complete integration into host cells. Regardless, even though these bacterial cell systems may not interact directly with host cellular functions or metabolism, their failure may cause symbionts to lose fitness. The failure of symbionts to function properly may then lead to negative host-level fitness impacts, host mitigated bacterial elimination, or eventual extinction of the entire symbiotic lineage (reviewed by Bennett and Moran 2015).

The genes retained by symbiont genomes generally experience elevated rates of molecular evolution, AT-biased mutations, and high rates of amino acid replacements (Moran, McCutcheon, and Nakabachi 2008; McCutcheon and Moran 2011; Wernegreen 2011). Over time, these mutations can impair protein function and catabolic efficiency (Lambert and Moran 1998; Huang et al. 2008). Mutations further render genes prone to environmental variation and stress (e.g., high temperatures), reducing their functional capabilities that may then lead to decline in host fitness (Dunbar et al. 2007; Kikuchi et al. 2016). In order to deal with this, bacterial symbionts universally retain certain chaperonins and heat-shock genes (e.g., *groESL* and *dnaK*) that can properly fold proteins compromised by accumulated mutations (Fares, Moya, and Barrio 2004; Sabater-Muñoz et al. 2015; Aguilar-Rodríguez et al. 2016). Chaperonin genes in hemipteran symbionts are among those under strong selection, presumably to preserve their central role (Herbeck et al. 2003; Fares, Moya, and Barrio 2005; Sabater-Muñoz et al. 2017). Remarkably, *groESL* and *dnaK* are also constitutively highly expressed in symbiotic bacteria, often orders of magnitude more than all other genes, including those involved in essential nutrition (Baumann, Baumann, and Clark 1996; Wilcox et al. 2003; McCutcheon, McDonald, and Moran 2009a,b; Poliakov et al. 2011; Bennett and Chong 2017).

Symbionts with tiny genomes also generally lose self-regulatory mechanism, and it has generally been assumed that the host may take over these roles. In particular, symbiotic bacteria tend to lose all or most of their transcription factors, and show attenuated abilities to regulate gene transcription in changing environmental and nutrient conditions (Wilcox et al. 2003; Moran, Dunbar, and Wilcox 2005). However, recent work has demonstrated that symbionts may instead be capable of regulating gene expression posttranscriptionally (reviewed by Thairu and Hansen 2019). Multiomic studies have revealed that *Buchnera* strains across anciently diverged aphid hosts maintain sRNAs that can regulate the translation of mRNA into proteins, particularly those involved in EAA synthesis (Hansen and Degnan 2014). Several identified sRNAs are differentially expressed during aphid developmental stages, leading to quantitative differences in the expression of the proteins they are predicted to regulate (Thairu, Cheng, and Hansen 2018). Similar patterns were also observed when pea aphids fed on host plants of different nutritive qualities (Kim et al. 2018). Beyond aphids, functional evidence of such self-regulatory mechanisms in other hemipteran symbionts is currently lacking. However, intriguing evidence from studies focusing on other aspects of bacterial gene expression (e.g., sharpshooter leafhoppers; Bennett and Chong 2017) have revealed that sRNAs do occur and are expressed in other bacterial symbionts.

11.5.3   Symbiont-symbiont support

In hemipteran systems where hosts rely on multiple obligate symbionts, partners often engage in cross feeding of essential metabolites and cellular resources. In the broadest sense, co-obligate microbes generally feed each other synthesized EAAs and B vitamins via the host, since they almost always evolve to provision these nutrients in a non-overlapping manor (McCutcheon and Moran 2010). However, partner microbes may also provide other essential metabolites for the complete synthesis of resources required by the whole system (Figure 11.4; Douglas 2016). For example, in sharpshooter leafhoppers, *Sulcia* is predicted to provide its bacterial partner, *Baumannia*, with essential metabolites for methionine (homoserine) and pantothenate (3-methyl-2-oxobutanoate) (McCutcheon and Moran 2007; Ankrah, Chouaia, and Douglas 2018). *Sulcia* is also predicted to provide *Baumannia* with resources for cell envelope synthesis (e.g., fatty acid and peptidoglycan synthesis). The latter case is remarkable since only *Baumannia*, and not the host, appears to require these resources. Thus, *Sulcia* seems to be helping stabilize the symbiosis with *Baumannia* since it also relies on it for essential nutrition.

The acquisition of novel symbionts has been proposed as an escape route from failing partners. In some cases, hosts appear to acquire additional symbionts to aid more ancient ones. For example, members of the planthopper family, the Cixiidae (Figure 11.1; *Oliarus* spp.), have acquired a third bacterial partner, *Purcelliella*, apparently to support EAA synthesis by its two other older ones, *Sulcia* and *Vidania* (Bennett and Mao 2018). *Purcelliella* does not provide EAAs, but rather encodes precursor metabolites required by its microbial partners. In the Sternorrhyncha, several aphid lineages have also gained various co-obligate symbionts to assist *Buchnera* in nutrition synthesis. The banana aphid (*Pentalonia nigronervosa*) has acquired a *Wolbachia*-like symbiont predicted to provide *Buchnera* with essential metabolites for the synthesis of lysine and riboflavin (De Clerck et al. 2015). Similarly, the cedar aphid has acquired a co-obligate bacterium, *Serratia symbiotica*, which has taken over tryptophan synthesis with metabolite-based aid from the existing *Buchnera* symbiont (Lamelas et al. 2011). In these examples, a younger symbiont is either supplanting, or assisting, the responsibilities of a more ancient one, providing a window into how hosts enlist additional partners to maintain their complex symbioses.

Finally, one of the more remarkable examples of microbe–microbe assistance occurs in the mealybugs (Figure 11.1; Coccoidea) that widely rely on the bacterial symbiont, *Tremblaya*. To sustain a symbiosis, *Tremblaya* has repeatedly taken up a diversity of intracellular symbionts of its own across different host insect lineages (von Dohlen et al. 2001; Husnik and McCutcheon 2016). In the citrus mealybug (*Planococcus citri*), *Tremblaya* is paired with the bacterium *Moranella* that directly contributes to synthesis of six of the ten EAAs required by the symbiosis (McCutcheon and von Dohlen 2011; Husnik et al. 2013). To access those resources, the host is predicted to use uniquely evolved mechanisms to destabilize *Moranella's* cell wall, releasing metabolites directly into *Tremblaya's* cytoplasma (Husnik et al. 2013). *Tremblaya* also appears to rely on *Moranella* for other basic cellular processes, including translation. Specifically, *Tremblaya* does not encode tRNA synthetases and instead likely relies on *Moranella*, which still encodes them all.

11.5.4   Host support and regulation of nutritional synthesis in symbionts

Although hemipteran symbionts variably encode intact EAA pathways, they generally require non-EAAs and other metabolites from their hosts in order to initiate and complete them (Figure 11.4). For example, hemipteran hosts widely upregulate genes involved in metabolite synthesis, including coenzyme A for leucine and lysine, phosphoribosyl pyrophosphate for histidine, and various metabolites for methionine (Sloan et al. 2014; Luan et al. 2015; Mao, Yang, and Bennett 2018). Symbionts also require non-EAAs, glutamine, and glutamate for most of the EAA pathways. In order to furnish these metabolites, a majority of hosts investigated so far upregulate the GS/GOGAT cycle that recycles ammonia waste into these products (e.g., Hansen and Moran 2011; Sloan et al. 2014; Ankrah, Chouaia, and Douglas 2018; Mao, Yang, and Bennett 2018). Furthermore, in pea aphids, it has been shown that the supply of glutamine is an important mechanism for regulating linked bacteriocyte-bacterial EAA metabolisms (Macdonald et al. 2012; Price et al. 2014). The provisioning of key metabolites may generally offer hosts a means of differentially regulating symbiont metabolic activities (discussed below).

In many cases, symbionts go on to lose genes from their core EAA and B vitamin synthesis pathways, despite being the primary reason hosts retain symbionts. In these systems, host-encoded enzymes appear to participate directly in the successful synthesis of these nutrients. For example, in the citrus mealybug, host-encoded genes are involved in the synthesis of eight of the ten EAAs that were formerly the provenance of *Tremblaya* (Husnik et al. 2013). Similar patterns are observed in nearly all other examined host systems (e.g., psyllids and aphids; Hansen and Moran 2011;

Sloan et al. 2014), except for the *Macrosteles quadrilineatus* leafhopper wherein *Sulcia* and *Nasuia* retain relatively complete EAA pathways (Mao, Yang, and Bennett 2018). In some cases, particular insect orthologs have been convergently used to replace commonly lost EAA transaminases. For example, many hemipteran symbionts have lost the branched-chain amino acid aminotransferase gene (*ilvE*; BCAT), which is repeatedly replaced by an insect-encoded BCAT gene (e.g., mealybugs and whiteflies; see Luan et al. 2015; Husnik and McCutcheon 2016). Recent work in *M. quadrilineatus* provides some insight into how such EAA genes may be replaced. In this system, although the *Sulcia* still retains *ilvE*, the insect BCAT ortholog is also selectively highly expressed in the *Sulcia* specific bacteriocytes, suggesting that it may be performing redundant functions and is in the process of replacing the bacterial copy (Mao, Yang, and Bennett 2018). Similar metabolic redundancy has been noted to occur for several EAA pathways in the whitefly (Luan et al. 2015). Thus, as hosts evolve such integrated complementarity, there is likely to be a period of evolutionary time where both host and symbiont genes are performing overlapping roles. The host gene, being less prone to the negative effects of drift, likely overtakes the metabolic role from the symbiont-encoded copy, which is then eventually lost.

Perhaps one of the more remarkable aspects of host support of EAA synthesis, is that insects often incorporate horizontally transferred genes (HTGs) from other bacteria to complete disrupted symbiont-encoded pathways. The incorporation of HTGs into the host genome for nutrition synthesis varies between host lineages. For example, the whitefly, *Bemisia tabaci*, relies on an HTG for the synthesis of arginine and lysine by the bacterium, *Portiera* (Luan et al. 2015). While psyllids also rely on an orthologous HTG for arginine synthesis, their bacterial symbiont *Carsonella* has further lost genes involved in phenylalanine synthesis that are complemented by additional HTG acquisitions (Sloan et al. 2014). Similarly, the citrus mealybug requires the same final two catabolic steps in lysine biosynthesis as do whiteflies (Husnik et al. 2013). However, the origin of these genes is a convergent adaptation to shore-up parallel losses from their distinct symbionts' genomes. In effect, the incorporation of HTGs provides the host with important catabolic capabilities that animals do not normally have (e.g., EAA synthesis), permitting them to avoid symbiotic failure and extinction.

Finally, recent work has identified several mechanisms by which hosts regulate the exchange and release of nutrients from their symbionts. For intracellular systems, host-encoded transporters may provide a key mechanism for regulating the exchange of EAAs and non-EAAs (Price et al. 2014; Feng et al. 2019). Symbionts are generally reliant on their hosts for non-EAAs and, in the pea aphid, a multi-non-EAA transporter, ApNEAAT1, has been identified that permits the bidirectional exchange of these resources (Feng et al. 2019). A separate transporter, ApGLNT1, in the outer bacteriocyte membrane regulates the influx of the non-EAA glutamine, which is broadly required for both non-EAA synthesis by the bacteriocyte and EAA synthesis by the symbiont *Buchnera* (Price et al. 2014). The ApGLNT1 transporter is competitively inhibited by high EAA arginine concentrations in the hemolymph, reducing the flow of glutamine into the bacteriocytes. Remarkably, changes in host epigenetic patterns are further observed in some of these genes and may be a means of further regulating nutrient exchange with symbionts (Kim, Thairu, and Hansen 2016; Kim et al. 2018).

In contrast, the picture of how nutrients are exchanged and regulated in extracellular symbionts is less clear. While these symbionts may simply excrete nutrients into the gut matrix, which can then be absorbed by the host, some evidence suggests host-encoded lysosomal pathways, or other antimicrobial mechanisms, lyse bacteria to release nutrients (Bauer et al. 2014). Hosts also express a number of B vitamin transporters and processing genes, likely allowing them to import and use these nutrients (Salem et al. 2014).

11.5.5   Host support and regulation of other symbiont cell functions

Host support of obligate symbionts extends far beyond just facilitating nutrition synthesis. This support also includes core bacterial cell processes, for example, provisioning of cellular energy, synthesis of cellular membranes, regulation of symbiont titer, and processing of cellular information, among others. The losses of these cellular capabilities raise the question of how symbionts function at all, or rather, if they should instead be considered fully integrated and regulated cellular entities akin to organelles (Husnik and Keeling 2019). Recent work across a diverse set of hosts has revealed that, however hemipteran symbionts are classified, they simply cannot exist without basic cellular resources from their hosts. This conundrum is highlighted by the fact that few intracellular symbionts ultimately retain the capability to synthesize their own lipid membranes. Hosts contribute these resources and the genes capable of furnishing phospholipid membranes, as demonstrated in the whitefly-*Portiera* bacterial symbiosis (Luan et al. 2015), are more highly expressed in the bacteriocytes. Symbionts are also generally missing some, or all, of the machinery required to generate cellular energy. In the divergently related pea aphid and leafhoppers, ADP/ATP translocases are highly expressed in bacteriocytes likely regulating the flow of ATP to resident symbionts (Nakabachi et al. 2005).

A critical aspect of supporting obligate symbionts is that they retain limited capabilities to control, or complete, cellular reproduction and the replication of cellular information. Although, symbiont cell reproduction occurs during host development (reviewed by Skidmore and Hansen 2017), how hemipteran hosts manage this process remains poorly understood (Wilson and Duncan 2015). In contrast, a clearer picture has emerged as to how hosts regulate the other end of the symbiont life cycle to control their population sizes. Several of the hemipteran hosts examined so far, highly express lysosomal genes, antimicrobial peptides (e.g., defensins), and various proteases (e.g., cathepsins) that target symbionts for lysis and recycling of their cellular components (e.g., Nakabachi et al. 2005; Nishikori et al. 2009; Shigenobu and Stern 2012; Futahashi et al. 2013; Mao, Yang, and Bennett 2018; Simonet et al. 2018). Genes underlying these mechanisms in host lineages often go through extensive duplication events in contrast to other insects that do not obligately rely on symbionts to the same degree (e.g., *Drosophila*) (Rispe et al. 2008; Shigenobu and Stern 2012; Futahashi et al. 2013; Mao, Yang, and Bennett 2018). Work in pea aphids has revealed that the use of these lysozyme and immune functions is selectively modulated during times of environmental stress and host development (Nishikori et al. 2009; Simonet et al. 2018). Similar patterns of selective regulation of these mechanisms have also been observed in the pyrrhocorids, suggesting that extracellular systems also use discrete immune-like functions to regulate their bacterial cell numbers and their nutritional contributions (Futahashi et al. 2013; Bauer et al. 2014).

Beyond managing symbiont cell numbers, it was recently illustrated that hosts may broadly facilitate bacterial DNA replication, transcription, and translation (Mao, Yang, and Bennett 2018). In the *M. quadrilineatus* leafhopper, bacteriocytes highly express host genes (some derived from horizontal transfer and mitochondrial reassignment) that could fill discrete gaps in bacterial DNA replication and RNA transcription holoenzymes. Comparison with other gene expression studies revealed that this level of compensation is likely common among hemipteran hosts (Mao, Yang, and Bennett 2018). For example, tRNA synthetases and processing genes are often lost from bacterial symbionts (McCutcheon, McDonald, and Moran 2009b; McCutcheon and von Dohlen 2011). Broadly across hemipterans, bacteriome tissues show direct compensatory expression of host genes capable of performing these roles (Mao, Yang, and Bennett 2018). Further work in cicadas demonstrates that despite such losses, bacterial tRNAs are still correctly processed, suggesting that host genes are involved (Van Leuven et al. 2019). Such informational processing genes, and particularly in replication and transcription protein complexes, would be required to participate directly in core bacterial enzyme complexes *within* the bacterial cytosol. This prediction is in contrast to nutrition synthesis where metabolites can be exchanged across membranes and integrated into separate host and symbiont metabolic pathways. Growing evidence indicates that indeed the import of a large number of host proteins into symbionts is likely widespread in hemipterans (McCutcheon and Keeling 2014; Husnik and Keeling 2019). Despite this critically important mechanism, only one insect-encoded protein as of the time of this writing, a horizontally transferred *rplA* gene in the pea aphid, has even been verified as expressed inside a bacterial cell at the time of this writing (Nakabachi et al. 2014).

11.6 Conclusion

Beneficial symbiosis in the Hemiptera have provided us with some of the clearest models of the coevolutionary process, particularly how disparate organisms become genomically, metabolically, and ecologically intertwined (Moran 2007; Douglas 2011; Feldhaar 2011; Bennett and Moran 2015). Essentially all animals interact with or directly depend on microbial processes. To this end, hemipteran symbioses have provided important insights into how hosts integrate and manage their microbiomes for nutritional benefits and other essential services (Douglas 2011; McFall-Ngai et al. 2013). Multiomic research in these groups has further informed our understanding of how animal and microbial (i.e., bacteria and fungi) genomes coevolve and function together in the service of maintaining a tightly integrated biological system for balanced health (Husnik et al. 2013; Wilson and Duncan 2015; Mao, Yang, and Bennett 2018).

Genomic data from the hemipteran symbionts themselves have also shed light on basic theories of molecular evolution, particularly regarding how drift and population bottlenecks can wreak havoc on genomes (Moran 1996; Wernegreen 2015; Sabater-Muñoz et al. 2017). And, although these interactions have long been held as models of mutualisms with reciprocal beneficial services handed out to both partners, new data and new theory are revealing that these coevolutionary interactions are far more complex (Garcia and Gerardo 2014; Bennett and Moran 2015; McCutcheon, Boyd, and Dale 2019). Beyond their importance in host health, degenerate symbionts with tiny genomes also pose significant environmental and metabolic challenges to their hosts (Bennett and Moran 2015). Becoming beholden to a symbiotic partner can trap formerly independent organisms into a spiraling rabbit hole of dependence, manipulation, and control.

Finally, where hemipteran symbiont genomes have shrunk to near organelle levels, they rely on extensive host resources and services. They are further intrinsically integrated into host cellular processes and required for systemic balance (McCutcheon 2016; Husnik and Keeling 2019). In these cases, study of these ancient symbioses can provide important insights into the processes that may have facilitated eukaryogenesis, organelle evolution, and host adaptation to maintain host-organelle interactions indefinitely. These are processes that have occurred rarely in the evolution of complex life and for which we have had few comparative examples to better understand them (Husnik and Keeling 2019). The repeated evolution of hemipteran symbioses may provide fertile ground for testing fundamental hypotheses about the origins of eukaryotes and biological complexity in general.

References

Aguilar-Rodríguez, J., B. Sabater-Muñoz, R. Montagud-Martínez, V. Berlanga, D. Alvarez-Ponce, A. Wagner, and M.A. Fares. 2016. The molecular chaperone DnaK is a source of mutational robustness. *Genome Biology and Evolution* 8:2979–91.

Anbutsu, H., M. Moriyama, N. Nikoh, T. Hosokawa, R. Futahashi, M. Tanahashi, X.Y. Meng et al. 2017. Small genome symbiont underlies cuticle hardness in beetles. *Proceedings of the National Academy of Sciences USA* 114:E8382–91.

Ankrah, N.Y.D., B. Chouaia, and A.E. Douglas. 2018. The cost of metabolic interactions in symbioses between insects and bacteria with reduced genomes. *mBio* 9:e01433–18.

Ankrah, N.Y.D., J. Luan, and A.E. Douglas. 2017. Cooperative metabolism in a three-partner insect-bacterial symbiosis revealed by metabolic modeling. *Journal of Bacteriology* 199:e00872–16.

Bauer, E., H. Salem, M. Marz, H. Vogel, and M. Kaltenpoth. 2014. Transcriptomic immune response of the cotton stainer *Dysdercus fasciatus* to experimental elimination of vitamin-supplementing intestinal symbionts. *PLOS ONE* 9:e114865.

Baumann, P. 2005. Biology bacteriocyte-associated endosymbionts of plant sap-sucking insects. *Annual Review of Microbiology* 59:155–89.

Baumann, P., L. Baumann, and M.A. Clark. 1996. Levels of *Buchnera aphidicola* chaperonin groEL during growth of the aphid *Schizaphis graminum*. *Current Microbiology* 32:279–85.

Beard, C.B., C. Cordon-Rosales, and R.V. Durvasula. 2002. Bacterial symbionts of the triatominae and their potential use in control of Chagas disease transmission. *Annual Review of Entomology* 47:123–41.

Bennett, G.M., and R.A. Chong. 2017. Genome-wide transcriptional dynamics in the companion bacterial symbionts of the glassy-winged sharpshooter (Cicadellidae: *Homalodisca vitripennis*) reveal differential gene expression in bacteria occupying multiple host organs. *G3* 7:3073–82.

Bennett, G.M., and M. Mao. 2018. Comparative genomics of a quadripartite symbiosis in a planthopper host reveals the origins and rearranged nutritional responsibilities of anciently diverged bacterial lineages. *Environmental Microbiology* 20:4461–72.

Bennett, G.M., J.P. McCutcheon, B.R. McDonald, and N.A. Moran. 2016. Lineage-specific patterns of genome deterioration in obligate symbionts of sharpshooter leafhoppers. *Genome Biology and Evolution* 8:296–301.

Bennett, G.M., and N.A. Moran. 2013. Small, smaller, smallest: The origins and evolution of ancient dual symbioses in a phloem-feeding insect. *Genome Biology and Evolution* 5:1675–88.

Bennett, G.M., and N.A. Moran 2015. Heritable symbiosis: The advantages and perils of an evolutionary rabbit hole. *Proceedings of the National Academy of Sciences USA* 112:10169–76.

Braendle, C., T. Miura, R. Bickel, A.W. Shingleton, S. Kambhampati, and D.L. Stern. 2003. Developmental origin and evolution of bacteriocytes in the aphid–*Buchnera* symbiosis. *PLoS Biology* 1:e1.

Buchner, P. 1965. *Endosymbiosis of Animals with Plant Microorganisms*. John Wiley & Sons.

Campbell, M.A., P. Łukasik, M.C. Meyer, M. Buckner, C. Simon, C. Veloso, A. Michalik, and J.P. McCutcheon. 2018. Changes in endosymbiont complexity drive host-level compensatory adaptations in cicadas. *mBio* 9:e02104–18.

Campbell, M.A., J.T. Van Leuven, R.C. Meister, K.M. Carey, C. Simon, and J.P. McCutcheon. 2015. Genome expansion via lineage splitting and genome reduction in the cicada endosymbiont *Hodgkinia*. *Proceedings of the National Academy of Sciences USA* 112:10192–99.

Chong, R.A., and N.A. Moran. 2016. Intraspecific genetic variation in hosts affects regulation of obligate heritable symbionts. *Proceedings of the National Academy of Sciences USA* 113:13114–19.

Chong, R.A., H. Park, and N.A. Moran. 2019. Genome evolution of the obligate endosymbiont *Buchnera aphidicola*. *Molecular Biology and Evolution* 36:1481–98.

Dan, H., N. Ikeda, M. Fujikami, and A. Nakabachi. 2017. Behavior of bacteriome symbionts during transovarial transmission and development of the asian citrus psyllid. *PLOS ONE* 12:e0189779.

De Clerck, C., A. Fujiwara, P. Joncour, S. Léonard, M.L. Félix, F. Francis, M.H. Jijakli, T. Tsuchida, and S. Massart. 2015. A metagenomic approach from aphid's hemolymph sheds light on the potential roles of co-existing endosymbionts. *Microbiome* 3:63.

Douglas, A.E. 2006. Phloem-sap feeding by animals: Problems and solutions. *Journal of Experimental Botany* 57:747–54.

Douglas, A.E. 2009. The microbial dimension in insect nutritional ecology. *Functional Ecology* 23:38–47.

Douglas, A.E. 2014. The molecular basis of bacterial–insect symbiosis. *Journal of Molecular Biology* 426:3830–7.

Douglas, A.E. 2011. Lessons from studying insect symbioses. *Cell Host and Microbe* 10:359–67.

Douglas, A.E. 2016. How multi-partner endosymbioses function. *Nature Reviews Microbiology* 14:731–43.

Dunbar, H.E., A.C.C. Wilson, N.R. Ferguson, and N.A. Moran. 2007. Aphid thermal tolerance is governed by a point mutation in bacterial symbionts. *PLoS Biology* 5:e96.

Engel, P., and N.A. Moran. 2013. The gut microbiota of insects – diversity in structure and function. *FEMS Microbiology Reviews* 37:699–735.

Fan, H.W., H. Noda, H.Q. Xie, Y. Suetsugu, Q.H. Zhu, and C.Z. Zhang. 2015. Genomic analysis of an ascomycete fungus from the rice planthopper reveals how it adapts to an endosymbiotic lifestyle. *Genome Biology and Evolution* 7:2623–34.

Fares, M.A., A. Moya, and E. Barrio. 2004. GroEL and the maintenance of bacterial endosymbiosis. *Trends in Genetics* 20:413–16.

Fares, M.A., A. Moya, and E. Barrio. 2005. Adaptive evolution in GroEL from distantly related endosymbiotic bacteria of insects. *Journal of Evolutionary Biology* 18:651–60.

Feldhaar, H. 2011. Bacterial symbionts as mediators of ecologically important traits of insect hosts. *Ecological Entomology* 36:533–43.

Feng, H., N. Edwards, C.M.H Anderson, M. Althaus, R.P. Duncan, Y.C. Hsu, C.W. Luetje, D.R.G. Price, A.C.C. Wilson, and D.T. Thwaites. 2019. Trading amino acids at the aphid-*Buchnera* symbiotic interface. *Proceedings of the National Academy of Sciences USA* 116:16003–11.

Futahashi, R., K. Tanaka, M. Tanahashi, N. Nikoh, Y. Kikuchi, B.L. Lee, and T. Fukatsu. 2013. Gene expression in gut symbiotic organ of stinkbug affected by extracellular bacterial symbiont. *PLOS ONE* 8:e64557.

Garcia, J.R., and N.M. Gerardo. 2014. The symbiont side of symbiosis: Do microbes really benefit? *Frontiers in Microbiology* 5:510.

Gibson, C.M., and M.S. Hunter. 2010. Extraordinarily widespread and fantastically complex: Comparative biology of endosymbiotic bacterial and fungal mutualists of insects. *Ecology Letters* 13:223–34.

Hacquard, S., R. Garrido-Oter, A. González, S. Spaepen, G. Ackermann, S. Lebeis, A.C. McHardy et al. 2015. Microbiota and host nutrition across plant and animal kingdoms. *Cell Host and Microbe* 17:603–16.

Hansen, A.K., and P.H. Degnan. 2014. Widespread expression of conserved small RNAs in small symbiont genomes. *The ISME Journal* 8 (12):2490–502.

Hansen, A.K., and N.A. Moran. 2011. Aphid genome expression reveals host-symbiont cooperation in the production of amino acids. *Proceedings of the National Academy of Sciences USA* 108:2849–54.

Hansen, A.K., and N.A. Moran. 2014. The impact of microbial symbionts on host plant utilization by herbivorous insects. *Molecular Ecology* 23:1473–96.

Herbeck, J.T., D.J. Funk, P.H. Degnan, and J.J. Wernegreen. 2003. A conservative test of genetic drift in the endosymbiotic bacterium *Buchnera*: Slightly deleterious mutations in the chaperonin groEL. *Genetics* 165:1651–60.

Hirota, B., G. Okude, H. Anbutsu, R. Futahashi, M. Moriyama, X.Y. Meng, N. Nikoh, R. Koga, and T. Fukatsu. 2017. A novel, extremely elongated, and endocellular bacterial symbiont supports cuticle formation of a grain pest beetle. *mBio* 8:e01482–17.

Hosokawa, T., Y. Kikuchi, X.Y. Meng, and T. Fukatsu. 2005. The making of symbiont capsule in the plataspid stinkbug *Megacopta Punctatissima*. *FEMS Microbiology Ecology* 54:471–77.

Huang, C.Y., C.Y. Lee, H.C. Wu, M.H. Kuo, and C.Y. Lai. 2008. Interactions of chaperonin with a weakly active anthranilate synthase from the aphid endosymbiont *Buchnera aphidicola*. *Microbial Ecology* 56:696–703.

Husnik, F., and P.J. Keeling. 2019. The fate of obligate endosymbionts: Reduction, integration, or extinction. *Current Opinion in Genetics & Development* 58–59:1–8.

Husnik, F., and J.P. McCutcheon. 2016. Repeated replacement of an intrabacterial symbiont in the tripartite nested mealybug symbiosis. *Proceedings of the National Academy of Sciences USA* 113:E5416–24.

Husnik, F., N. Nikoh, R. Koga, L. Ross, R.P. Duncan, M. Fujie, M. Tanaka et al. 2013. Horizontal gene transfer from diverse bacteria to an insect genome enables a tripartite nested mealybug symbiosis. *Cell* 153:1567–78.

Ioannidis, P., Y. Lu, N. Kumar, T. Creasy, S. Daugherty, M.C. Chibucos, J. Orvis et al. 2014. Rapid transcriptome sequencing of an invasive pest, the brown marmorated stink bug *Halyomorpha halys*. *BMC Genomics* 15:738–22.

Johnson, K.P., C.H. Dietrich, F. Friedrich, R.G. Beutel, B. Wipfler, R.S. Peters, J.M. Allen et al. 2018. Phylogenomics and the evolution of hemipteroid insects. *Proceedings of the National Academy of Sciences USA* 115:12775–80.

Kaiser, B. 1980. Licht-und elektronenmikroskopische untersuchung der symbionten von *Graphocephala Coccinea* Forstier (Homoptera: Jassidae). *International Journal of Insect Morphology and Embryology* 9:79–88.

Kaiwa, N., T. Hosokawa, N. Nikoh, M. Tanahashi, M. Moriyama, X.Y. Meng, T. Maeda et al. 2014. Symbiont-supplemented maternal investment underpinning host's ecological adaptation. *Current Biology* 24:2465–70.

Kaltenpoth, M., S.A. Winter, and A. Kleinhammer. 2009. Localization and transmission route of *Coriobacterium glomerans*, the endosymbiont of pyrrhocorid bugs. *FEMS Microbiology Ecology* 69:373–83.

Kellner, R.L.L., and K. Dettner. 1996. Differential efficacy of toxic pederin in deterring potential arthropod predators of *Paederus* (Coleoptera: Staphylinidae) offspring. *Oecologia* 107:293–300.

Kenyon L.J., T. Meulia, and Z.L. Sabree. 2015. Habitat visualization and genomic analysis of 'Candidatus Pantoea carbekii,' the primary symbiont of the brown marmorated stink bug. *Genome Biology and Evolution* 7:620–35.

Kikuchi, Y., T. Hosokawa, and T. Fukatsu. 2007. Insect-microbe mutualism without vertical transmission: A stinkbug acquires a beneficial gut symbiont from the environment every generation. *Applied and Environmental Microbiology* 73:4308–16.

Kikuchi, Y., T. Hosokawa, and T. Fukatsu. 2011a. An ancient but promiscuous host-symbiont association between *Burkholderia* gut symbionts and their heteropteran hosts. *The ISME Journal* 5:446–60.

Kikuchi, Y., T.T. Hosokawa, and T. Fukatsu. 2011b. Specific developmental window for establishment of an insect-microbe gut symbiosis. *Applied and Environmental Microbiology* 77:4075–81.

Kikuchi, Y., A. Tada, D.L. Musolin, N. Hari, T. Hosokawa, K. Fujisaki, and T. Fukatsu. 2016. Collapse of insect gut symbiosis under simulated climate change. *mBio* 7:e01578–16.

Kikuchi, Y., and I. Yumoto. 2013. Efficient colonization of the bean bug *Riptortus pedestris* by an environmentally transmitted *Burkholderia* symbiont. *Applied and Environmental Microbiology* 79:2088–91.

Kim, D., B.F. Minhas, H.L. Byarlay, and A.K. Hansen. 2018. Key transport and ammonia recycling genes involved in aphid symbiosis respond to host-plant specialization. *G3* 8(7):2433–43.

Kim, D., M.W. Thairu, and A.K. Hansen. 2016. Novel insights into insect-microbe interactions-role of epigenomics and small RNAs. *Frontiers in Plant Science* 7:1164.

Kim, J.K., N.H. Kim, H.A. Jang, Y. Kikuchi, C.H. Kim, T. Fukatsu, and B.L. Lee. 2013a. Specific midgut region controlling the symbiont population in an insect-microbe gut symbiotic association. *Applied and Environmental Microbiology* 79:7229–33.

Kim, J.K., Y.J. Won, N. Nikoh, H. Nakayama, S.H. Han, Y. Kitkuchi, Y.H. Rhee, H.Y. Park, J.Y. Kwon, K. Kurokawa, N. Dohmae, T. Fukatsu, and B.L. Lee. 2013b. Polyester synthesis genes associated with stress resistance are involved in an insect–bacterium symbiosis. *Proceedings of the National Academy of Sciences* 110:E2381–89.

Kim, J.K., H.A. Jang, Y.J. Won, Y. Kikuchi, S.H. Han, C.H. Kim, N. Nikoh, T. Fukatsu, and B.L. Lee. 2014. Purine biosynthesis-deficient Burkholderia mutants are incapable of symbiotic accommodation in the stinkbug. *The ISME journal* 8:552–63.

Kim, J.K., J.B. Lee, Y.R. Huh, H.A. Jang, C.H. Kim, J.W. Yoo, and B.L. Lee. 2015. *Burkholderia* gut symbionts enhance the innate immunity of host *Riptortus Pedestris*. *Developmental and Comparative Immunology* 53:265–69.

Kobiałka, M., A. Michalik, J. Szwedo, and T. Szklarzewicz. 2018. Diversity of symbiotic microbiota in Deltocephalinae leafhoppers (Insecta, Hemiptera, Cicadellidae). *Arthropod Structure and Development* 47:268–78.

Kobiałka, M., A. Michalik, M. Walczak, Ł. Junkiert, and T. Szklarzewicz. 2016. *Sulcia* symbiont of the leafhopper *Macrosteles Laevis* (Ribaut, 1927) (Insecta, Hemiptera, Cicadellidae: Deltocephalinae) harbors *Arsenophonus* Bacteria. *Protoplasma* 253:903–12.

Koga, R., G.M. Bennett, J.R. Cryan, and N.A. Moran. 2013. Evolutionary replacement of obligate symbionts in an ancient and diverse insect lineage. *Environmental Microbiology* 15:2073–81.

Koga, R., X.Y. Meng, T. Tsuchida, and T. Fukatsu. 2012. Cellular mechanism for selective vertical transmission of an obligate insect symbiont at the bacteriocyte-embryo interface. *Proceedings of the National Academy of Sciences USA* 109:E1230–37.

Koga, R., and N.A. Moran. 2014. Swapping symbionts in spittlebugs: Evolutionary replacement of a reduced genome symbiont. *The ISME Journal* 8:1237–46.

Lambert, J.D., and N.A. Moran. 1998. Deleterious mutations destabilize ribosomal RNA in endosymbiotic bacteria. *Proceedings of the National Academy of Sciences USA* 95:4458–62.

Lamelas, A., M.J. Gosalbes, A. Manzano-Marín, J. Peretó, A. Moya, and A. Latorre. 2011. *Serratia symbiotica* from the aphid *Cinara cedri*: A missing link from facultative to obligate insect endosymbiont. *PLoS Genetics* 7:e1002357.

Lee, J.B., K.E. Park, S.A. Lee, S.H. Jang, H.J. Eo, H.A. Jang, C.H. Kim et al. 2017. Gut symbiotic bacteria stimulate insect growth and egg production by modulating hexamerin and vitellogenin gene expression. *Developmental and Comparative Immunology* 69:12–22.

Luan, J.B., W. Chen, D.K. Hasegawa, A.M. Simmons, W.M. Wintermantel, K.S. Ling, Z. Fei, S.S. Liu, and A.E. Douglas. 2015. Metabolic coevolution in the bacterial symbiosis of whiteflies and related plant sap-feeding insects. *Genome Biology and Evolution* 7:2635–47.

Luan, J., X. Sun, Z. Fei, and A.E. Douglas. 2018. Maternal inheritance of a single somatic animal cell displayed by the bacteriocyte in the whitefly *Bemisia tabaci*. *Current Biology* 28:459–65.

Łukasik, P., K. Nazario, J.T. Van Leuven, M.A. Campbell, M. Meyer, A. Michalik, P. Pessacq, C. Simon, C. Veloso, and J.P. McCutcheon. 2018. Multiple origins of interdependent endosymbiotic complexes in a genus of cicadas. *Proceedings of the National Academy of Sciences USA* 115:E226–235.

Macdonald, S.J., G.G. Lin, C.W. Russell, G.H. Thomas, and A.E. Douglas. 2012. The central role of the host cell in symbiotic nitrogen metabolism. *Proceedings of the Royal Society B: Biological Sciences* 279:2965–73.

Mao, M., X. Yang, and G.M. Bennett. 2018. Evolution of host support for two ancient bacterial symbionts with differentially degraded genomes in a leafhopper host. *Proceedings of the National Academy of Sciences USA* 115 (50):E11691–700.

Mao, M., X. Yang, K. Poff, and G.M. Bennett. 2017. Comparative genomics of the dual-obligate symbionts from the treehopper, *Entylia carinata* (Hemiptera: Membracidae), provide insight into the origins and evolution of an ancient symbiosis. *Genome Biology and Evolution* 9:1803–15.

Matsuura, Y., Y. Kikuchi, T. Miura, and T. Fukatsu. 2015. Ultrabithorax is essential for bacteriocyte development. *Proceedings of the National Academy of Sciences USA* 112:9376–81.

McCutcheon, J.P. 2016. From microbiology to cell biology: When an intracellular bacterium becomes part of its host cell. *Current Opinion in Cell Biology* 41:132–36.

McCutcheon, J.P., B.M. Boyd, and C. Dale. 2019. The life of an insect endosymbiont from the cradle to the grave. *Current Biology* 29:R485–95.

McCutcheon, J.P., and P.J. Keeling. 2014. Endosymbiosis: Protein targeting further erodes the organelle/symbiont distinction. *Current Biology* 24:R654–55.

McCutcheon, J.P., B.R. McDonald, and N.A. Moran. 2009a. Origin of an alternative genetic code in the extremely small and GC-rich genome of a bacterial symbiont. *PLoS Genetics* 5:e1000565.

McCutcheon, J.P., B.R. McDonald, and N.A. Moran. 2009b. Convergent evolution of metabolic roles in bacterial co-symbionts of insects. *Proceedings of the National Academy of Sciences USA* 106:15394–99.

McCutcheon, J.P., and N.A. Moran. 2007. Parallel genomic evolution and metabolic interdependence in an ancient symbiosis. *Proceedings of the National Academy of Sciences USA* 104:19392–97.

McCutcheon, J.P., and N.A. Moran. 2010. Functional convergence in reduced genomes of bacterial symbionts spanning 200 million years of evolution. *Genome Biology and Evolution* 2:708–18.

McCutcheon, J.P., and N.A. Moran. 2011. Extreme genome reduction in symbiotic bacteria. *Nature Reviews Microbiology* 10:13–26.

McCutcheon, J.P., and C.D. von Dohlen. 2011. An interdependent metabolic patchwork in the nested symbiosis of mealybugs. *Current Biology* 21:1366–72.

McFall-Ngai, M.J. 2015. Giving microbes their due – Animal life in a microbially dominant world. *The Journal of Experimental Biology* 218:1968–73.

McFall-Ngai, M., M.G. Hadfield, T.C.G. Bosch, H.V. Carey, T. Domazet-Lošo, A.E. Douglas, N. Dubilier et al. 2013. Animals in a bacterial world, a new imperative for the life sciences. *Proceedings of the National Academy of Sciences USA* 110:3229–36.

Mira, A., and N.A. Moran. 2002. Estimating population size and transmission bottlenecks in maternally transmitted endosymbiotic bacteria. *Microbial Ecology* 44:137–43.

Moran, N.A. 1996. Accelerated evolution and Muller's rachet in endosymbiotic bacteria. *Proceedings of the National Academy of Sciences USA* 93:2873–78.

Moran, N.A. 2007. Symbiosis as an adaptive process and source of phenotypic complexity. *Proceedings of the National Academy of Sciences USA* 104:8627–33.

Moran, N.A., and P. Baumann. 2000. Bacterial endosymbionts in animals. *Current Opinion in Microbiology* 3:270–75.

Moran, N.A., and G.M. Bennett. 2014. The tiniest tiny genomes. *Annual Review of Microbiology* 68:195–215.

Moran, N.A., H.E. Dunbar, and J.L. Wilcox. 2005. Regulation of transcription in a reduced bacterial genome: Nutrient-provisioning genes of the obligate symbiont *Buchnera aphidicola*. *Journal of Bacteriology* 187:4229–37.

Moran, N.A., J.P. McCutcheon, and A. Nakabachi. 2008. Genomics and evolution of heritable bacterial symbionts. *Annual Review of Genetics* 42:165–90.

Moran, N.A., H.J. McLaughlin, and R. Sorek. 2009. The dynamics and time scale of ongoing genomic erosion in symbiotic bacteria. *Science* 323:379–82.

Moran, N.A., P. Tran, and N.M. Gerardo. 2005. Symbiosis and insect diversification: An ancient symbiont of sap-feeding insects from the bacterial phylum Bacteroidetes. *Applied and Environmental Microbiology* 71:8802–10.

Nakabachi, A., K. Ishida, Y. Hongoh, M. Ohkuma, and S. Miyagishima. 2014. Aphid gene of bacterial origin encodes a protein transported to an obligate endosymbiont. *Current Biology* 24:R640–41.

Nakabachi, A., S. Shigenobu, N. Sakazume, T. Shiraki, Y. Hayashizaki, P. Carninci, H. Ishikawa, T. Kudo, and T. Fukatsu. 2005. Transcriptome analysis of the aphid bacteriocyte, the symbiotic host cell that harbors an endocellular mutualistic bacterium, Buchnera. *Proceedings of the National Academy of Sciences USA* 102: 5477–82.

Nakabachi, A., R. Ueoka, K. Oshima, R. Teta, A. Mangoni, M. Gurgui, N.J. Oldham et al. 2013. Defensive bacteriome symbiont with a drastically reduced genome. *Current Biology* 23:1478–84.

Nakabachi, A., A. Yamashita, H. Toh, H. Ishikawa, H.E. Dunbar, N.A. Moran, and M. Hattori. 2006. The 160-Kilobase genome of the bacterial endosymbiont *Carsonella*. *Science* 314:267.

Nikoh, N, T. Hosokawa, M. Moriyama, K. Oshima, M. Hattori, and T. Fukatsu. 2014. Evolutionary origin of insect-*Wolbachia* nutritional mutualism. *Proceedings of the National Academy of Sciences USA* 111:10257–62.

Nikoh, N., T. Hosokawa, K. Oshima, M. Hattori, and T. Fukatsu. 2011. Reductive evolution of bacterial genome in insect gut environment. *Genome Biology and Evolution* 3:702–14.

Nishikori, K., K. Morioka, T. Kubo, and M. Morioka. 2009. Age- and morph-dependent activation of the lysosomal system and *Buchnera* degradation in aphid endosymbiosis. *Journal of Insect Physiology* 55:351–57.

Noda, H., and Y. Koizumi. 2003. Sterol biosynthesis by symbiotes: Cytochrome P450 sterol C-22 desaturase genes from yeast-like symbiotes of rice planthoppers and anobiid beetles. *Insect Biochemistry and Molecular Biology* 33:649–58.

Normark, B.B. 2003. The evolution of alternative genetic systems in insects. *Annual Review of Entomology* 48:397–423.

Ohbayashi, T., K. Takeshita, W. Kitagawa, N. Nikoh, R. Koga, X.Y. Meng, K. Tago et al. 2015. Insect's intestinal organ for symbiont sorting. *Proceedings of the National Academy of Sciences USA* 112:E5179–88.

Otero-Bravo, A., and Z.L. Sabree. 2015. Inside or out? possible genomic consequences of extracellular transmission of crypt-dwelling stinkbug mutualists. *Frontiers in Ecology and Evolution* 3:64.

Otero-Bravo, A., S. Goffredi, and Z.L. Sabree. 2018. Cladogenesis and genomic streamlining in extracellular endosymbionts of tropical stink bugs. *Genome Biology and Evolution* 10:680–93.

Park, K.E., S.H. Jang, J. Lee, S.A. Lee, Y. Kikuchi, Y.S. Seo, and B. Luel Lee. 2018. The roles of antimicrobial peptide, rip-thanatin, in the midgut of *Riptortus Pedestris*. *Developmental and Comparative Immunology* 78:83–90.

Pérez-Brocal, V., R. Gil, S. Ramos, A. Lamelas, M. Postigo, J.M. Michelena, F.J. Silva, A. Moya, and A. Latorre. 2006. A small microbial genome: The end of a long symbiotic relationship? *Science* 314:312–13.

Poliakov, A., C.W. Russell, L. Ponnala, H.J. Hoops, Q. Sun, A.E. Douglas, and K.J. van Wijk. 2011. Large-scale label-free quantitative proteomics of the pea aphid-*Buchnera* symbiosis. *Molecular and Cellular Proteomics* 10:M110.007039.

Price, D.R.G., H. Feng, J.D. Baker, S. Bavan, C.W. Luetje, and A.C.C. Wilson. 2014. Aphid amino acid transporter regulates glutamine supply to intracellular bacterial symbionts. *Proceedings of the National Academy of Sciences USA* 111:320–25.

Rabatel, A., G. Febvay, K. Gaget, G. Duport, P. Baa-Puyoulet, P. Sapountzis, N. Bendridi et al. 2013. Tyrosine pathway regulation is host-mediated in the pea aphid symbiosis during late embryonic and early larval development. *BMC Genomics* 14:235.

Rispe, C., M. Kutsukake, V. Doublet, S. Hudaverdian, F. Legeai, J.C. Simon, D. Tagu, and T. Fukatsu. 2008. Large gene family expansion and variable selective pressures for Cathepsin B in aphids. *Molecular Biology and Evolution* 25:5–17.

Sabater-Muñoz, B., M. Prats-Escriche, R. Montagud-Martínez, A. López-Cerdán, C. Toft, J. Aguilar-Rodríguez, A. Wagner, and M.A. Fares. 2015. Fitness trade-offs determine the role of the molecular chaperonin GroEL in buffering mutations. *Molecular Biology and Evolution* 32:2681–93.

Sabater-Muñoz, B., C. Toft, D. Alvarez-Ponce, and M.A. Fares. 2017. Chance and necessity in the genome evolution of endosymbiotic bacteria of insects. *The ISME Journal* 11:1291–304.

Salem, H., E. Bauer, A.S. Strauss, H. Vogel, M. Marz, and M. Kaltenpoth. 2014. Vitamin supplementation by gut symbionts ensures metabolic homeostasis in an insect host. *Proceedings of the Royal Society B: Biological Sciences* 281:20141838.

Salem, H., L. Florez, N. Gerardo, and M. Kaltenpoth. 2015. An out-of-body experience: The extracellular dimension for the transmission of mutualistic bacteria in insects. *Proceedings of the Royal Society B: Biological Sciences* 282:20142957.

Salem, H., E. Kreutzer, S. Sudakaran, and M. Kaltenpoth. 2013. Actinobacteria as essential symbionts in firebugs and cotton stainers (Hemiptera, Pyrrhocoridae). *Environmental Microbiology* 15:1956–68.

Santos-Garcia, D., A. Latorre, A. Moya, G. Gibbs, V. Hartung, K. Dettner, S.M. Kuechler, and F.J. Silva. 2014. Small but powerful, the primary endosymbiont of moss bugs, *Candidatus* Evansia Muelleri, holds a reduced genome with large biosynthetic capabilities. *Genome Biology and Evolution* 6:1875–93.

Santos-Garcia, D., F.J. Silva, S. Morin, K. Dettner, and S.M. Kuechler. 2017. The all-rounder *Sodalis*: A new bacteriome-associated endosymbiont of the lygaeoid bug *Henestaris Halophilus* (Heteroptera: Henestarinae) and a critical examination of its evolution. *Genome Biology and Evolution* 9:2893–910.

Shigenobu, S., and D.L. Stern. 2012. Aphids evolved novel secreted proteins for symbiosis with bacterial endosymbiont. *Proceedings of the Royal Society B: Biological Sciences* 280:20121952-52.

Shigenobu, S., H. Watanabe, M. Hattori, Y. Sakaki, and H. Ishikawa. 2000. Genome sequence of the endocellular bacterial symbiont of aphids *Buchnera* Sp. APS. *Nature* 407:81–6.

Simonet, P., G. Duport, K. Gaget, M. Weiss-Gayet, S. Colella, G. Febvay, H. Charles, J. Viñuelas, A. Heddi, and F. Calevro. 2016a. Direct flow cytometry measurements reveal a fine-tuning of symbiotic cell dynamics according to the host developmental needs in aphid symbiosis. *Scientific Reports* 6:19967.

Simonet, P., K. Gaget, S. Balmand, M.R. Lopes, N. Parisot, K. Buhler, G. Duport et al. 2018. Bacteriocyte cell death in the Pea Aphid/Buchnera symbiotic system. *Proceedings of the National Academy of Sciences USA* 115: E1819–28.

Simonet, P., K. Gaget, N. Parisot, G. Duport, M. Rey, G. Febvay, H. Charles, P. Callaerts, S. Colella, and F. Calevro. 2016b. Disruption of phenylalanine hydroxylase reduces adult lifespan and fecundity, and impairs embryonic development in parthenogenetic pea aphids. *Scientific Reports* 6:34321.

Skidmore, I.H., and A.K. Hansen. 2017. The evolutionary development of plant-feeding insects and their nutritional endosymbionts. *Insect Science* 24:910–28.

Sloan, D.B., and N.A. Moran. 2012a. Endosymbiotic bacteria as a source of carotenoids in whiteflies. *Biology Letters* 8:986–89.

Sloan, D.B., and N.A. Moran. 2012b. Genome reduction and co-evolution between the primary and secondary bacterial symbionts of psyllids. *Molecular Biology and Evolution* 29:3781–92.

Sloan, D.B., A. Nakabachi, S. Richards, J. Qu, S.C. Murali, R.A. Gibbs, and N.A. Moran. 2014. Parallel histories of horizontal gene transfer facilitated extreme reduction of endosymbiont genomes in sap-feeding insects. *Molecular Biology and Evolution* 31:857–71.

Sudakaran, S., C. Kost, and M. Kaltenpoth. 2017. Symbiont acquisition and replacement as a source of ecological innovation. *Trends in Microbiology* 25:375–90.

Sudakaran, S., F. Retz, Y. Kikuchi, C. Kost, and M. Kaltenpoth. 2015. Evolutionary transition in symbiotic syndromes enabled diversification of phytophagous insects on an imbalanced diet. *The ISME Journal* 9:2581–604.

Szklarzewicz, T., B. Grzywacz, J. Szwedo, and A. Michalik. 2015. Bacterial symbionts of the leafhopper *Evacanthus interruptus* (Linnaeus, 1758) (Insecta, Hemiptera, Cicadellidae: Evacanthinae). *Protoplasma* 253:379–91.

Thairu, M.W., S. Cheng, and A.K. Hansen. 2018. A sRNA in a reduced mutualistic symbiont genome regulates its own gene expression. *Molecular Ecology* 27:1766–76.

Thairu, M.W., and A.K. Hansen. 2019. It's a small, small world: Unravelling the role and evolution of small RNAs in organelle and endosymbiont genomes. *FEMS Microbiology Letters* 366:fnz049.

True, J.R. 2003. Insect melanism: The molecules matter. *Trends in Ecology and Evolution* 18:640–47.

van Ham, R.C.H., J. Kamerbeek, C. Palacios, C. Rausell, F. Abascal, U. Bastolla, J.M. Fernández et al. 2003. Reductive genome evolution in *Buchnera Aphidicola*. *Proceedings of the National Academy of Sciences USA* 100:581–86.

Van Leuven, J.T., M. Mao, D.D. Xing, G.M. Bennett, and J.P. McCutcheon. 2019. Cicada endosymbionts have tRNAs that are correctly processed despite having genomes that do not encode all of the tRNA processing machinery. *mBio* 10:e01950–18.

Van Leuven, J.T., R.C. Meister, C. Simon, and J.P. McCutcheon. 2014. Sympatric speciation in a bacterial endosymbiont results in two genomes with the functionality of one. *Cell* 158:1270–80.

Vogel, K.J., and N.A. Moran. 2011. Effect of host genotype on symbiont titer in the aphid— *Buchnera* symbiosis. *Insects* 2:423–34.

Vogel, K.J., and N.A. Moran. 2013. Functional and evolutionary analysis of the genome of an obligate fungal symbiont. *Genome Biology and Evolution* 5:891–904.

Von Dohlen, C.D., S. Kohler, S.T. Alsop, and W.R. McManus. 2001. Mealybug beta-proteobacterial endosymbionts contain gamma-proteobacterial symbionts. *Nature* 412:433–36.

Weglarz, K.M., N.G. Havill, G.R. Burke, and C.D. von Dohlen. 2018. Partnering with a pest: Genomes of hemlock woolly adelgid symbionts reveal atypical nutritional provisioning patterns in dual-obligate bacteria. *Genome Biology and Evolution* 10:1607–21.

Wernegreen, J.J. 2002. Genome evolution in bacterial endosymbionts of insects. *Nature Reviews Genetics* 3:850–61.

Wernegreen, J.J. 2011. Reduced selective constraint in endosymbionts: Elevation in radical amino acid replacements occurs genome-wide. *PLOS ONE* 6:e28905.

Wernegreen, J.J. 2015. Endosymbiont evolution: Predictions from theory and surprises from genomes. *Annals of the New York Academy of Sciences* 1360:16–35.

Wilcox, J.L., H.E. Dunbar, R.D. Wolfinger, and N.A. Moran. 2003. Consequences of reductive evolution for gene expression in an obligate endosymbiont. *Molecular Microbiology* 48:1491–500.

Wilson, A.C.C., P.D. Ashton, F. Calevro, H. Charles, S. Colella, G. Febvay, G. Jander et al. 2010. Genomic insight into the amino acid relations of the pea aphid, *Acyrthosiphon pisum*, with its symbiotic bacterium *Buchnera Aphidicola*. *Insect Molecular Biology* 19:249–58.

Wilson, A.C.C., and R.P. Duncan. 2015. Signatures of host/symbiont genome coevolution in insect nutritional endosymbioses. *Proceedings of the National Academy of Sciences USA* 112:10255–61.

12 Symbiosis for insect cuticle formation

Hisashi Anbutsu and Takema Fukatsu

Contents

12.1 Introduction

Insects account for the majority of all the species described so far, constituting the most dominant animal group in the terrestrial ecosystem. By contrast, the number of described bacterial species is incomparably smaller than the number of described insect species (Grimaldi and Engel 2005). However, microbiologists know that the described bacterial species are actually the tip of an iceberg, plausibly representing far less than 1% of all bacterial species on Earth, and thus there is no doubt that an enormous number of unknown bacterial species remain to be discovered (Gasc et al. 2015; Solden et al. 2016). Accordingly, insect–bacterium symbiotic associations are ubiquitously observed, and interactions between them are extremely diverse (Buchner 1965; Bourtzis and Miller 2003; Zchori-Fein and Miller 2011).

Beetles, representing the insect order Coleoptera, account for the majority of the biodiversity described so far (Grimaldi and Engel 2005; Hunt et al. 2007; Stork et al. 2015). The beetles are characterized by their sclerotized exoskeleton. Our common knowledge is that "beetles are hard insects." Needless to say, the hard cuticle of beetles is formed by an array of molecular and cellular machineries that are encoded on the beetle's genome (Noh et al. 2016). Recently, however, it has turned out that symbiotic bacteria also significantly contribute to cuticle pigmentation and hardening of weevils (Kuriwada et al. 2010; Vigneron et al. 2014; Anbutsu et al. 2017). Here, we review the mechanisms underlying the symbiont-mediated cuticle hardening in weevils and suggest the possibility that the symbiont-mediated cuticle hardening may be found in a greater variety of insects than previously envisaged.

12.2   Weevil–*Nardonella* endosymbiosis

Among beetles, weevils represent the most species-rich group, embracing over 70,000 described species in the world, and include such highly sclerotized lineages as *Pachyrhynchus* spp., *Eupholus* spp., *Rhynchophorus* spp., *Trigonopterus* spp., and others (Faleiro 2006; Hunt et al. 2007; Oberprieler et al. 2007; McKenna et al. 2009; Seago et al. 2009; Riedel et al. 2013). The weevils' hard cuticles are important for their survival and adaptation as they confer mechanical strength, antipredator effects, desiccation tolerance, and other beneficial qualities (Crowson 1981; Weissling and Giblin-Davis 1993; Tseng et al. 2014; van de Kamp et al. 2016).

Many weevils are associated with a specific γ-proteobacterial lineage, *Nardonella*, within the cytoplasm of the bacteriome. The size and shape of the bacteriomes and the bacterial cells may differ considerably among weevil species (Figure 12.1) (Anbutsu

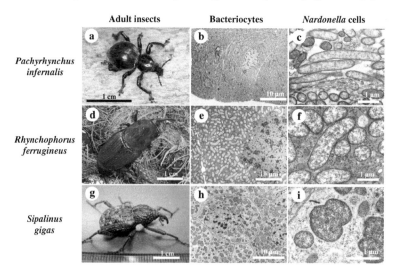

FIGURE 12.1   Transmission electron microscopic images of *Nardonella* in larval bacteriomes. (a–c) Black hard weevil *P. infernalis*. (d–f) Red palm weevil *R. ferrugineus*. (g–i) Giant weevil *S. gigas*. (a, d, g) Images of adult insects. (b, e, h) Bacteriocytes. (c, f, i) *Nardonella* cells. (Modified from Anbutsu, H. et al. 2017. *Proc. Natl. Acad. Sci. USA* 114: E8382–E8391.)

et al. 2017). The evolutionary origin of this endosymbiotic relationship is traced to the common ancestor of the weevils, which is older than 100 million years (Lefèvre et al. 2004; Conord et al. 2008; Hosokawa and Fukatsu 2010; Kuriwada et al. 2010; Rinke et al. 2011; Hirsch et al. 2012; Hosokawa et al. 2015; White et al. 2015; Huang et al. 2016). Despite the long-lasting symbiotic relationship, little has been known about the biological function of *Nardonella* for the host weevils (Kuriwada et al. 2010).

12.3   *Nardonella* genome is extremely reduced and specialized for tyrosine synthesis

We sequenced the entire genome sequences of *Nardonella* endosymbionts, which are associated with four weevil species,—the red palm weevil *Rhynchophorus ferrugineus*, the giant weevil *Sipalinus gigas*, the West Indian sweet potato weevil *Euscepes postfasciatus*, and the black hard weevil *Pachyrhynchus infernalis*. The genomes of all four *Nardonella* endosymbionts were highly reduced, ranging from 0.20 Mb to 0.23 Mb in size, and the number of protein-coding open reading frames was estimated as 196–226 (Figure 12.2) (Anbutsu et al. 2017). Reflecting this

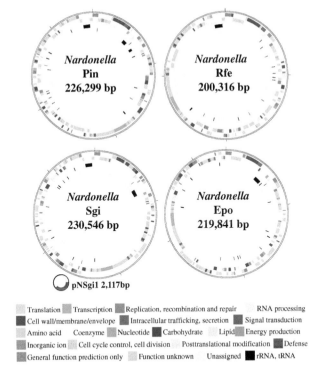

FIGURE 12.2   *Nardonella* genomes identified from four weevil species: *Nardonella* Pin of the black hard weevil *P. infernalis*, *Nardonella* Rfe of the red palm weevil *R. ferrugineus*, *Nardonella* Sgi of the giant weevil *S. gigas*, and *Nardonella* Epo of the West Indian sweet potato weevil *E. postfasciatus*. (Modified from Anbutsu, H. et al. 2017. *Proc. Natl. Acad. Sci. USA* 114: E8382–E8391.)

extreme genome reduction, almost all metabolic genes conserved in the majority of γ-proteobacteria have been lost (Anbutsu et al. 2017). Exceptionally, only the tyrosine synthesis pathway and the peptidoglycan synthesis pathway were present. Considering that peptidoglycan synthesis is necessary for the construction of bacterial cell walls, strikingly, *Nardonella*'s genome seems to be substantially specialized for the production of only one amino acid, tyrosine. Insects require tyrosine for sclerotization and melanization of the cuticle layer in the exoskeleton (True 2003; Noh et al. 2016). Therefore, we suspected that tyrosine synthesized by *Nardonella* might contribute to the formation and coloration of the hard exoskeleton of the host weevils.

12.4   *Nardonella* endosymbiotic system in *Pachyrhynchus infernalis*

The black hard weevil *P. infernalis* (Figure 12.3a and b) inhabits the Yaeyama Islands, Okinawa, Japan. As the name suggests, it has a highly sclerotized black exoskeleton. The weevil is so hard that it is not easy to penetrate the beetle body to make a pinned specimen. In a Japanese popular science fiction comic *Terra Formers*, *P. infernalis* is caricatured as an armored alien monster ("Pachyrhynchus Infernalis Terraformar," *Terra Formers* Wiki, accessed December 10, 2019, https:// terraformars.fandom.com/wiki/Pachyrhynchus_Infernalis_Terraformar), which makes the tiny beetle famous not only for insect enthusiasts but also for comic lovers. In the larvae, symbiotic bacteria are localized in a bacteriome that surrounds the foregut-midgut junction like a rosary (Figure 12.3c). In adult females, the symbiotic bacteria are concentrated in a tip region of the ovarioles and developing oocytes (Figure 12.3d). In bacteriomes, symbionts are packed tightly in the cytoplasm of numerous bacteriocytes (Figure 12.3e). Similar patterns of *Nardonella* localization have been observed in *R. ferrugineus*, *S. gigas*, *E. postfasciatus*, and other weevil species (Hosokawa and Fukatsu 2010; Hosokawa et al. 2015; Anbutsu et al. 2017).

12.5   *Nardonella*-harboring bacteriome as a tyrosine-producing organ

First, we conducted a tracer experiment to confirm the ability of *Nardonella* to synthesize tyrosine by using an *in vitro* assay system (Figure 12.3f–h). The *Nardonella*-harboring bacteriomes were dissected from mature weevil larvae and cultured for 2 h in a medium containing ^{15}N-labeled glutamine. As a result, essential amino acids were scarcely labeled with ^{15}N, reflecting the fact that both the host and the symbiont lack the synthetic pathways for essential amino acids. Some nonessential amino acids were labeled with ^{15}N, likely by the host's synthetic pathways for these amino acids. Notably, among semiessential amino acids, only tyrosine exhibited a high labeling rate (about 50%) (Figure 12.3i). These results support the idea that the bacteriome of *P. infernalis* preferentially synthesize a large amount of tyrosine (Anbutsu et al. 2017).

In a variety of insect–microbial symbiotic systems, it has been reported that symbiotic bacteria are often so heat-sensitive that high-temperature treatment of host insects causes suppression or removal of the symbiont infections (Wernegreen 2012; Kikuchi et al. 2016). As for *P. infernalis*, rearing of larvae at 30°C resulted in significantly reduced symbiont titers in the bacteriome compared to control larvae

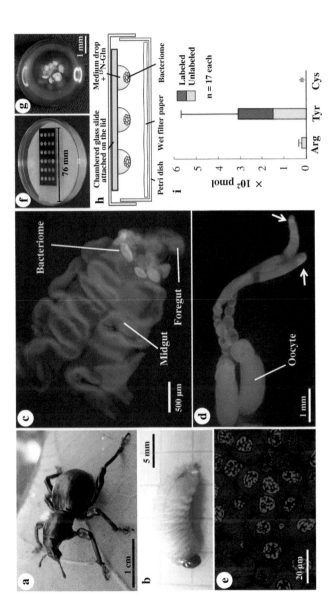

FIGURE 12.3 *P. infernalis–Nardonella* endosymbiotic system and tyrosine synthesis in larval bacteriome. (a) Adult insect. (b) Final instar larva. (c–e) FISH imaging of *Nardonella* endosymbiont. *Nardonella*'s 16S rRNA and host's nuclear DNA are visualized in red and blue, respectively. (c) Dissected larval alimentary tract with the bacteriome surrounding the foregut–midgut junction. (d) Ovarioles dissected from an adult female. Arrows indicate the *Nardonella* localization at the ovariole tips. (e) *Nardonella* in the cytoplasm of larval bacteriocytes. (f–i) *In vitro* assay of *Nardonella*'s tyrosine synthesis in *P. infernalis*. (f–h) Experimental system for hanging drop culture. (f) External view of the culture system. (g) Isolated larval bacteriome of *P. infernalis* in hanging drop medium. (h) Schematic view of the culture system. (i) Quantification of semiessential amino acids released from an isolated larval bacteriome of *P. infernalis* in the medium containing ^{15}N-glutamine. Columns and bars show means and standard deviations, in which ^{15}N-labeled and unlabeled amino acid fractions are represented by dark green and light green, respectively. Asterisks indicate undetected amino acid, cysteine. (Modified from Anbutsu, H. et al. 2017. *Proc. Natl. Acad. Sci. USA* 114: E8382–E8391.)

reared at 25°C. When the bacteriomes dissected from these larvae were subjected to the tracer experiment, the larvae reared at 30°C exhibited drastically reduced tyrosine synthesis in comparison with the larvae reared at 25°C (Anbutsu et al. 2017). These results confirm that *Nardonella* is certainly involved in tyrosine synthesis in the bacteriome, and that the bacteriome of *P. infernalis* functions as a tyrosine-producing organ.

12.6   Suppression of *Nardonella* by antibiotic and its effects on tyrosine and DOPA provisioning

We developed an agar-based artificial diet rearing system for *P. infernalis*. On the control diet without antibiotic, larvae developed normally, whose bacteriomes were full of *Nardonella* (e.g., Figures 12.1b and 12.3e). On the antibiotic-supplemented diet with 0.003% rifampicin, larvae also developed, but their bacteriomes were *Nardonella*-depleted (Figure 12.4a). *Nardonella* titers were around ten times lower in the antibiotic-treated larvae than in the control larvae.

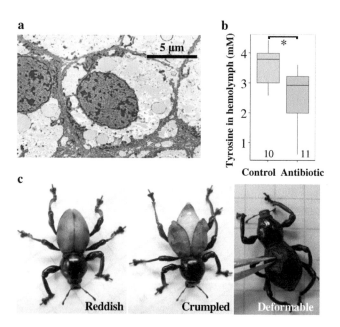

FIGURE 12.4 Effects of antibiotic on *Nardonella* infection, tyrosine production, and adult color and cuticle formation in *P. infernalis*. (a) Transmission electron microscopic images of the bacteriocytes in the larva reared on the artificial diet supplemented with 0.003% rifampicin. (b) Tyrosine levels in the hemolymph of mature larvae. Asterisks indicate statistically significant differences (*t*-test; *, $P < 0.05$). Tukey box plots indicate the median (bold line), 25th and 75th percentiles (box edges), range (whiskers), with sample sizes represented at the bottom. (c) Antibiotic-treated adult insects with reddish elytra, crumpled fragile elytra, and soft and deformable elytra. (Modified from Anbutsu, H. et al. 2017. *Proc. Natl. Acad. Sci. USA* 114: E8382–E8391.)

In the process of cuticle pigmentation and sclerotization in beetles and other insects, tyrosine and its derivative, L-3,4-dihydroxyphenylalanine (L-DOPA), are primary substrates for initiating a series of chemical reactions for cuticle tanning, polymerization, and melanization (True 2003; Noh et al. 2016). In *P. infernalis*, hemolymphal levels of both compounds were the highest at the pupal stage, suggesting that these metabolites are stored in pupae for adult cuticle formation (Anbutsu et al. 2017). At the mature larval stage, symbiotic larvae synthesize and accumulate tyrosine more actively than symbiont-depleted larvae (Figure 12.4b). At the pupal stage, L-DOPA levels suggested more active recruitment of L-DOPA in symbiotic pupae than in symbiosis-deficient pupae, which is presumably consumed for adult cuticle construction (Anbutsu et al. 2017).

12.7 Contribution of *Nardonella* to adult cuticle formation in *Pachyrhynchus infernalis*

Most of the symbiotic adult weevils that emerged from the control diet exhibited normal morphology, with hard and black elytra (e.g., Figures 12.1a and 12.3a). By contrast, the *Nardonella*-suppressed adult insects that emerged from the antibiotic-supplemented diet frequently showed remarkable morphological abnormalities such as reddish, crumpled, and deformable elytra (Figure 12.4c). Elytra from the antibiotic-treated adult insects were significantly more reddish than those from the control adult insects. Physical property analysis of the dissected elytra using a viscoelastometer revealed that the elastic modulus, a qualitative index of physical hardness, did not differ significantly between the control insects and the antibiotic-treated insects. Notably, however, the elytra of the symbiosis-deficient insects were remarkably thinner than those of the symbiotic insects (Anbutsu et al. 2017).

All these results described above indicate that, through the long-lasting and intimate host–symbiont association, the ancient endosymbiont *Nardonella* has evolved an extremely streamlined genome, which is specialized for a specific biological function, namely tyrosine provisioning. The *Nardonella*-provisioned tyrosine underpins the highly sclerotized exoskeleton of weevils, confers physical strength, desiccation resistance, and other adaptive advantages and potentially contributes to their diversity and prosperity in the terrestrial ecosystem.

12.8 Incomplete tyrosine synthesis pathway of *Nardonella* and complementation by host genes

Notably, *Nardonella*'s tyrosine synthesis pathway was incomplete at the final step, lacking the *tyrB* gene encoding tyrosine aminotransferase (Anbutsu et al. 2017). Hence, we suspected that some host gene(s) expressed in the bacteriome might catalyze the final step reaction and convert 4-hydroxyphenylpyruvate (4-HPP) to tyrosine. In the genome of *Drosophila melanogaster* (Adams et al. 2000), three aminotransferases, namely glutamate oxaloacetate transaminase 1 (*GOT1*), glutamate oxaloacetate transaminase 2 (*GOT2*), and tyrosine aminotransferase (*TAT*) can potentially convert 4-HPP to tyrosine. For a comprehensive survey of these aminotransferase genes, we conducted RNA-sequencing analysis of RNA samples prepared from the bacteriome and the midgut of mature larvae of *P. infernalis* reared on the control diet and the antibiotic-supplemented diet. We identified three *GOT1* (designated as *GOT1A*, *GOT1B*, and *GOT1C*), two *GOT2* (*GOT2A* and *GOT2B*), and one *TAT* gene sequences. Molecular phylogenetic analyses showed that these genes are orthologous to the corresponding genes identified from *D. melanogaster* and other insects (Anbutsu et al. 2017). Among them, *GOT1A* and *GOT2A* were very highly expressed in the bacteriomes of the *Nardonella*-infected control insects. Quantitative RT-PCR analysis of the dissected tissues verified that these two genes were very highly expressed in the larval bacteriomes (Anbutsu et al. 2017).

Based on these results, we attempted RNA interference (RNAi) experiments to knockdown the expression levels of *GOT2A*, *GOT1A*, and *TAT*. By injection of dsRNA into mature larvae, expression levels of *GOT2A* and *GOT1A* were significantly suppressed, whereas *TAT* did not respond to RNAi, probably because its original expression level was very low (Figure 12.5a). When the mature larvae were injected with dsRNAs and subsequently subjected to the dissection of bacteriomes for *in vitro* culturing experiments, tyrosine production by the bacteriomes was suppressed in association with the drastically reduced *GOT2A* and *GOT1A* expression levels (Figure 12.5a) (Anbutsu et al. 2017). In another experiment, even when *GOT2A* was solely suppressed by RNAi, the dsRNA-injected insects frequently exhibited morphological abnormalities such as reddish, crumpled, and/or deformed elytra (Figure 12.5b) (Anbutsu et al. 2017).

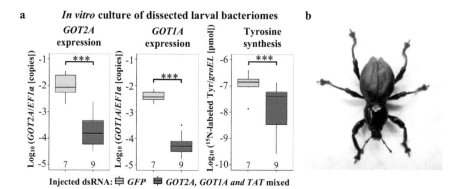

FIGURE 12.5 *GOT1* and *GOT2* genes of *P. infernalis* are involved in tyrosine synthesis and adult cuticle formation. (a) Tyrosine synthesis activity of larval bacteriomes suppressed by RNA interference targeting *GOT2A*, *GOT1A*, and *TAT*. Double-strand RNAs were injected into mature larvae, their bacteriomes were dissected seven days after the injection and cultured in a medium containing ^{15}N-labeled glutamine for 2 h, and the bacteriomes were subjected to quantitative RT-PCR while the culture media were analyzed by LC-MS for quantification of ^{15}N-labeled tyrosine. Asterisks indicate statistically significant differences (likelihood-ratio test of GLM assuming a Gamma error distribution; ***, $P < 0.001$). Tukey box plots are as shown in Figure 12.4, with outliers, which are larger or smaller than 1.5 times the interquartile range from the box edge (dots). (b) Image of an adult insect on the day of emergence, which was subjected to larval injection with *GOT2A* dsRNA. (Modified from Anbutsu, H. et al. 2017. *Proc. Natl. Acad. Sci.* USA 114: E8382–E8391.)

These results indicate that, in the bacteriome of *P. infernalis*, *Nardonella* with the near-complete shikimate pathway synthesizes 4-HPP, and the missing final step reaction from 4-HPP to tyrosine is catalyzed by the host-derived aminotransferases. In this way, the incomplete synthesis pathway for the symbiont metabolite is functionally complemented by the host genes.

12.9   Insights from weevil-*Nardonella* symbiosis: Host's final step control over symbiont's metabolic pathway

Taken together, these results strongly suggest that *Nardonella*'s sole and most crucial role, tyrosine synthesis, is regulated at the final step by the host's aminotransferase genes. These genes are upregulated in the bacteriome and involved in the formation and pigmentation of the adult cuticle. Plausibly, the control from the host side may be evolutionarily relevant to, on the one hand, the extremely reduced *Nardonella* genome incapable of transcriptional and translational regulations, and on the other hand, the drastic change of tyrosine demand upon metamorphosis from nonsclerotized larvae to highly sclerotized adults.

Here, we suggest that the final step regulation of symbiont metabolism by host genes may be found not only in the weevil–*Nardonella* endosymbiosis but also in other insect–microbe endosymbiotic systems, on the ground that such a metabolic configuration readily enables the host's control over the symbiont's metabolic activities. For example, in the aphid–*Buchnera* endosymbiosis, such patterns are observed in the synthesis pathways of essential amino acids isoleucine, leucine, valine and phenylalanine, and also tyrosine. In these pathways, the final step enzyme genes are lacking in the symbiont genome and complemented by the following host genes: branched-chain aminotransferase for isoleucine, leucine and valine; aspartate aminotransferase for phenylalanine; and phenylalanine 4-monooxygenase for tyrosine (Shigenobu et al. 2000; International Aphid Genomics Consortium 2010; Wilson et al. 2010; Hansen and Moran 2011; Shigenobu and Wilson 2011; Rabatel et al. 2013; Russell et al. 2013; Simonet et al. 2016). In the ant *Cardiocondyla obscurior*, its endosymbiont *Westeberhardia* was reported to exhibit some metabolic properties reminiscent of *Nardonella*. The 0.53 Mb reduced genome of *Westeberhardia* has lost many metabolic capabilities but retains a near-complete shikimate pathway for synthesizing 4-HPP, which is presumably converted to tyrosine by host's tyrosine aminotransferase (Klein et al. 2016). Whether this host–symbiont metabolic complementarity actually works—and contributes to cuticle formation in the ant species—deserves future experimental verification.

12.10   Insights from weevil-*Nardonella* symbiosis: How do symbiont replacements proceed?

Despite the long-lasting and intimate symbiotic relationship between weevils and *Nardonella*, previous studies have identified many weevil lineages in which *Nardonella* had been either lost or replaced by novel bacterial symbionts (Lefèvre et al. 2004; Heddi and Nardon 2005; Conord et al. 2008; Toju et al. 2010 2013; Toju and Fukatsu 2011).

The highly specific function of *Nardonella* may have facilitated symbiont losses and replacements. For example, if some weevil lineages had evolved to live on food sources containing sufficient tyrosine, evolutionary losses of *Nardonella* would readily occur in these lineages. Considering that many bacteria are able to synthesize tyrosine and other amino acids, any non-*Nardonella* bacterial associates, such as secondary facultative symbionts and gut microbial associates, may be potentially capable of compensating for *Nardonella*'s biological function. In *Sitophilus* grain weevils, a γ-proteobacterial lineage *Sodalis pierantonius* has taken over the original *Nardonella* endosymbiont (Lefèvre et al. 2004; Heddi and Nardon 2005; Conord et al. 2008; Oakeson et al. 2014). The weevil-associated *Sodalis* genome was determined as 4.5 Mb in size, retaining many metabolic pathways intact (Oakeson et al. 2014). Several classic and recent studies have documented various biological roles of the *Sodalis* symbiont for the grain weevils. At phenotypic levels, it has been reported to enhance growth, survival, and fecundity (Nardon 1973; Grenier et al. 1986; Vigneron et al. 2014), to improve flight activity (Nardon 1973; Grenier et al. 1994), and to facilitate cuticular tanning and hardening (Nardon 1973; Wicker and Nardon 1982; Vigneron et al. 2014). At biochemical and metabolic levels, it has been shown to provision B vitamins (Wicker 1983), to supply aromatic amino acids phenylalanine and tyrosine (Wicker and Nardon 1982; Vigneron et al. 2014), and to metabolize methionine and sarcosine (Gasnier-Fauchet et al. 1986; Gasnier-Fauchet and Nardon 1986). Also, it was suggested to be involved in mitochondrial energy metabolism (Heddi et al. 1993; Heddi and Nardon 1993). Our results strongly suggest that tyrosine provisioning is the primary essential role of the weevil–bacterium endosymbiosis, while the other biological functions were probably acquired secondarily in association with the symbiont replacement from *Nardonella* to *Sodalis* in the lineage of grain weevils.

12.11   Symbiosis for insect cuticle formation: General phenomena across diverse insect taxa

Finally, we suggest that symbiosis-assisted cuticle formation, pigmentation, and sclerotization may not be restricted to weevils (Kuriwada et al. 2010; Vigneron et al. 2014; Anbutsu et al. 2017) but also found in other insect groups. Previous studies reported that, when obligate bacterial symbionts are experimentally removed, host insects with developed exoskeleton—including beetles, stinkbugs, and others,—become pale and soft: for example, the saw-toothed grain beetle *Oryzaephilus surinamensis* (Coleoptera: Sylvanidae) (Hirota et al. 2017; Engl et al. 2018), the kudzu bugs *Megacopta* spp. (Hemiptera: Plataspidae) (Hosokawa et al. 2006), the burrowing bugs *Adomerus* spp. (Hemiptera: Cydnidae) (Hosokawa et al. 2013), the stinkbugs *Nezara viridula* and *Plautia stali* (Hemiptera: Pentatomidae) (Hosokawa et al. 2016; Kikuchi et al. 2016), and so on (Figure 12.6). Genomic analysis suggested the possibility of symbiont-mediated tyrosine provisioning for cuticle formation in the ant *C. obscurior* (Hymenoptera: Formicidae) (Klein et al. 2016). Considering that the formation of thick and hard cuticle requires much tyrosine, it seems plausible that, if not all, some beetles, stinkbugs, and other insects with more or less sclerotized cuticle may generally be dependent on symbiont-provisioned tyrosine supply. Future

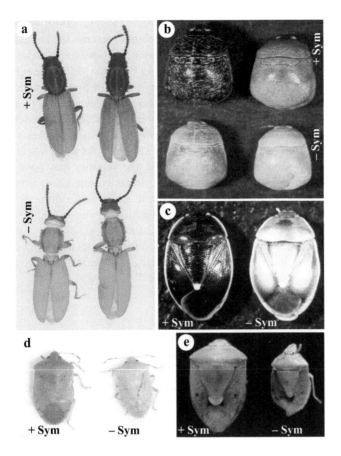

FIGURE 12.6   Pale and soft cuticle observed with aposymbiosis in diverse insects. (a) Sylvanid grain beetle *O. surinamensis.* Top, symbiotic adult insects; Bottom, aposymbiotic adult insects. Young adult insects of the same age (1–2-day-old) are shown (see Hirota et al. 2017). (b) Plataspid stinkbugs *M. punctatissima* (left) and *M. cribraria* (right). Top, symbiotic adult insects; Bottom, aposymbiotic adult insects (see Hosokawa et al. 2006). (c) Cydnid stinkbug *Adomerus triguttulus*. Left, symbiotic adult insect; Right, aposymbiotic adult insect (see Hosokawa et al. 2013). (d) Southern green stinkbug *Nezara viridula*. Left, symbiotic adult insect; Right, aposymbiotic adult insect (see Kikuchi et al. 2016). (e) Brown-winged green stinkbug *Plautia stali*. Left, symbiotic adult insect; Right, aposymbiotic adult insect (see Hosokawa et al. 2016). (Images by courtesy of Takahiro Hosokawa (b, c, e) and Yoshitomo Kikuchi (d).)

studies should focus on the genomic, metabolic, and physiological aspects of these insect–microbe symbiotic systems.

12.12   Conclusion and perspective

Many symbiotic bacteria are indispensable for the growth, survival, and reproduction of their host by supplying vital nutrients, such as essential amino acids and vitamins. The symbiotic partners are often integrated into an almost inseparable biological entity (Moran et al. 2008; Douglas 2009). In such close and long-lasting symbiotic relationships, the symbiont genomes tend to exhibit peculiar characteristics such as structural degeneration, massive gene losses, and drastic size reduction (Wernegreen 2002; Moran et al. 2008; McCutcheon and Moran 2011). The weevil-*Nardonella* endosymbiotic system provides another example of an extremely reduced symbiont genome in a different insect group through a different evolutionary trajectory, which represents an unprecedented form of nutritional symbiosis. The shikimate pathway for tyrosine synthesis is present in most microorganisms and plants, but not in animals, including insects. The stiffer exoskeleton provided by the symbiotic tyrosine supplier may have contributed, at least to some extent, to the adaptation, diversity, and prosperity of weevils. Our finding presents an impressive example as to how symbiotic bacteria, or their gene(s), can influence the evolution of a host organism, and how far such an intimate and ancient host–symbiont association can go over evolutionary time. On the other hand, our finding sheds light on the reasons as to why such a highly sophisticated symbiotic system entails ironic evolutionary consequences such as instability, losses, and replacements.

Acknowledgments

We thank the following collaborators: Minoru Moriyama, Naruo Nikoh, Takahiro Hosokawa, Ryo Futahashi, Masahiko Tanahashi, Xian-Ying Meng, Takashi Kuriwada, Naoki Mori, Kenshiro Oshima, Masahira Hattori, Manabu Fujie, Noriyuki Satoh, Taro Maeda, Shuji Shigenobu, Ryuichi Koga, Takuya Aikawa, Katsunori Nakamura, Takuma Takanashi, Wataru Toki, Yosuke Usui, Norikuni Kumano, Munetoshi Maruyama, Akira Oyafuso, Katsushi Yamaguchi, Tomoko F. Shibata, Hiroshi Shimizu, Hideyuki Tsukada, and Yoshitomo Kikuchi. This study was supported by the JSPS KAKENHI Grant Numbers JP25221107 to T.F., N.N., T.H., and R.K.; JP22128001 and JP22128007 to T.F., N.N., and S.S.; the Program for Promotion of Basic and Applied Researches for Innovations in Bio-Oriented Industry to T.F., and NIBB Collaborative Research Projects for Integrative Genomics to T.F. and S.S.

References

Adams, M. D., S. E. Celniker, R. A. Holt et al. 2000. The genome sequence of *Drosophila melanogaster*. *Science* 287: 2185–2195.
Anbutsu, H., M. Moriyama, N. Nikoh et al. 2017. Small genome symbiont underlies cuticle hardness in beetles. *Proc. Natl. Acad. Sci. USA.* 114: E8382–E8391.
Bourtzis, K., and T. A. Miller. 2003. *Insect Symbiosis*. Boca Raton: CRC Press/Taylor & Francis.
Buchner, P. 1965. *Endosymbiosis of Animals with Plant Microorganisms*. New York: Interscience.
Conord, C., L. Despres, A. Vallier et al. 2008. Long-term evolutionary stability of bacterial endosymbiosis in Curculionoidea: Additional evidence of symbiont replacement in the Dryophthoridae family. *Mol. Biol. Evol.* 25: 859–868.
Crowson, R. A. 1981. *The Biology of the Coleoptera*. London: Academic Press.
Douglas, A. E. 2009. The microbial dimension in insect nutritional ecology. *Funct. Ecol.* 23: 38–47.

Engl, T., N. Eberl, C. Gorse et al. 2018. Ancient symbiosis confers desiccation resistance to stored grain pest beetles. *Mol. Ecol.* 27: 2095–2108.

Faleiro, J. R. 2006. A review of the issues and management of the red palm weevil *Rhynchophorus ferrugineus* (Coleoptera: Rhynchophoridae) in coconut and date palm during the last one hundred years. *Int. J. Trop. Insect Sci.* 26: 135–154.

Gasc, C., C. Ribiere, N. Parisot et al. 2015. Capturing prokaryotic dark matter genomes. *Res. Microbiol.* 166: 814–830.

Gasnier-Fauchet, F., A. Gharib, and P. Nardon. 1986. Comparison of methionine metabolism in symbiotic and aposymbiotic larvae of *Sitophilus oryzae* L. (Coleoptera: Curculionidae). I. Evidence for a glycine N-methyltransferase-like activity in the aposymbiotic larvae. *Comp. Biochem. Physiol. B* 85: 245–250.

Gasnier-Fauchet, F., and P. Nardon. 1986. Comparison of methionine metabolism in symbiotic and aposymbiotic larvae of *Sitophilus oryzae* L. (Coleoptera: Curculionidae). II. Involvement of the symbiotic bacteria in the oxidation of methionine. *Comp. Biochem. Physiol. B* 85: 251–254.

Grenier, A. M., P. Nardon, and G. Bonnot. 1986. Importance de la symbiose dans la croissance des populations de *Sitophilus oryzae* L. (Coleoptère Curculionidae). *Oecol. Appl.* 7: 93–110.

Grenier, A. M., C. Nardon, and P. Nardon. 1994. The role of symbiotes in flight activity of *Sitophilus* weevils. *Entomol. Exp. Appl.* 70: 201–208.

Grimaldi, D., and M. S. Engel. 2005. *Evolution of the Insects.* New York: Cambridge Univ. Press.

Hansen, A. K., and N. A. Moran. 2011. Aphid genome expression reveals host-symbiont cooperation in the production of amino acids. *Proc. Natl. Acad. Sci. USA.* 108: 2849–2854.

Heddi, A., F. Lefebvre, and P. Nardon. 1993. Effect of endocytobiotic bacteria on mitochondrial enzymatic activities in the weevil *Sitophilus oryzae* (Coleoptera: Curculionidae). *Insect Biochem. Mol. Biol.* 23: 403–411.

Heddi, A., and P. Nardon. 1993. Mitochondrial DNA expression in symbiotic and aposymbiotic strains of *Sitophilus oryzae. J. Stored Prod. Res.* 29: 243–252.

Heddi, A., and P. Nardon. 2005. *Sitophilus oryzae* L.: A model for intracellular symbiosis in the Dryophthoridae weevils (Coleoptera). *Symbiosis* 39: 1–11.

Hirota, B., G. Okude, H. Anbutsu et al. 2017. A novel, extremely elongated, and endocellular bacterial symbiont supports cuticle formation of a grain pest beetle. *mBio* 8: e01482-17.

Hirsch, J., S. Strohmeier, M. Pfannkuchen, and A. Reineke. 2012. Assessment of bacterial endosymbiont diversity in *Otiorhynchus* spp. (Coleoptera: Curculionidae) larvae using a multitag 454 pyrosequencing approach. *BMC Microbiol.* 12: S6.

Hosokawa, T., and T. Fukatsu. 2010. *Nardonella* endosymbiont in the West Indian sweet potato weevil *Euscepes postfasciatus* (Coleoptera: Curculionidae). *Appl. Entomol. Zool.* 45: 115–120.

Hosokawa, T., M. Hironaka, K. Inadomi, H. Mukai, N. Nikoh, and T. Fukatsu. 2013. Diverse strategies for vertical symbiont transmission among subsocial stinkbugs. *PLOS ONE* 8: e65081.

Hosokawa, T., Y. Ishii, N. Nikoh, M. Fujie, N. Satoh, and T. Fukatsu. 2016. Obligate bacterial mutualists evolving from environmental bacteria in natural insect populations. *Nat. Microbiol.* 1: 15011.

Hosokawa, T., Y. Kikuchi, N. Nikoh, M. Shimada, and T. Fukatsu. 2006. Strict host-symbiont cospeciation and reductive genome evolution in insect gut bacteria. *PLoS Biol.* 4: e337.

Hosokawa, T., R. Koga, K. Tanaka, M. Moriyama, H. Anbutsu, and T. Fukatsu. 2015. *Nardonella* endosymbionts of Japanese pest and non-pest weevils (Coleoptera: Curculionidae). *Appl. Entomol. Zool.* 50: 223–229.

Huang, X., Y. S. Huang, J. Y. Zhang, F. Lu, J. Wei, and M. X. Jiang. 2016. The symbiotic bacteria *Nardonella* in rice water weevil (Coleoptera: Curculionidae): Diversity, density, and associations with host reproduction. *Ann. Entomol. Soc. Am.* 109: 415–423.

Hunt, T., J. Bergsten, Z. Levkanicova et al. 2007. A comprehensive phylogeny of beetles reveals the evolutionary origins of a superradiation. *Science* 318: 1913–1916.

International Aphid Genomics Consortium. 2010. Genome sequence of the pea aphid *Acyrthosiphon pisum*. *PLoS Biol.* 8: e1000313.

Kikuchi, Y., A. Tada, D. L. Musolin et al. 2016. Collapse of insect gut symbiosis under simulated climate change. *mBio* 7: e01578-16.

Klein, A., L. Schrader, R. Gil et al. 2016. A novel intracellular mutualistic bacterium in the invasive ant *Cardiocondyla obscurior*. *ISME J.* 10: 376–388.

Kuriwada, T., T. Hosokawa, N. Kumano, K. Shiromoto, D. Haraguchi, and T. Fukatsu. 2010. Biological role of *Nardonella* endosymbiont in its weevil host. *PLOS ONE.* 5: e13101.

Lefèvre, C., H. Charles, A. Vallier, B. Delobel, B. Farrell, and A. Heddi. 2004. Endosymbiont phylogenesis in the Dryophthoridae weevils: Evidence for bacterial replacement. *Mol. Biol. Evol.* 21: 965–973.

McCutcheon, J. P., and N. A. Moran. 2011. Extreme genome reduction in symbiotic bacteria. *Nat. Rev. Microbiol.* 10: 13–26.

McKenna, D. D., A. S. Sequeira, A. E. Marvaldi, and B. D. Farrell. 2009. Temporal lags and overlap in the diversification of weevils and flowering plants. *Proc. Natl. Acad. Sci. USA.* 106: 7083–7088.

Moran, N. A., J. P. McCutcheon, and A. Nakabachi. 2008. Genomics and evolution of heritable bacterial symbionts. *Annu. Rev. Genet.* 42: 165–190.

Nardon, P. 1973. Obtention d'une souche asymbiotique chez le charançon *Sitophilus sasakii*: Différentes méthodes et comparaison avec la souche symbiotique d'origine. *C. R. Acad. Sci. Paris D.* 3: 65–67.

Noh, M. Y., S. Muthukrishnan, K. J. Kramer, and Y. Arakane. 2016. Cuticle formation and pigmentation in beetles. *Curr. Opin. Insect Sci.* 17: 1–9.

Oakeson, K. F., R. Gil, A. L. Clayton et al. 2014. Genome degeneration and adaptation in a nascent stage of symbiosis. *Genome Biol. Evol.* 6: 76–93.

Oberprieler, R. G., A. E. Marvaldi, and R. S. Anderson. 2007. Weevils, weevils, weevils everywhere. *Zootaxa* 1668: 491–520.

Rabatel, A., G. Febvay, K. Gaget et al. 2013. Tyrosine pathway regulation is host-mediated in the pea aphid symbiosis during late embryonic and early larval development. *BMC Genomics.* 14: 235.

Riedel, A., K. Sagata, S. Surbakti, R. Tänzler, and M. Balke. 2013. One hundred and one new species of *Trigonopterus* weevils from New Guinea. *ZooKeys.* 280: 1–150.

Rinke, R., A. S. Costa, F. P. Fonseca, L. C. Almeida, I. Delalibera Júnior, and F. Henrique-Silva. 2011. Microbial diversity in the larval gut of field and laboratory populations of the sugarcane weevil *Sphenophorus levis* (Coleoptera, Curculionidae). *Genet. Mol. Res.* 10: 2679–2691.

Russell, C. W., S. Bouvaine, P. D. Newell, and A. E. Douglas. 2013. Shared metabolic pathways in a coevolved insect-bacterial symbiosis. *Appl. Environ. Microbiol.* 79: 6117–6123.

Seago, A. E., P. Brady, J. P. Vigneron, and T. D. Schultz. 2009. Gold bugs and beyond: A review of iridescence and structural colour mechanisms in beetles (Coleoptera). *J. R. Soc. Interface.* 6: S165–S184.

Shigenobu, S., H. Watanabe, M. Hattori, Y. Sakaki, and H. Ishikawa. 2000. Genome sequence of the endocellular bacterial symbiont of aphids *Buchnera* sp. APS. *Nature.* 407: 81–86.

Shigenobu, S., and A. C. C. Wilson. 2011. Genomic revelations of a mutualism: The pea aphid and its obligate bacterial symbiont. *Cell Mol. Life Sci.* 68: 1297–1309.

Simonet, P., K. Gaget, N. Parisot et al. 2016. Disruption of phenylalanine hydroxylase reduces adult lifespan and fecundity, and impairs embryonic development in parthenogenetic pea aphids. *Sci. Rep.* 6: 34321.

Solden, L., K. Lloyd, and K. Wrighton. 2016. The bright side of microbial dark matter: Lessons learned from the uncultivated majority. *Curr. Opin. Microbiol.* 31: 217–226.

Stork, N. E., J. McBroom, C. Gely, and A. J. Hamilton. 2015. New approaches narrow global species estimates for beetles, insects, and terrestrial arthropods. *Proc. Natl. Acad. Sci. USA.* 112: 7519–7523.

Toju, H., and T. Fukatsu. 2011. Diversity and infection prevalence of endosymbionts in natural populations of the chestnut weevil: Relevance of local climate and host plants. *Mol. Ecol.* 20: 853–868.

Toju, H., T. Hosokawa, R. Koga et al. 2010. "*Candidatus* Curculioniphilus buchneri," a novel clade of bacterial endocellular symbionts from weevils of the genus *Curculio. Appl. Environ. Microbiol.* 76: 275–282.

Toju, H., A. S. Tanabe, Y. Notsu, T. Sota, and T. Fukatsu. 2013. Diversification of endosymbiosis: Replacements, co-speciation and promiscuity of bacteriocyte symbionts in weevils. *ISME J.* 7: 1378–1390.

True, J. R. 2003. Insect melanism: The molecules matter. *Trends Ecol. Evol.* 18: 640–647.

Tseng, H. Y., C. P. Lin, J. Y. Hsu, D. A. Pike, and W. S. Huang. 2014. The functional significance of aposematic signals: Geographic variation in the responses of widespread lizard predators to colourful invertebrate prey. *PLOS ONE.* 9: e91777.

van de Kamp, T., A. Riedel, and H. Greven. 2016. Micromorphology of the elytral cuticle of beetles, with an emphasis on weevils (Coleoptera: Curculionoidea) *Arthropod Str. Dev.* 45: 14–22.

Vigneron, A., F. Masson, A. Vallier et al. 2014. Insects recycle endosymbionts when the benefit is over. *Curr. Biol.* 24: 2267–2273.

Weissling, T. J., and R. M. Giblin-Davis. 1993. Water loss dynamics and humidity preference of *Rhynchophorus cruentatus* (Coleoptera: Curculionidae) adults. *Environ. Entomol.* 22: 93–98.

Wernegreen, J. J. 2002. Genome evolution in bacterial endosymbionts of insects. *Nat. Rev. Genet.* 3: 850–861.

Wernegreen, J. J. 2012. Mutualism meltdown in insects: Bacteria constrain thermal adaptation. *Curr. Opin. Microbiol.* 15: 255–262.

White, J. A., N. K. Richards, A. Laugraud, A. Saeed, M. M. Curry, and M. R. McNeill. 2015. Endosymbiotic candidates for parasitoid defense in exotic and native New Zealand weevils. *Microb. Ecol.* 70: 274–286.

Wicker, C. 1983. Differential vitamin and choline requirements of symbiotic and aposymbiotic *S. oryzae* (Coleoptera: Curculionidae). *Comp. Biochem. Physiol. A* 76: 177–182.

Wicker, C., and P. Nardon. 1982. Development responses of symbiotic and aposymbiotic weevils *Sitophilus oryzae* L. (Coleoptera, Curculionidae) to a diet supplemented with aromatic amino acids. *J. Insect Physiol.* 28: 1021–1024.

Wilson, A. C. C., P. D. Ashton, F. Calevro et al. 2010. Genomic insight into the amino acid relations of the pea aphid, *Acyrthosiphon pisum*, with its symbiotic bacterium *Buchnera aphidicola. Insect Mol. Biol.* 19: 249–258.

Zchori-Fein, E., and T. A. Miller. 2011. *Manipulative Tenants: Bacteria Associated with Arthropods.* Boca Raton: CRC Press/Taylor & Francis.

13 Microbial determinants of folivory in insects

Aileen Berasategui and Hassan Salem

Contents

13.1 Introduction

A centerpiece to ecosystem productivity, land plants are key drivers of energy conversion and carbon fixation (Zelitch 1975; Kroth 2015). As the primary photosynthetic organs of plants, the bulk of Earth's carbon reservoirs are fixed and stored by leaves (Vogelman, Nishio, and Smith 1996). Abundant in energy-rich sugars, and endowed with a steady supply of essential nutrients, peptides, lipids, and cofactors, leaves share many features of a nutritionally balanced diet. Obligate folivory, however, evolved a limited number of times throughout the metazoan tree of life (McNab 1988; Chivers 1989; Rand et al. 1990; Currano, Labandeira, and Wilf 2010).

Despite the extensive radiation of land plants across every major continent and the ubiquity of foliage as a seemingly accessible resource, most animals lack the metabolic and physiological adaptations necessary to subsist on leaves as a sole source of nutrition. First, and most prominently, animals largely lack the enzymatic circuitry necessary to maximize the dietary value of ingested foliage. The highly recalcitrant polysaccharides that define the fibrous features of leaves can only be hydrolyzed by a specific range of enzymes that are ancestrally encoded by plants and their specialized pathogens (Walton 1994; Kubicek, Starr, and Glass 2014). Towards accessing the nutritionally rich cytosol, folivores must first contend with the main structural polysaccharide components of the plant cell wall, despite generally lacking an endogenous repertoire of essential digestive enzymes. Second, leaves are typically enriched with a range of constitutive and induced plant secondary metabolites evolved to mitigate the incidence and impact of herbivory (Levin 1976; Stotz et al. 2000; Wittstock and Gershenzon 2002). Ranging from terpenoids to alkaloids, these defensive compounds disrupt the integrity of the digestive epithelial lining of animals and compromise the functionality of neuronal networks when ingested. Finally, foliage can be transient, as with deciduous trees and shrubs spanning Earth's temperate and polar regions (Jackson 1967). The total abscission of leaves poses as a considerable hurdle for folivores during the winter season (Giron et al. 2007), necessitating the evolution of strategies to survive in the absence of a specialized diet for months on end. Towards mitigating these challenges, strikingly convergent adaptations arose in independent folivorous lineages, many of which are mediated through symbioses with metabolically dynamic microbial communities.

With the advent of sequencing technologies and metabolic modeling, coupled with the development of conceptual frameworks to study the functionality of the metaorganism (McFall-Ngai et al. 2013), microbes are increasingly recognized as an essential source of adaptations for animals (Douglas 2015; Chomicki et al. 2019). By contributing complementary metabolic profiles, symbioses can upgrade the phenotypic complexity of both partners (Moran 2007), spurring their radiation into novel ecological niches and triggering their diversification. Microbial symbionts are especially recognized for fueling their hosts' specializations on nutritionally challenging diets (Douglas 2015). Sap-feeding invertebrates are consistently demonstrated to partner with symbionts that supplement the essential amino acids lacking in their diet (Hansen and Moran 2014; Baumann 2005). Haematophagous animals contend with the B-vitamin deficiency of their bloodmeals through nutritional partnerships with endosymbionts contributing these cofactors (Akman et al. 2002; Rio, Attardo, and Weiss 2016; Duron et al. 2018). Wood feeding is made possible through the digestive range of their lignocellulolytic symbionts across a number of metazoan taxa (Brune 2014; Brune and Dietrich 2015). Here, we argue that leaves constitute an equally specialized diet, necessitating innovations that extend well beyond the metabolic range of most animals. In outlining the convergent metabolic features of folivore microbiomes, we point towards the outsized role microbes play towards upgrading the dietary value of ingested leaves. Given the compositionally simple, experimentally tractable partnerships folivorous insects form with their microbial partners, we emphasize the unique suitability of these study systems to pursue hypothesis-driven research into the adaptive impact of symbiosis in its intersection with the evolution of leaf-feeding behavior (Figure 13.1).

13.2   Deconstructing the plant cell wall

Serving as the largest reservoir of organic carbon on Earth, plant cell walls are metabolically inaccessible to most animals (Rose 2003). Composed primarily of cellulose, hemicellulose, lignin, and pectin, these polysaccharides define the mechanical properties and endow the wall with its characteristic rigidity (Mohnen 2008; Burton, Gidley, and Fincher 2010). Instrumental towards ensuring the structural integrity of the plant cell, the polysaccharidic matrix also contributes towards adhesion and signal transduction (Burton, Gidley, and Fincher 2010). As the foremost barrier separating the nutritionally rich cytosol from the extracellular matrix, the plant cell wall safeguards against intracellular infection by pathogens and parasites (Underwood 2012). For folivores, the benefits of degrading the plant cell wall are thus twofold: (1) an upgraded carbon economy by tapping into a recalcitrant source of energetically valuable sugars, and (2) mediating an efficient extraction of limiting nitrogenous content and lipids from the cytosol for a more balanced diet.

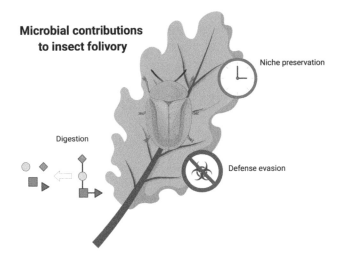

FIGURE 13.1 Microbial contributions to insect folivory include the degradation of recalcitrant plant cell wall polysaccharides (e.g. pectin, cellulose, and hemicellulose), the avoidance of noxious secondary compounds, and the preservation of leaves as their ecological niche. Overcoming these obstacles allow folivores to exploit an otherwise imbalanced nutritional resource.

Towards deconstructing the plant cell wall, a conserved battery of enzymes (e.g. glucoside hydrolases, polysaccharide lyases, etc.) are necessary to transform representative polysaccharidic sequences into simple sugars that can be metabolized throughout the digestive tract (Kubicek, Starr, and Glass 2014). Complementary to a range of endogenous digestive enzymes (Calderón-Cortés et al. 2012; McKenna et al. 2019), functional descriptions across a myriad of herbivore gut microbiomes also revealed the importance of the resident community in mediating plant biomass degradation (Martens et al. 2011; Engel, Martinson, and Moran 2012; Pope et al. 2012; Patel et al. 2014). In ruminants, relatively stable gut microbiomes serve as bioreactors for plant cell wall degradation and fermentation (Dai et al. 2015). Koalas and wombats—generalist and specialist herbivores, respectively—harbor highly convergent microbial communities enriched for the production of cellulases and xylanases to process a leafy diet (Shiffman et al. 2017). While the consistent annotation of carbohydrate-active enzymes may highlight the adaptive potential of gut microbiomes to their herbivorous hosts, assigning specific functions to individual taxa within these communities remains challenging given their complexity. As highlighted earlier, these limitations are less pronounced in the partnerships insects form with their symbionts (Douglas 2015). Stabilized through millions of years of coevolution between host and microbe, the symbioses that foliage-feeding ants and beetles engage in with microorganisms serve as some of the most streamlined partnerships both compositionally and functionally (Figure 13.2).

Leaf-cutting ants (Hymenoptera: Formicidae: Attini) are among the most prolific defoliators in tropical forests (Mueller et al. 2001; Wirth et al. 2003). Conspicuous and widespread in the New World tropics, members of the Attini tribe form

FIGURE 13.2 Microbial symbionts of folivores mediate host plant use across a diversity of insect hosts. (a) *Ca* Stammera capleta (above) produces two pectinolytic enzymes that facilitate degradation of plant cell walls, making the cytosolic content available for its tortoise beetle hosts (below). (b) The fungal partner of leaf-cutting ants functions as an external digestive system, both degrading most types of recalcitrant sugars present in leaves, as well as detoxifying plant phenolics. (c) Bacteria in the oral secretions of the Colorado potato beetle deceive tomato plants into perceiving herbivore attack as microbial, diverting plant resources into mounting an incorrect defense strategy (credit Tavo Romann). (d) The velvet bean caterpillar seems to avoid plant protease inhibitors by associating with gut bacteria whose proteases are unaffected by plant inhibitors. (e) Gypsy moth caterpillars feeding on quaking aspen acquire a gut microbiome able to degrade their host plant salicinoids. (f) *Wolbachia* symbionts allow leaf miners to maintain a green island in yellowing leaves, preserving their food source despite a decaying environment (credit Dr. Susannah Lydon).

enormous colonies that can host millions of workers and feature nests that subsist for decades (Wirth et al. 2003). Displaying a division of labor characteristic of other social insects, colonies host reproductively active queens, and differentially sized morphological worker castes divided to fulfill an array of specialized tasks (Weber 1966; Currie and Stuart 2001). This includes defending the colony, clearing the nest of debris and waste, and foraging for freshly cut leaves (Currie and Stuart 2001). As individual colonies are capable of foraging hundreds of kilograms of leaves each year, leaf-cutting ants drastically alter forest ecosystems and drive nutrient cycling (Wirth et al. 2003). But at the center of the ants' remarkable capacity to deconstruct massive amounts of plant biomass is a specialized, cultivated leucocoprineous fungus that the colony depends on for nutrition (Weber 1966; Mueller et al. 2001). A partnership dated at ~60 million years old, attini ants and their cultivars have diversified rapidly since the origin of the symbiosis, totaling 220 known species (Mueller et al. 2001). Provisioning their cultivars with fresh leaves, the ants manage their fungiculure in ways that parallel human agricultural practices (Mueller et al. 2005). Here, the ants inoculate, manipulate and transplant their cultivars to maximize growth. Carefully tended and continuously manured through fecal fluids, the fungus is cultivated for several weeks ahead of harvesting the mature mycelium and its associated gongylidia (specialized nodules) (Mueller et al. 2005). Rich in nutrients, the gongylidia serve as vacuolized bundles that nourish the colony. Larval development depends entirely on consuming the cultivated fungus (Quinlan and Cherrett 1978), while adult workers supplement their cultivar meals with sugary extracts from floral nectaries and other plant juices (Murakami and Higashi 1997). Given the elemental role of the cultivar, attine ants transmit the fungus vertically through trophophoresy (Wirth et al. 2003). Acquiring inocula from their natal gardens, reproductive females transfer the fungal cultivar via specialized pockets towards establishing a starter culture (Mueller et al. 2005).

As the main food source for the colony, the cultivar's primary function is to convert plant biomass into nutrients for the ants (Suen et al. 2010). Serving as an external digestive system, the central metabolic features of the fungal gardens within the nest functionally resemble the gut microbiomes of bovines and other folivorous ruminants (Suen et al. 2010; Aylward et al. 2015). Structurally differentiated into distinct strata, fresh foliar material is introduced only to the top layer ahead of stepwise degradation process that lasts for six weeks and concludes with the removal and expulsion of degraded biomass by worker ants into refuse dumps (Wirth et al. 2003). Throughout this process, the quantification of plant cell wall polysaccharides revealed marked reduction in cellulose, hemicellulose and pectin. Most prominently, cellulose content decreased by 30% following passage through the fungal garden (Suen et al. 2010). Genome annotation of cultivars isolated from *Atta cephalotes* and *Acromyrmex echinatior* identified 145 predicted plant biomass-degrading enzymes, including 81 glycoside hydrolases, 6 polysaccharide lyases and nine carbohydrate esterases (Aylward et al. 2013). Complementary metaproteomic characterization predicted many of these enzymes hydrolyze the complete spectrum of polysaccharides that comprise the plant cell wall. Gene expression analyses of the cultivar revealed that genes encoding plant cell wall degrading enzymes reached their highest expression in the bottom section of the fungal garden (Grell et al. 2013). This is consistent with the observed physical transformation of plant substrates from green, leafy material on the top layer, to an amalgam of degraded biomass and mature fungal cells. Strikingly, many of these glycolytic enzymes are characterized in the fecal droplets the ants deposit to manure the garden throughout the cultivation process (Schiøtt et al. 2010). This suggests that the proteins survive the ants' digestive system ahead of reapplication to fresh plant material, possibly as a secondary adaptation to prolong the deconstructive efficacy of enzymes within the gardens (Schiøtt et al. 2010).

While enormous colony sizes and a clear division of labor allows for leaf-cutting ants to engage in an ancient agricultural practice that culminates with outsourcing essential digestive processes to an external partner, other insects coopt microbial metabolic diversity by engaging in highly intimate nutritional symbioses. One clear example involves tortoise leaf beetles (Coleoptera: Chrysomelidae: Cassidinae) and their symbiont, *Candidatus* Stammera capleta (henceforth *Stammera*). With ~3000 described species arranged in 170 genera and 24 tribes, cassidines are an exceptionally diverse group of herbivorous beetles (Chaboo 2007). Despite their cosmopolitan distribution, cassidines nonetheless diversified most rapidly in the tropics where their biogeography closely aligns with their host plants (Windsor 1987; Windsor, Riley, and Stockwell 1992; Morrison and Windsor 2018). Towards acquiring essential digestive enzymes to process a strictly folivorous diet, tortoise leaf beetles engage in a highly streamlined symbiosis with *Stammera*.

Localized extracellularly in symbiotic organs connected to the foregut, *Stammera* populations are maintained as monocultures resulting in low strain diversity across host populations (Salem et al. 2017). Strikingly, given the microbe's localization, *Stammera* possesses a drastically reduced genome (0.27 Mb) that is largely dedicated to informational processing (transcription, translation), replication, and the production essential digestive enzymes. Encoded within *Stammera*'s limited metabolism are two pectinolytic enzymes, polygalacturonase and rhamnogalacturonan lyase (Salem et al. 2017). The former is an endo-active glycoside hydrolase (family 28) that cleaves homogalacturonan, nature's most abundant pectic class; while the latter is a polysaccharide lyase (family 4) that hydrolyzes the heteropolymeric backbone of pectin through a beta-elimination reaction (Salem et al. 2017). Collectively, both symbiont-derived enzymes are responsible for the insect's pectinolytic phenotype, an essential adaptation given the high abundance of pectin in foliage (Burton, Gidley, and Fincher 2010) and the recalcitrant complexity of the polysaccharide (Mohnen 2008). Consistent with the specialized role of *Stammera*, symbiont elimination diminishes the digestive capacity of cassidines, notably in relation to the two pectic classes (Salem et al. 2017). This corresponds with low larval survivorship and slow developmental times relative to symbiotic insects. Given the mutualistic impact of *Stammera*, female leaf beetles ensure a stable continuum of the symbiosis through strict vertical transmission by packaging the microbe into caplet-like structures deposited individually over the anterior pole of each egg. Upon hatching, emerging larva consume the caplets, and in the process, acquire their starting inoculum of a pectinase-producing partner.

In line with the assessment that *Stammera* possesses the smallest known genome of any extracellular microbe (Salem et al. 2017), transcriptional profiling revealed a tightly regulated and metabolically integrated symbiosis, expanding our view of the minimal metabolism required to sustain life outside of a host cell (Bauer et al. 2020). In contrast to endosymbionts with highly reduced genomes, *Stammera* does not utilize aerobic respiration for energy generation. Rather, energy production and the recovery of reducing equivalents are achieved through the oxidation of sugars to pyruvate via glycolysis, followed by the fermentation of pyruvate to lactate through the activity of lactate dehydrogenase (Bauer et al. 2020). Aerobic respiration is typically conserved in most nutritional endosymbionts since amino acid and vitamin biosynthesis relies on precursors produced through the citric acid cycle, tying respiratory energy generation with the mutualistic factors that underlie the host–symbiont partnership (McCutcheon and Moran 2012). But since *Stammera*'s mutualistic role does not lie in supplementing micronutrients, selection to maintain a complex respiratory apparatus is relaxed in favor of fermentative one. While less efficient in terms of ATP output per unit of glucose, the symbiont's obligate reliance on fermentation for energy generation is unlikely to be costly given the host's carbohydrate-rich diet. This is supported by the upregulation of sugar transport into the symbiotic organ to fuel the minimal fermentative metabolism of an essential digestive symbiont (Bauer et al. 2020).

Broadscale genome sequencing of representative *Stammera* strains revealed that the differential distribution of symbiont-encoded pectinolytic enzymes can drastically shape the digestive physiology of Cassidinae beetles with direct implications on host plant use (Salem et al. In press). Conserved across the *Stammera* pangenome is the ability to produce and supplement polygalacturonase, highlighting the homogalacturonan-targeting pectinase as a foundational enzyme for the stability of the symbiosis with cassidines. In contrast, the annotation of rhamnogalacturonan lyase is limited to a subset of *Stammera* strains, as is the ability to deconstruct the heteropolymeric sequence of pectin (Salem et al. In press). Consistent with *in silico* predictions, beetles harboring *Stammera* encoding polygalacturonase and rhamnogalacturonan lyase display a greater pectinolytic range relative to cassidines whose symbionts only supplement the former of the two digestive enzymes (Salem et al. In press). Matching an ability to metabolize a greater diversity of universal plant polysaccharides, cassidines deploying both pectinases have radiated to exploit a wider range of host plants. In reconciling detailed records of life history traits with comparative genomics, transcriptomics and biochemical assays, the symbiosis between tortoise leaf beetles and *Stammera* serves as an example of how small changes to a symbiont's metabolic range can drastically impact the phenotypic complexity and the adaptive potential of its metazoan host.

13.3   Symbiont-mediated evasion of plant defenses

Plants counter challenges from herbivores and pathogens through the production of noxious secondary metabolites; some are constitutively expressed (Wittstock and Gershenzon 2002) while others can be induced upon attack (Stotz et al. 2000). Enriched within leaves, plant toxins include metabolites as chemically diverse as alkaloids, cyanogenic glycosides, phenolics, terpenes, benzoxazinoids, and glucosinolates, among others (Fürstenberg-Hägg, Zagrobelny, and Bak 2013). Their mode of action remains elusive in many cases, but these compounds are often involved in the disruption of gut membranes, hindering metabolism and preventing normal molecular signaling, ion and nutrient transport, as well as triggering the interruption of hormone-controlled physiological processes (Mithöfer and Boland 2012). Thus, folivores are under a strong selective pressure to evolve strategies to overcome these compounds. While many of these adaptations are endogenously encoded in a folivore's metabolic repertoire (Zhu-Salzman, Bi, and Liu 2005; Després, David, and Gallet 2007), it is now evident that microbial symbionts play a central role in mediating host plant use by interfering with and degrading plant chemical defenses.

Contingent on the nature of the attack, plants induce different defense responses. Following microbial infection, plants activate salicylic acid (SA)-dependent defenses, whereas herbivory induces jasmonic acid (JA) synthesis. Often, these pathways negatively cross talk. Towards feeding on tomato plant (Solanaceae) leaves, the Colorado potato beetle (*Leptinotarsa decemlineata*) secretes bacteria-containing saliva into the wound. Detecting the threat as microbial instead of herbivorous, the plant mounts a defense response based on SA instead of JA. This deception benefits the insect, which avoids antiherbivore defenses and experiences higher larval growth (Chung et al. 2013). Among the several bacterial taxa present in the oral secretion, only three isolates, *Pseudomonas, Enterobacter,* and *Stenotrophomonas*, suppress JA-dependent defenses. In particular, among the different bacterial components in these three isolates, flagellin isolated from *Pseudomonas*, was identified as one of the effectors that downregulate JA synthesis (Chung et al. 2013). These strategies are conserved in other herbivorous insect lineages. For instance, Aster Yellows (AY) phytoplasmas, a plant pathogen, could be considered a facultative symbiont of its vector leafhopper *Macrosteles quadrilineatus*. AY excretes an effector protein (SAP11) that modulates host plant (*Arabidopsis*) defense responses, downregulating the production of JA-derived metabolites to the benefit of the insect, which experiences higher fecundity (Sugio et al. 2011).

While the aforementioned examples feature symbionts protecting their insect host against plant allelochemicals prior to their synthesis, most known examples involve protection against metabolites that have already been produced, either by symbiont-mediated avoidance or breakdown of these compounds. Upon herbivorous attack, soybean plants induce the production of protein inhibitors targeting folivore proteinases, consequently hindering insect digestion (Carlini and Grossi-de-Sá 2002). Despite this, the velvet-bean caterpillar, *Anticarsia gemmatalis*, represents a major pest of soybeans. Antibiotic treatment of these insects disrupts its gut bacterial community and results in lower caterpillar growth and survival, suggesting a beneficial role of microbes. Subsequent enzymatic assays demonstrate that proteolytic and lipolytic activities are significantly affected by symbiont loss (Visôtto et al. 2009). Interestingly, some members of the gut microbiome are able to synthesize proteinases that are immune to soybean proteinase inhibitors (Pilon et al. 2013), suggesting that microbial proteases may serve as a secondary set of enzymes that mediate the insect host circumventing plant defenses.

The gypsy moth (*Lymantria dispar*), a generalist folivore and pest, is well known for population outbreaks that drastically alter forest ecosystems (Liebhold et al. 1994). Given its broad host plant range, *L. dispar* caterpillars tolerates an equally diverse assortment of plant secondary metabolites. The gypsy moth's preferred host plant, the quaking aspen (*Populus tremuloides*), is rich in terpenes, phenolic glycosides (salicinoids), and tannins (Lindroth and St. Clair 2013). Earlier descriptions of *L. dispar* gut bacterial community found members of the *Rhodococus* genus; noted for their production of terpene-degrading enzymes (van der Vlugt-Bergmans and van der Werf 2001; Broderick et al. 2004). Salicinoids lower growth and development in gypsy moth caterpillars (Hemming and Lindroth 1995). While *L. dispar* has evolved some adaptations to overcome plant toxins, such as a highly alkaline midgut, and a battery of detoxification enzymes, these strategies can be overwhelmed by high concentrations of phenolic glycosides. However, gypsy moth larvae harbor gut bacterial communities that can degrade salicinoids, leading to increased larval growth following chemical challenges by the secondary metabolite (Mason, Couture, and Raffa 2014). As described in other lepidopterans, the gut microbiome composition in this species is largely shaped by the microbial community present in ingested leaves, whereas its structure is determined by insect physiology (Mason and Raffa 2014).

Accordingly, caterpillars harboring an aspen-derived community are better adapted to tolerate aspen-derived defenses than those harboring nonaspen associated bacteria (Mason, Couture, and Raffa 2014). It appears that the benefit of these gut symbionts to their insect hosts may be derived from their ability to defend themselves against plant secondary metabolites in their primary environment as leaf-associated microbes.

Among fungus-growing ants, the transition to folivory corresponded with the necessity to contend with a wide diversity of plant secondary metabolites (Berenbaum 1988). Although leaf cutting ants can exploit nearly 75% of all plants present in the New World tropical forests, they avoid foraging on some species (Howard, Cazin, and Wiemer 1988), presumably those that contain toxins with detrimental effect to themselves and their symbiotic fungal partner (Seaman 1984). The metabolic repertoire of the fungal cultivar, which includes the ability to detoxify some plant toxins, may have been a key factor for leaf-cutting ants in becoming dominant folivores in the New World. Towards degrading phenols, the fungal cultivar of attini ants produces several phenol-oxidizing enzymes of the laccase family. One of them, LgLcc1 is highly expressed in the gongylidia and is ingested by ant workers, surviving digestion before being defecated on top of the garden. This mechanism ensures that laccase activity is highest in the garden top layers where new leaf material is being deposited and detoxification is most needed (De Fine Licht et al. 2013). Additionally, the presence of the laccase in the gut of foragers may aid in detoxifying phenolics ants may drink when cutting and chewing leaves to manure the fungal garden. Analyses of orthologous genes encoding this phenol-oxidizing enzyme demonstrate that LgLcc1 has been selected for in gongylidia-producing fungal gardens. Consistent with this observation, laccase activity is highest in the nests of leaf-cutting ants than in their noncutting counterparts (De Fine Licht et al. 2013).

Microbial degradation of plant noxious compounds is not unique to folivorous insects. Folivorous vertebrates such as cows, sheep and rats, as well as birds, harbor bacterial communities with the ability to degrade plant secondary metabolites (García-Amado et al. 2007; Kohl and Dearing 2012; Kohl et al. 2016). Similarly, other herbivorous insects besides folivores also outsource some of their adaptations against plant toxins to microbial symbionts (Barr et al. 2010; Boone et al. 2013; Hammerbacher et al. 2013; Ceja-Navarro et al. 2015; Welte et al. 2016; Berasategui et al. 2017). Thus, there is increasing evidence that symbiotic microbes can mediate host plant use through the manipulation, degradation and inactivation of plant defenses, and is possibly representative of a widespread occurance (Shen and Dowd 1990).

13.4   Niche preservation

Dietary specialization carries the risk of temporal instability. This is most evident in the challenges faced by folivores in their coevolution with deciduous plants, where resource quality and accessibility are directly influenced by seasonality and abiotic conditions (Giron et al. 2007; Kaiser et al. 2010; Gutzwiller et al. 2015; Zhang et al. 2017). Despite a leaf's inevitable developmental progression, from initiation to senescence (Bar and Ori 2014), numerous herbivorous lineages have nonetheless evolved strategies to stall that process to continue exploiting foliage nutritionally well beyond seasonal fall. Endophagous herbivores, such leaf miners and insect gallers, can achieve this by stalling morphogenesis and coopting plant architecture to generate structures that buffer against environmental change, protect from natural enemies, and, ultimately, extend the lifespan of a dietary niche. By coopting their host plant's metabolism, endophagous herbivores can actively trigger the differentiation of their microhabitat to shape its chemical composition and dietary value *in situ*. Among leaf-miners, this process is mediated by the bacterial symbiont, *Wolbachia*.

Responsible for their spectacular induction of "green islands" on yellowing leaves during autumn, the leaf-miner moth *Phyllonorycter* (Lepidoptera: Gracillariidae) is able to preserve a photosynthetically active patch embedded within an otherwise decaying leaf (Giron et al. 2007). Defined by elevated cytokinins, these patches reflect the continued maintenance of chlorophyll and the inhibited progression of senescence. Infection by *Wolbachia* directly impacts the ability of *Phyllonorycter* to induce green islands, since symbiont loss corresponds with the absence of the phenotype in yellowing leaves (Kaiser et al. 2010). Responding to a deteriorating ecological niche, aposymbiotic insects exhibit compensatory feeding and higher levels of mortality relative to *Phyllonorycter* infected by *Wolbachia* (Kaiser et al. 2010). By mediating its host's manipulation leaf tissues vis-à-vis cytokinin production, *Wolbachia* ensures that the insect's access to sugar-rich, metabolically active habitat despite a rapidly decaying enviornment (Zhang et al. 2017). While symbiont-induced nutritional homeostasis is estimated to be widespread across the Gracillariidae, the identification of two separate *Wolbachia* strains suggests several independent origins of green-island induction within this insect family (Gutzwiller et al. 2015).

13.5   Conclusions

Symbiont acquisition and replacement are essential processes, coinciding with the integration of novel metabolic features and the ability to exploit previously inaccessible niches (Moran 2007). Among insects exploiting highly specialized diets, microbes serve as an important source of metabolites and supplements towards upgrading their hosts' nutritional ecology. This is best documented in the evolutionary independent associations between sap-feeding hemipterans and nutritional endosymbionts that supplement the essential amino acids lacking in their diet (Bennett and Moran 2015). Similarly, convergent mutualistic factors are enriched in the primary endosymbionts of blood-feeding insects. Across ticks (Duron et al. 2018), bedbugs (Hosokawa et al. 2010), and tsetse flies (Akman et al. 2002), haematophagy is made possible through symbiont-encoded B vitamin contributions to balance a diet that is highly deficient in these cofactors. This chapter outlines that leaf feeding poses an inherently different set of challenges for obligate folivores. Challenges largely countered by animals engaging in stable symbioses with microbes that contribute towards the degradation of complex polymers, detoxification of noxious compounds, and niche preservation. While many of the examples highlighted here feature taxonomically diverse microbes, the range of services endowed to the insect host are nonetheless conserved. Given that conservation, and the experimental tractability of insect symbioses to pursue novel and emerging questions within the field of microbiome research, we emphasize the suitability of folivorous insects as dynamic study systems to illustrate the metaorganismal basis of adaptation in animals.

References

Akman, L., Yamashita, A., Watanabe, H., Oshima, K., Shiba, T., Hattori, M., Aksoy, S. 2002. Genome sequence of the endocellular obligate symbiont of tsetse flies, Wigglesworthia glossinidia. *Nature Genetics* 32(3):402-407. doi:10.1038/ng986.

Aylward, F.O., Burnum-Johnson, K.E., Tringe, S.G., Teiling, C., Tremmel, D.M., Moeller, J.A., Scott, J.J. et al. 2013. Leucoagaricus gongylophorus produces diverse enzymes for the degradation of recalcitrant plant polymers in leaf-cutter ant fungus gardens. *Applied Environmental Microbiology* 79: 3770–3778. doi: 10.1128/AEM.03833-12.

Aylward, F. O., Khadempour, L., Tremmel, D. M., McDonald, B. R., Nicora, C. D., Wu, S., Moore, R.J. et al. 2015. Enrichment and broad representation of plant biomass-degrading enzymes in the specialized hyphal swellings of Leucoagaricus gongylophorus, the fungal symbiont of leaf-cutter ants. *PLoS ONE* 10(8): e0134752. doi:10.1371/journal.pone.0134752.

Bar, M., and Ori, N. 2014. Leaf development and morphogenesis. *Development*, 141:4219-4230. doi:10.1242/dev.106195.

Barr, K.L., Hearne, L.B., Briesacher, S., Clark, T.L., Davis, G.E. 2010. Microbial symbionts in insects influence down-regulation of defense genes in maize. *PLoS ONE* 5(6): e11339. doi:10.1371/journal.pone.0011339

Bauman, P. 2005. Biology bacteriocyte-associated endosymbionts of plant sap-sucking insects. *Annual Reviews of Microbiology* 59:155–189. doi: 10.1146/annurev.micro.59.030804.121041.

Bauer, E., Kaltenpoth, M., Salem, H. 2020. Minimal fermentative metabolism fuels extracellular symbiont in a leaf beetle. *The ISME Journal* 14:866–870. doi.org/10.1038/s41396-019-0562-1.

Bennett, G.M., and Moran, N.A. 2015. Heritable symbiosis: The advantages and perils of an evolutionary rabbit hole. *Proceedings of the National Academy of Sciences USA* 112(33):10169-10176. doi: 10.1073/pnas.1421388112.

Berasategui, A., Salem, H., Paetz, C., Santoro, M., Gershenzon, J., Kaltenpoth, M., Schmidt, A. 2017. Gut microbiota of the pine weevil degrades conifer diterpenes and increases insect fitness. *Molecular Ecology* 26(15):4099–4110. doi:10.1111/mec.14186.

Berenbaum, M.R. 1988. Allelochemicals in insect-microbe-plant interactions; agents provocateurs in the coevolutionary arms race. In: *Novel Aspects of Insect-Plant Interactions* (Barbosa P, Letourneau DK), pp. 97–123. John Wiley & Sons, New York.

Boone, C.K., Keefover-Ring, K., Mapes, A.C., Adams, A.S., Bohlmann, J., Raffa, K.F. 2013. Bacteria associated with a tree-killing insect reduce concentrations of plant defense compounds. *Journal of Chemical Ecology* 39: 1003–1006. doi: 10.1007/s10886-013-0313-0.

Broderick, N.A., Raffa, K.F., Goodman, R.M., Handelsman, J. 2004. Census of the bacterial community of the gypsy moth larval midgut by using culturing and culture-independent methods. *Applied and Environmental Microbiology* 70: 293-300. doi: 10.1128/AEM.70.1.293-300.2004

Brune, A. 2014. Symbiotic digestion of lignocellulose in termite guts. *Natural Reviews Microbiology* 12(3):168-180. doi: 0.1038/nrmicro3182.

Brune, A., Dietrich C. 2015. The gut microbiota of termites: digesting the diversity in the light of ecology and evolution. *Annual Reviews Microbiology* 69:145-166. doi: 10.1146/annurev-micro-092412-155715

Burton, R. A., Gidley, M. J., and Fincher, G. B. 2010. Heterogeneity in the chemistry, structure and function of plant cell walls. *Nature Chemical Biology* 6(10): 724–32. doi:10.1038/nchembio.439.

Calderón-Cortés, N., Quesada, M., Watanabe, H., Cano-Camacho, H. and Oyama, K. 2012. Endogenous Plant Cell Wall Digestion: A key mechanism in insect evolution. *Annual Review of Ecology, Evolution, and Systematics* 43(1): 45–71. doi:10.1146/annurev-ecolsys-110411-160312.

Carlini, C. R., and Grossi-de-Sá, M. F. 2002. Plant toxic proteins with insecticidal properties. A review on their potentialities as bioinsecticides. *Toxicon: Official Journal of the International Society on Toxinology* 40(11): 1515–39. doi:10.1016/s0041-0101(02)00240-4.

Ceja-Navarro, J. A., Vega, F. E., Karaoz, U., Hao, Z., Jenkins, S., Lim, H. C., Kosina P., Infante, F., Northen, T. R., and Brodie, E. L. 2015. Gut microbiota mediate caffeine detoxification in the primary insect pest of coffee. *Nature Communications* 6(July): 7618. doi:10.1038/ncomms8618.

Chaboo, C. S. 2007. Biology and phylogeny of the *Cassidinae gyllenhal* sensu lato (tortoise and leaf-mining beetles) (Coleoptera, Chrysomelidae). *Bulletin of the American Museum of Natural History* 305(June): 1–250.

Chivers, D. J. 1989. Adaptations of digestive systems in non-ruminant herbivores. *Proceedings of the Nutrition Society* 48(1): 59–67. doi:10.1079/PNS19890010.

Chomicki, G., Weber, M., Antonelli, A., Bascompte, J., and Kiers, E. T. 2019. The Impact of Mutualisms on Species Richness. *Trends in Ecology & Evolution* 34(8): 698–711. doi:10.1016/j.tree.2019.03.003

Chung, S. H., Rosa C., Scully, E. D., Peiffer, M., Tooker, J. F., Hoover, K., Luthe, D. S., and Felton, G. W. 2013. Herbivore exploits orally secreted bacteria to suppress plant defenses. *Proceedings of the National Academy of Sciences of the United States of America* 110(39): 15728–33. doi:10.1073/pnas.1308867110.

Currano, E. D., Labandeira, C. C., and Wilf, P. 2010. Fossil Insect folivory tracks paleotemperature for six million years. *Ecological Monographs* 80(4): 547–67. doi:10.1890/09-2138.1.

Currie, C. R., and Stuart, A. E. 2001. Weeding and grooming of pathogens in agriculture by ants. *Proceedings of the Royal Society B: Biological Sciences* 268(1471): 1033–39. doi:10.1098/rspb.2001.1605.

Dai, X., Tian, Y., Li, J., Luo, Y., Liu, D., Zheng, H., Wang, J., Dong, Z., Hu, S., and Huang, Li. 2015. Metatranscriptomic analyses of plant cell wall polysaccharide degradation by microorganisms in the cow rumen. *Applied and Environmental Microbiology* 81(4): 1375–86. doi:10.1128/AEM.03682-14.

De Fine Licht, H. H., Schiøtt, M., Rogowska-Wrzesinska, A., Nygaard, S., Roepstorff, P., and Boomsma, J. J. 2013. Laccase detoxification mediates the nutritional alliance between leaf-cutting ants and fungus-garden symbionts. *Proceedings of the National Academy of Sciences of the United States of America* 110(2): 583–87. doi:10.1073/pnas.1212709110.

Després, L., David, J.-P., and Gallet, C. 2007. The Evolutionary ecology of insect resistance to plant chemicals. *Trends in Ecology & Evolution* 22(6): 298–307. doi:10.1016/j.tree.2007.02.010.

Douglas, A. E. 2015. Multiorganismal insects: Diversity and function of resident microorganisms. *Annual Review of Entomology* 60(1): 17–34. doi:10.1146/annurev-ento-010814-020822.

Duron, O., Morel, O., Noël, V., Buysse, M., Binetruy, F., Lancelot, R., Loire, E. et al. 2018. Tick-bacteria mutualism depends on B vitamin synthesis pathways. *Current Biology: CB* 28(12): 1896–1902.e5. doi:10.1016/j.cub.2018.04.038.

Engel, P., Martinson, V. G., and Moran, N. A. 2012. Functional diversity within the simple gut microbiota of the honey bee. *Proceedings of the National Academy of Sciences* 109(27): 11002–7. doi:10.1073/pnas.1202970109.

Fürstenberg-Hägg, J., Zagrobelny, M., and Bak, S. 2013. Plant Defense against insect herbivores. *International Journal of Molecular Sciences* 14(5): 10242–97. doi:10.3390/ijms140510242.

García-Amado, M. A., Michelangeli, F., Gueneau, P., Perez, M. E., and Domínguez-Bello, M. E. 2007. Bacterial detoxification of saponins in the crop of the avian foregut fermenter *Opisthocomus hoazin*. *Journal of Animal and Feed Sciences* 16(Suppl 2): 82–5. doi:10.22358/jafs/74460/2007.

Giron, D., Kaiser, W., Imbault, N., and Casas, J. 2007. Cytokinin-mediated leaf manipulation by a leafminer caterpillar. *Biology Letters* 3(3): 340–43. doi:10.1098/rsbl.2007.0051.

Grell, M. N., Linde, T., Nygaard, S., Nielsen, K. L., Boomsma, J. J., and Lange, L. 2013. The fungal fymbiont of *Acromyrmex* leaf-cutting ants expresses the full spectrum of genes to degrade cellulose and other plant cell wall polysaccharides. *BMC Genomics* 14(1): 928. doi:10.1186/1471-2164-14-928.

Gutzwiller, F., Dedeine, F., Kaiser, W., Giron, D., and Lopez-Vaamonde, C. 2015. Correlation between the Green-Island phenotype and *Wolbachia* infections during the evolutionary diversification of Gracillariidae leaf-mining moths. *Ecology and Evolution* 5(18): 4049–62. doi:10.1002/ece3.1580.

Hammerbacher, A., Schmidt, A., Wadke, N., Wright, L. P., Schneider, B., Bohlmann, J., Brand, W. A., Fenning, T. M., Gershenzon, J., and Paetz, C. 2013. A common fungal associate of the spruce bark beetle metabolizes the stilbene defenses of Norway spruce. *Plant Physiology* 162(3): 1324–36. doi:10.1104/pp.113.218610.

Hansen, A. K., and Moran, N. A. 2014. The impact of microbial symbionts on host plant utilization by herbivorous insects. *Molecular Ecology* 23(6): 1473–96. doi:10.1111/mec.12421.

Hemming, J. D. C., and Lindroth, R. L. 1995. Intraspecific variation in aspen phytochemistry: Effects on performance of gypsy moths and forest tent caterpillars. *Oecologia* 103(1): 79–88. doi:10.1007/BF00328428.

Hosokawa, T., Koga, R., Kikuchi, R., Meng, X.-Y., and Fukatsu, T. 2010. *Wolbachia* as a bacteriocyte-associated nutritional mutualist. *Proceedings of the National Academy of Sciences* 107(2): 769–74. doi:10.1073/pnas.0911476107.

Howard, J. J., Cazin, J., and Wiemer, D. F. 1988. Toxicity of terpenoid deterrents to the leaf-cutting ant *Atta cephalotes* and its mutualistic fungus. *Journal of Chemical Ecology* 14(1): 59–69. doi:10.1007/BF01022531.

Jackson, L. W. R. 1967. Effect of shade on leaf structure of deciduous tree species. *Ecology* 48(3): 498–99. doi:10.2307/1932686.

Kaiser, W., Huguet, E., Casas, J., Commin, C., and Giron, D. 2010. Plant green-island phenotype induced by leaf miners is mediated by bacterial symbionts. *Proceedings. Biological Sciences* 277(1692): 2311–19. doi:10.1098/rspb.2010.0214

Kohl, K. D., Connelly, J. W., Dearing, M. D., and Forbey, J. S. 2016. Microbial detoxification in the gut of a specialist avian herbivore, the Greater Sage-Grouse. *FEMS Microbiology Letters* 363(14). doi:10.1093/femsle/fnw144.

Kohl, K. D., and Dearing, M. D. 2012. Experience matters: Prior exposure to plant toxins enhances diversity of gut microbes in herbivores. *Ecology Letters* 15(9): 1008–15. doi:10.1111/j.1461-0248.2012.01822.x.

Kroth, P. G. 2015. The biodiversity of carbon assimilation. *Journal of Plant Physiology* 172(January): 76–81. doi:10.1016/j.jplph.2014.07.021.

Kubicek, C. P., Starr, T. L., and Glass, N. L. 2014. Plant cell wall-degrading enzymes and their secretion in plant-pathogenic fungi. *Annual Review of Phytopathology* 52: 427–51. doi:10.1146/annurev-phyto-102313-045831.

Levin, D. A. 1976. The chemical defenses of plants to pathogens and herbivores. *Annual Review of Ecology and Systematics* 7: 121–59.

Liebhold, A. M., Elmes, G. A., Halverson, J. A., and Quimby, J. 1994. Landscape characterization of forest susceptibility to gypsy moth defoliation. *Forest Science* 40(1): 18–29. doi:10.1093/forestscience/40.1.18.

Lindroth, R. L., and St. Clair, S. B. 2013. Adaptations of quaking aspen *(Populus tremuloides* Michx.) for defense against herbivores. *Forest Ecology and Management* 299: 14–21. doi:10.1016/j.foreco.2012.11.018.

Martens, E. C., Lowe, E. C., Chiang, H., Pudlo, N. A., Wu, M., McNulty, N. P., Abbott, D. W. et al. 2011. Recognition and degradation of plant cell wall polysaccharides by two human gut symbionts. *PLoS Biology* 9(12): e1001221. doi:10.1371/journal.pbio.1001221.

Mason, C. J., Couture, J. J., and Raffa, K. F. 2014. Plant-associated bacteria degrade defense chemicals and reduce their adverse effects on an insect defoliator. *Oecologia* 175(3): 901–10. doi:10.1007/s00442-014-2950-6.

Mason, C. J., and Raffa, K. F. 2014. Acquisition and structuring of midgut bacterial communities in gypsy moth (Lepidoptera: Erebidae) larvae. *Environmental Entomology* 43(3): 595–604. doi:10.1603/EN14031.

McCutcheon, J. P., and Moran, N. A. 2012. Extreme genome reduction in symbiotic bacteria. *Nature Reviews Microbiology* 10(1): 13–26. doi:10.1038/nrmicro2670.

McFall-Ngai, M., Hadfield, M.G., Bosch, T. C. G., Carey, H. V., Domazet-Lošo, T., Douglas, E. E., Dubilier, N. et al. 2013. Animals in a bacterial world, a new imperative for the life sciences. *Proceedings of the National Academy of Sciences* 110(9): 3229–36. doi:10.1073/pnas.1218525110.

McKenna, D. D., Shin, S., Ahrens, D., Balke, M., Beza-Beza, C., Clarke, D. J., Donath, A. et al. 2019. The evolution and genomic basis of beetle diversity. *Proceedings of the National Academy of Sciences* 116(49): 24729–24737. doi:10.1073/pnas.1909655116

McNab, B. K. 1988. Food habits and the basal rate of metabolism in birds. *Oecologia* 77(3): 343–49. doi:10.1007/BF00378040.

Mithöfer, A., and Boland, W. 2012. Plant defense against Herbivores: Chemical aspects. *Annual Review of Plant Biology* 63: 431–50. doi:10.1146/annurev-arplant-042110-103854.

Mohnen, D. 2008. Pectin structure and biosynthesis. *Current Opinion in Plant Biology, Physiology and Metabolism* - Edited by Markus P. and Kenneth K., 11(3): 266–77. doi:10.1016/j.pbi.2008.03.006.

Moran, N. A. 2007. Symbiosis as an adaptive process and source of phenotypic complexity. *Proceedings of the National Academy of Sciences* 104(suppl 1): 8627–33. doi:10.1073/pnas.0611659104.

Morrison, C. R., and Windsor, D. M. 2018. The Life history of *Chelymorpha alternans* (Coleoptera: Chrysomelidae: Cassidinae) in Panamá. *Annals of the Entomological Society of America* 111(1): 31–41. doi:10.1093/aesa/sax075.

Mueller, U. G., Gerardo, N. M., Aanen, D. K., Six, D. L., and Schultz, T. R. 2005. The evolution of agriculture in insects. *Annual Review of Ecology, Evolution, and Systematics* 36: 563–95.

Mueller, U. G., Schultz, T. R., Currie, C. R., Adams, R. M., and Malloch, D. 2001. The origin of the attine ant-fungus mutualism. *The Quarterly Review of Biology* 76(2): 169–97.

Murakami, T., and Higashi, S. 1997. Social organization in two primitive attine ants, *Cyphomyrmex rimosus* and *Myrmicocrypta ednaella*, with reference to their fungus substrates and food sources. *Journal of Ethology* 15(1): 17–25. doi:10.1007/BF02767322.

Patel, D. D., Patel A. K., Parmar, N. R., Shah, T. M, Patel, J. B, Pandya, P. R., and Joshi, C. G. 2014. Microbial and carbohydrate active enzyme profile of buffalo rumen metagenome and their alteration in response to variation in the diet. *Gene* 545(1): 88–94. doi:10.1016/j.gene.2014.05.003.

Pilon, F. M., Visôtto, L. E., Guedes, R. N. C., and Oliveira, M. G. A. 2013. Proteolytic activity of gut bacteria Isolated from the Velvet Bean Caterpillar *Anticarsia gemmatalis. Journal of Comparative Physiology. B, Biochemical, Systemic, and Environmental Physiology* 183(6): 735–47. doi:10.1007/s00360-013-0744-5.

Pope, P. B., Mackenzie, A. K., Gregor, I., Smith, WQ., Sundset, M. A., McHardy, A. C., Morrison, M., and Eijsink, V. G. H. 2012. Metagenomics of the Svalbard Reindeer rumen microbiome reveals abundance of polysaccharide utilizationlLoci. *PLOS ONE* 7(6): e38571. doi:10.1371/journal.pone.0038571.

Quinlan, R. J., and Cherrett, J. M. 1978. Aspects of the symbiosis of the leaf-cutting ant *Acromyrmex Octospinosus* (Reich) and Its Food Fungus. *Ecological Entomology* 3(3): 221–30. doi:10.1111/j.1365-2311.1978.tb00922.x.

Rand, A. S., Dugan, B. A., Monteza, H., and Vianda D. 1990. The Diet of a Generalized Folivore: Iguana Iguana in Panama". *Journal of Herpetology* 24(2): 211–14. doi:10.2307/1564235.

Rio, R. V. M., Attardo, G. M., and Weiss, B. L. 2016. Grandeur Alliances: Symbiont Metabolic Integration and Obligate Arthropod Hematophagy". *Trends in Parasitology* 32(9): 739–49. doi:10.1016/j.pt.2016.05.002.

Rose, J. K. C. 2003. *The Plant Cell Wall.* CRC Press.

Salem, H., Bauer, E., Kirsch, R., Berasategui, A., Cripps, M., Weiss, B., Koga, R. et al. 2017. Drastic Genome reduction in an herbivore's pectinolytic symbiont". *Cell* 171(7): 1520–1531.e13. doi:10.1016/j.cell.2017.10.029.

Salem, H., Kirsch, R., Pauchet, Y., Berasategui, A., Fukumori, K., Cripps, M., Windsor, D. M., Fukatsu, T., and Gerardo, N. M. Symbiont digestive range reflects host plant breadth in herbivorous beetles. *Current Biology,* In press.

Schiøtt, M., Rogowska-Wrzesinska, A., Roepstorff, P. and Boomsma, J. J. 2010. Leaf-cutting ant fungi produce cell wall degrading pectinase complexes reminiscent of phytopathogenic fungi". *BMC Biology* 8(1): 156. doi:10.1186/1741-7007-8-156.

Seaman, F. C. 1984. The effects of tannic acid and other phenolics on the growth of the fungus cultivated by the leaf-cutting ant, *Myrmicocrypta buenzlii. Biochemical Systematics and Ecology* 12(2): 155–58. doi:10.1016/0305-1978(84)90028-0.

Shen, S. K. and Dowd, P. F. 1990. Insect Symbionts: A Promising Source of Detoxifying Enzymes. *ACS Symposium Series - American Chemical Society (USA).* http://agris.fao.org/agris-search/search.do?recordID=US9305809.

Shiffman, M. E., Soo, R. M., Dennis, P. G., Morrison, M., Tyson, G. W., and Hugenholtz, P. 2017. Gene and genome-centric analyses of koala and wombat fecal microbiomes point to metabolic specialization for eucalyptus digestion. *Peer Journal* 5(November): e4075. doi:10.7717/peerj.4075.

Stotz, H. U., Pittendrigh, B. R., Kroymann, J., Weniger, K., Fritsche, J., Bauke, A. and Mitchell-Olds, T. 2000. Induced plant defense responses against chewing insects. Ethylene signaling reduces resistance of Arabidopsis against Egyptian Cotton Worm but not Diamondback Moth. *Plant Physiology* 124(3): 1007–18. doi:10.1104/pp.124.3.1007.

Suen, G., Scott, J. J., Aylward, F. O., Adams, S. M., Tringe, S. G., Pinto-Tomás, A. A., Foster, C. E. et al. 2010. An insect herbivore microbiome with high plant biomass-degrading capacity. *PLOS Genetics* 6(9): e1001129. doi:10.1371/journal.pgen.1001129

Sugio, A., Kingdom, H. N., MacLean, A. M., Grieve, V. M., and Hogenhout, S. A. 2011. Phytoplasma protein effector SAP11 enhances insect vector reproduction by manipulating plant development and defense hormone biosynthesis. *Proceedings of the National Academy of Sciences* 108(48): E1254–63. doi:10.1073/pnas.1105664108.

Underwood, W. 2012. The plant cell wall: A dynamic barrier against pathogen invasion. *Frontiers in Plant Science* 3: 85. doi:10.3389/fpls.2012.00085.

Visôtto, L. E., Oliveira, M. G. A., Guedes, R. N. C., Ribon, A. O. B., and Good-God, P. I. V. 2009. Contribution of gut bacteria to digestion and development of the Velvetbean Caterpillar, *Anticarsia gemmatalis. Journal of Insect Physiology* 55(3): 185–91. doi:10.1016/j.jinsphys.2008.10.017.

van der Vlugt-Bergmans, C. J., and van der Werf, M. J. 2001. Genetic and biochemical characterization of a novel monoterpene epsilon-lactone hydrolase from *Rhodococcus erythropolis* DCL14. *Applied and Environmental Microbiology* 67(2): 733–41. doi:10.1128/AEM.67.2.733-741.2001.

Vogelman, T. C., Nishio, J. N., and Smith, W. K. 1996. Leaves and light capture: Light propagation and gradients of carbon fixation within leaves. *Trends in Plant Science* 1(2): 65–70. doi:10.1016/S1360-1385(96)80031-8.

Walton, J. D. 1994. Deconstructing the cell wall. *Plant Physiology* 104(4): 1113–18.

Weber, N. A. 1966. Fungus-growing ants. *Science* 153(3736): 587–604. doi:10.1126/science.153.3736.587.

Welte, C. U., de Graaf, R. M., van den Bosch, T. J. M., Op den Camp, H. J. M., van Dam, N. M., and Jetten, M. S. M. 2016. Plasmids from the gut microbiome of cabbage root fly larvae encode SaxA that catalyses the conversion of the plant toxin 2-phenylethyl isothiocyanate. *Environmental Microbiology* 18(5): 1379–90. doi:10.1111/1462-2920.12997

Windsor, D. M. 1987. Natural history of a subsocial tortoise beetle, *Acromis sparsa* Boheman (Chrysomelidae, Cassidinae) in Panama. *Psyche: A Journal of Entomology.* 94(1-2), 127–50.

Windsor, D. M., Riley, E. G., and Stockwell, H. P. 1992. An introduction to the biology and systematics of Panamanian tortoise beetles (Coleoptera: Chrysomelidae: Cassidinae). In *Insects of Panama and Mesoamerica. Selected Studies.* (Ed.) by Quintero, D., Aiello, A. Selected studies. Oxford University Press, New York. 372–91.

Wirth, R., Herz, H., Ryel, R. J., Beyschlag, W., and Hölldobler, B. 2003. The natural history of leaf-cutting ants. In *Herbivory of Leaf-Cutting Ants: A Case Study on Atta colombica in the Tropical Rainforest of Panama*, (Ed.) by R. Wirth, H. Herz, R. J. Ryel, W. Beyschlag and B. Hölldobler, 5–48. Ecological Studies. Berlin, Heidelberg: Springer Berlin Heidelberg. doi:10.1007/978-3-662-05259-4_2.

Wittstock, U., and Gershenzon, J. 2002. Constitutive plant toxins and their role in defense against herbivores and pathogens. *Current Opinion in Plant Biology* 5(4): 300–7.

Zelitch, I. 1975. Pathways of carbon fixation in green plants. *Annual Review of Biochemistry* 44: 123–45. doi:10.1146/annurev.bi.44.070175.001011.

Zhang, H., Guiguet, A., Dubreuil, G., Kisiala, A., Andreas, P., Emery, P. J. R., Huguet, E., Body, M., and Giron, D. 2017. Dynamics and origin of cytokinins involved in plant manipulation by a leaf-mining insect. *Insect Science* 24(6): 1065–78. doi:10.1111/1744-7917.12500.

Zhu-Salzman, K., Bi, J.-L., and Liu, T.-X. 2005. Molecular strategies of plant defense and Insect counter-defense. *Insect Science* 12(1): 3–15. doi:10.1111/j.1672-9609.2005.00002.x.

14 Right on cue
Microbiota promote plasticity of zebrafish digestive tract

Michelle S. Massaquoi and Karen J. Guillemin

Contents

14.1 Introduction

All animals on earth have evolved within a world teeming with microscopic life. The genesis of bacteria dates back 2.6 billion years, whereas primitive humans evolved only 315 thousand years ago from ancestors who had always coexisted with microbes. We refer to the community of bacteria, viruses, fungi, and archaea inhabiting a multicellular host as its microbiota. Only recently have biologists considered an organism as an ecosystem of many, rather than an isolated individual. Although microbes were initially vilified as pathogens, just a small fraction of the thousands of microbial species cause disease in animals. Microbial life is now being appreciated for its multitude of roles in host homeostasis. As our knowledge of the intricate interactions between host organisms and resident microbiota increases, it is fundamentally changing how we view many aspects of animal biology, including animal development.

Viewed from an evolutionary perspective, microbes have shaped animal history by influencing their fitness throughout their lifespans. As described by the Modern Synthesis, natural selection dictates how organisms that are better adapted to their environment will pass down their genes to the next generation. Resident microbes can shift that fitness landscape for mature organisms, for example by supplying degradative enzymatic capacities to allow hosts to access new sources of nutrition. Additionally, microbes can influence which specific genotypes survive and reproduce by shaping the developmental programs that determine how an organism forms from a single-cell embryo to a mature multicellular adult. Polyphenism is a biological phenomenon in which distinct phenotypes can arise from a single clonal genotype, demonstrating the plasticity of developmental biology. Predation, temperature, and nutrient availability are all direct environmental factors that induce polyphenism.

Resident microbes can also be added to this list of factors that influence a host's developmental trajectories. For example, the presence or absence of the bacterium *Wolbachia* significantly impacts ovary and oocyte development in the parasitic wasp, *Aosbara tabida Nees* (Dedeine et al. 2001).

Vertebrates harbor dense and complex microbiota, especially in their digestive tracts (Ley et al. 2008). The impact of the microbiota can be studied through the use of microbiologically sterile, or "germ-free" animals and with gnotobiology experiments, using biological systems in which all members of a community are known. In germ-free mice, not only does the absence of the microbiota impair maturation of the gut and associated mucosal immune system (Belkaid and Harrison 2017), but many distal organ systems are also impaired (Schroeder and Bäckhed 2016). For example, germ-free mice have stunted development of their intestinal villi capillary network (Stappenbeck et al. 2002) as well as incomplete bone formation (Sjögren et al. 2012). Although the majority of germ-free and gnotobiotic studies of vertebrates have been conducted in laboratory mice, comparisons across other vertebrate models are invaluable for understanding which host responses to microbiota are conserved across multiple host lineages.

In this chapter, we will discuss insights about the developmental impacts of the microbiota in the model vertebrate *Danio rerio*, the zebrafish. George Streisinger at the University of Oregon pioneered the use of zebrafish as a model system for developmental and genetic research (Grunwald and Eisen 2002). There are many advantages to working with zebrafish as a model vertebrate. In addition to genome conservation with mammals, zebrafish are genetically tractable, with many mutant and transgenic lines readily available. They also have high fecundity and are optically transparent during embryonic to larval stages, making this model organism ideal for studying developmental processes in real time. Additionally, because zebrafish develop *ex utero*, they can easily be derived germ-free by surface sterilization of the outer chorion for experiments aiming to understand host–microbe interactions (Melancon et al. 2017). Culture collections of zebrafish-associated bacterial isolates, with draft genome sequences, are available, enabling gnotobiotic experimentation (Stephens et al. 2016). Together, the transparent properties of larval zebrafish with the use of transgenic lines— that allow tracking of specific cell types and gnotobiological experiments—enable a high-resolution perspective of how the microbiota influence host development. Below, we discuss the ways in which the gut microbiota impact different aspects of zebrafish larval development, following the animals' first exposure to environmental microbes upon hatching out of their chorions. We use insights gleaned from gnotobiotic zebrafish studies to speculate how host–microbe interactions evolved to modulate developmental program plasticity to optimize the organisms' fitness for different environments.

14.2   Development under immune surveillance

The vertebrate intestine serves multiple roles, both as an organ for food digestion and nutrients absorption and as an immunological organ for harboring the body's most abundant microbial population. The intestinal epithelium can be thought of as the "inner skin" of the body because like the epidermis, it interacts with the outside environment, not only interfacing directly with microbial cells inhabiting the intestinal lumen, but also the rich source of bioactive molecules they secrete (Fischbach and

Segre 2016). With the multitudes of microbial life that populate any given vertebrate, a healthy immune system continuously monitors the intestinal lumen and senses threats posed by resident or invading microbes. Appropriately balanced responses to the microbiota are critical for symbiosis because on the one hand, lack of defense can lead to microbial growth and on the other hand, excessive inflammatory responses can be detrimental to both host and microbe cell populations. As the host develops, the resident microbes help train the immune system to achieve this appropriately balanced response (Belkaid and Harrison 2017).

The gnotobiotic zebrafish model has allowed a detailed characterization of the different immune responses elicited by individual members of the microbiota (Murdoch and Rawls 2019). For example, zebrafish mono-associated with different zebrafish-derived bacterial isolates will exhibit different levels of immune gene expression (Rawls et al. 2006) and accumulate different numbers of neutrophils, a type of white blood cell that lead the immune system's inflammatory response (Rolig et al. 2015). These types of data inspire the question of how different bacterial residents elicit different immunological responses in the host. Two possible and not mutually exclusive mechanisms are that host immune sensors are differentially stimulated by different bacteria, and that different bacteria produce immunomodulatory factors altering host immune responses.

During larval development, zebrafish rely on their innate immune system for microbial sensing prior to the maturation of their adaptive immune system in juvenile stages. The best characterized of their innate immune sensors are the Toll-like family receptors (TLRs), which are part of an ancient pattern recognition family of receptors (Jault, Pichon, and Chluba 2004; Deguine and Barton 2014). TLR activation is mediated by the sensing of generic microbial products termed microbial-associated molecular patterns (MAMPs), such as cell wall components and flagellin, which subsequently regulates the appropriate immune response (Deguine and Barton 2014). The specificity and downstream response of TLR signaling is partly dictated by the differential recruitment of intracellular adaptor molecules. Myeloid differentiation primary response 88 (MyD88) (Hall et al. 2009) is a common adaptor of TLRs and Interleukin-1 receptor that regulates the expression of pro- or antiinflammatory cytokines and mitogen-activated protein kinase (MAPK) signaling for cell survival or proliferation (Akira 2003; Larsson et al. 2012). The zebrafish genome has duplicated *tlr* genes with a single copy of *myd88,* that has conserved function in modulating innate immune responses (Jault et al. 2004; Meijer et al. 2004; Van Der Sar et al. 2006; Bates et al. 2007; Hall et al. 2009; Burns et al. 2017). Of note, Myd88-deficient zebrafish have a completely attenuated intestinal neutrophil influx, indistinguishable from germ-free animals (Bates et al. 2007; Burns et al. 2017). This indicates that much of the immunological responses to the intestinal microbiota are mediated through Myd88. One trait that varies dramatically across zebrafish bacterial isolates is their capacity for motility within the intestine (Schlomann et al. 2018; Wiles et al. 2018). Variation in bacterial *in vivo* production of motility machinery, such as flagellin subunits, could account for different capacities of different bacterial isolates to activate Myd88-dependent immune responses.

Resident bacteria also modulate host immune responses through specific secreted factors. For example, Rolig and colleagues demonstrated that a *Shewanella* isolate was an especially potent suppressor of neutrophil intestinal influx, a response that could be recapitulated with *Shewanella* secreted factors (Rolig et al. 2015). More recently, Rolig and colleagues showed that several *Aeromonas* strains secrete a protein, *Aeromonas* immune modulator A (AimA), that dampens neutrophil influx and proinflammatory cytokine expression and also confers a colonization advantage to the bacteria (Rolig et al. 2018). The crystal structure of AimA revealed that it has two distinct domains with related folds, both individually retaining the capacity to regulate neutrophils. To investigate whether AimA confers a colonization advantage to *Aeromonas* through attenuating inflammation, the authors measured the abundance of *Aeromonas* strains with and without AimA in *myd88* mutant zebrafish. They found that in these immunocompromised hosts with a limited immune response, *Aeromonas* no longer required AimA for maximal intestinal colonization, suggesting that in the face of a normal host immune response, the bacteria benefit from AimA's ability to dampen inflammation (Rolig et al. 2018). This study illustrates that Myd88-mediated host responses to the microbiota modulate features of the host environment that impact the fitness of resident bacteria. This finding is corroborated by the fact that isolates of *Aeromonas* experimentally evolved to colonize wild type zebrafish intestines are less fit when introduced into *myd88* deficient hosts (Robinson et al. 2018). Thus, the innate immune system acts as a conduit to intercept and respond to microbial cues and may underlie microbiota-mediated developmental plasticity.

14.3   Developmental plasticity at the luminal interface

In addition to the immune cells of the intestinal mucosa, the cells of the intestinal epithelium are situated to interact with the most abundant microbial population in the vertebrate body. The embryonic stages of zebrafish development happen within the sterile environment of the chorion, but once they hatch from their chorions into the larval stage (~2–3 days post fertilization (dpf)), they are inoculated with their first resident microbes from the surrounding environment. The load of microbes dramatically expands within the zebrafish intestine from 4–6 dpf (Bates et al. 2006; Jemielita et al. 2014; Stephens et al. 2016), which coincides with important developmental time points within the digestive system's intestine, pancreas, and liver. By 6 dpf the intestine exhibits clear compartmentalization into morphologically and transcriptionally distinct regions: the intestinal bulb (duodenum and jejunum-like), mid-intestine (ileum-like), and posterior intestine (colon-like) (Ng et al. 2005; Lickwar et al. 2017).

Raised under germ-free conditions, larval (6dpf) zebrafish resemble their conventionally reared counterparts at the level of gross morphology of their digestive organs, but the epithelium is characterized by a paucity of proliferation (Rawls, Samuel, and Gordon 2004) and of secretory cells, including mucus-secreting goblet cells and hormone secreting enteroendocrine cells (Bates et al. 2006) (Figure 14.1). These traits cannot be restored by exposure of germ-free animals to a generic activator of TLR signaling, the Gram-negative bacterial cell wall component lipopolysaccharide (LPS), but can be restored by mono-association with certain bacterial isolates (Bates et al. 2006; Cheesman et al. 2011).

A prominent and highly conserved molecular pathway that modulates cell proliferation of many tissues including the intestinal epithelium is Wnt signaling. Within the canonical pathway, Wnt binds to the cell receptor Frizzled and activates

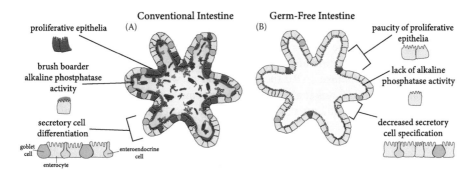

FIGURE 14.1 Intestinal microbiota induce development of host intestinal epithelia. In a cross-sectional view of a conventional intestine (A), the microbiota stimulate proliferation of the intestinal epithelia, expression of brush boarder intestinal alkaline phosphatase activity (AP, a marker of enterocyte maturation), and specification of secretory cell fates. In a germ-free intestine (B), epithelia exhibit less proliferation, decreased expression of markers of enterocyte maturity and a paucity of secretory cells.

the transcriptional regulator Beta-catenin by preventing its proteasome-mediated degradation and allowing it to translocate to the nucleus to turn on Wnt target genes. In the larval zebrafish intestine, both Wnt signaling and resident microbiota promote epithelial cell proliferation and accumulation of Beta-catenin (Cheesman et al. 2011). Mono-associated zebrafish with an *Aeromonas* strain, or exposure to this strain's secreted products, was sufficient to rescue intestinal epithelial proliferation. This study also discovered that Myd88 was required for intestinal epithelial cell proliferation, with *myd88*-deficient animals resembling the germ-free state. Together, the data indicate that host sensing of microbial products stimulates intestinal epithelial proliferation through canonical Wnt signaling, directly linking the innate immune sensing of the microbiota with a canonical animal developmental program (Figure 14.2).

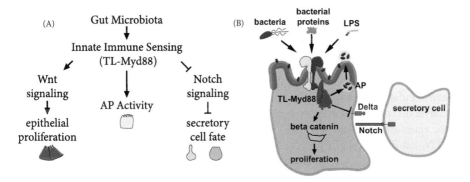

FIGURE 14.2 Innate immune sensing by Myd88 drives intestinal epithelial development in response to microbial cues. Epistasis diagram (A) showing pathways downstream of Myd88 impacting epithelial proliferation, alkaline phosphatase (AP) activity, and secretory cell fate. Microbial cues from bacteria or their secreted products differentially impact the Myd88-dependent downstream responses (B). Lipopolysaccharide (LPS) is only sufficient to stimulate AP activity, but does not rescue epithelia proliferation and secretory cell fates.

The continually renewing intestinal epithelial cells differentiate into distinct functional cell types to carry out diverse functions such as absorbing nutrients, secreting a protective mucus barrier, and relaying cell-to-cell communication. One marker of the mature absorptive enterocyte lineage is the enzymatic activity of alkaline phosphatase (AP), the expression of which requires Myd88 and is stimulated by LPS (Bates et al. 2007). This enzyme in turn dephosphorylates LPS, rendering it less proinflammatory and promoting host tolerance of the microbiota (Bates et al., 2007). Another indication of intestinal epithelial maturation is the expression of glycoconjugates in different patterns along the length of the digestive tract (Falk, Roth, and Gordon 1994; Wu et al. 2009). In the absence of microbiota, larval zebrafish show a striking decrease in AP activity (Figure 14.1) and levels of certain glucoconjugates, including N-acetylgalactosamine and galactosea1,3galactosyl moieties (Bates et al. 2006). As the intestinal epithelium matures, its distinct functional regions become more prominent. For example, in larval zebrafish an ileum-like region of the midgut becomes apparent, where cells containing large lysosomes engage in luminal protein absorption and degradation (Lickwar et al. 2017; Park et al. 2019). The microbiota are necessary for maturation of this region of the gut (Rawls et al. 2004; Bates et al. 2006). In addition to absorbing molecules from the lumen, intestinal cells also secrete molecules into the lumen and bloodstream. Secretory cells within the intestinal epithelial play many critical roles for intestinal and vertebrate homeostasis. Mucus-secreting goblet cells maintain a protective mucosal barrier between the luminal microbiota and epithelium. Enteroendocrine cells sample the luminal contents and communicate with the extraintestinal cells by hormone secretion. In the absence of the microbiota, the specification and differentiation of goblet and serotonin-secreting enteroendocrine cells is reduced, with older germ-free larvae having secretory cell numbers similar to younger conventionally reared animals (Bates et al. 2006).

A prominent, conserved developmental pathway known to regulate cell-fate decisions in many tissues, including the intestinal epithelium, is Notch-Delta signaling. Within intestinal secretory cell development, appropriate regulation of Notch-Delta signaling is essential for the decision between absorptive versus secretory cell types. Notch receptors are transmembrane proteins that form a juxtracrine signaling response when bound by the membrane protein Delta on a neighboring cell. Notch-Delta binding on the outside of the two neighboring cells induces cleavage of the Notch receptor, releasing the intracellular domain to translocate to the nucleus and regulate gene expression. Inhibiting Notch signaling within zebrafish causes a conversion of the intestinal epithelium to a secretory cell fate (Crosnier et al. 2005). In the zebrafish intestinal epithelium, absence of the microbiota phenocopies activation of Notch-Delta signaling with an expansion of absorptive cells relative to secretory cells (Bates et al. 2006; Troll et al. 2018). Inhibiting Notch-Delta signaling specifically within the intestinal epithelium significantly increases the number of secretory cells in both conventional and germ free reared zebrafish, indicating that the microbiota acts upstream of Notch-Delta signaling to specify intestinal secretory cell fate (Troll et al. 2018). Echoing the microbiota's impact on host intestinal epithelial proliferation, Myd88-dependent host sensing of the microbiota is necessary for specification of appropriate proportions of secretory cells (Troll et al. 2018). Double mutant animals lacking Myd88 and Notch signaling have high numbers of secretory cells, placing Myd88 upstream of Notch signaling (Troll et al. 2018). In this model, the microbiota are perceived by the innate immune system, which inhibits Notch-Delta signaling to promote secretory cell differentiation (Figure 14.2).

Developmental biologists typically study processes of organ growth, differentiation, and functionalization under highly controlled conditions, allowing them to focus on the intrinsic programs of gene regulation through canonical pathways, such as the Wnt and Notch-Delta signaling that balance intestinal epithelial cell proliferation and differentiation (Sancho, Cremona, and Behrens 2015). Studies in the gnotobiotic larval zebrafish demonstrate that the presence and composition of the microbiota modulate these canonical signaling pathways and modify intestinal developmental programs. In the case of both of Wnt and Notch signaling, their modulation by microbiota is mediated through Myd88-dependent processes. Since LPS is insufficient to activate Wnt and Notch signaling-mediated changes in the germ-free intestinal epithelium, although it is sufficient to upregulate intestinal AP expression, this suggest that the microbiota produce multiple factors that nonredundantly stimulate innate immune sensing pathways and impact diverse aspects of host development.

14.4   Beyond the lumen: A secreted bacterial protein impacts pancreas development

Branching from the direct host–microbe interface of the intestinal epithelium is the hepato-pancreatic ductal system. This ductal system does not harbor appreciable numbers of microbes, but it supplies the intestinal lumen with digestive enzymes from the pancreas, and bile from the liver and gall bladder for digestion. The pancreas is a dual gland composed of exocrine tissue that secretes enzymes and endocrine tissue that secretes hormones and regulates blood sugar levels. The beta cell population is an essential host cell type as it is the sole provider of insulin for the body. After sensing blood glucose, beta cells release insulin into the blood stream allowing all host cell types to uptake blood-glucose for cellular respiration. As with the intestine, important developmental time points of the pancreas coincides with the assembly of the microbiota; beta cell mass dramatically expands during 4–6 dpf simultaneously with the increase of intestinal bacterial abundance (Kimmel and Meyer 2010). In the absence the microbiota, the insulin-producing beta cells of the zebrafish pancreas fail to undergo their normal developmental expansion (Hill et al. 2016).

By mono-associating with zebrafish-associated bacterial isolates, Hill et al. (2016) found that only a few members of the zebrafish microbiota (several *Aeromonas* and *Shewanella spp.*) were sufficient to rescue the normal expansion of beta cell mass in germ-free zebrafish, suggesting involvement of a specific cue rather than a generic product of all microbiota members. From the known genome sequences of zebrafish bacterial isolates, Hill et al. (2016) compared the genomes of the mono-associated bacteria that rescued beta cell development to those that did not and discovered an uncharacterized bacterial secreted protein, Beta Cell Expansion Factor A (BefA), that was necessary within *Aeromonas* and sufficient alone to rescue beta cell expansion within germ-free derived zebrafish. Further, BefA facilitates expansion of beta cells by inducing their proliferation, and homologs of BefA produced by members of the human gut microbiota have conserved function in facilitating zebrafish beta cell development. Luminal proteins have been shown to traffic to distant organs (Park et al. 2019), suggesting a possible mechanism by which BefA could influence the developmental of an extraintestinal tissue. The long-term consequences on host metabolism and fitness of this reduced beta cell expansion in larval life have not yet been characterized, but mice with reduced perinatal beta cell proliferation have reduced beta cell mass and impaired glucose homeostasis as adults (Berger et al. 2015). More broadly, BefA provides an example of how the presence or absence of specific microbial members during developmental stages can impact organ development, further supporting the idea of microbiota composition modulating developmental plasticity.

14.5   Conclusions

The gnotobiotic zebrafish model reveals how the presence of microbiota modulate developmental programs of the digestive tract and how individual microbial isolates and community combinations can elicit unique host responses at the tissue and cellular levels (Rolig et al. 2015; Burns and Guillemin 2017; Burns et al. 2017; Rolig et al. 2017). We postulate that host developmental programs use microbial cues to interpret their environments and adapt tissue development accordingly (Massaquoi and Guillemin 2018). Viewed through this lens, microbiota-induced changes in the intestinal epithelium can be understood as accommodations for coexisting with an abundant microbial community. Upregulation of the LPS-detoxifying enzyme AP prevents excessive intestinal inflammation in response to resident bacteria. Increasing epithelial cell turnover is a protective adaptation for removing damaged cells, while also providing resident microbes with increased nutrients on shed cellular material. Increasing the proportion of secretory cells would also promote coexistence with microbes. Through the secretion of over thirty hormones, enteroendocrine cells dictate many aspects of the luminal environment from glucose metabolism, gut motility, to digestion and absorption (Sikander, Rana, and Prasad 2009; Vincent, Sharp, and Raybould 2011; Tolhurst, Reimann, and Gribble 2012; Pais, Gribble, and Reimann 2016). The mucus secreted from goblet cells not only maintains a protective barrier between luminal contents and epithelia, it also provides nutrients and habitats for the microbiota (Kashyap et al. 2013; Desai et al. 2016; Johansson and Hansson 2016). Host sensing of specific microbiota membership may also be a mechanism for interpreting host ecology and specifically nutritional landscapes, which will foster different environmental microbes that serve as source pools for host colonization. Modulation of beta cell mass in response to resident microbiota may represent an adaptation in anticipation of metabolic demands based on nutrient scarcity or abundance.

An important tenant in evolution is that natural selection acts on the genetic variation within a species population. We would argue that resident microbiota, which are highly variable between individuals, add another layer of phenotypic variation, beyond that encoded in genomes, that has shaped animal evolution. In particular, if certain developmental programs are especially sensitized to microbial cues, then the microbiota of particular developmental stages may dictate functional adaptations that impact host fitness in given environments. We can think of this as a form of microbiota-induced polyphenism. Results from the zebrafish model provide molecular examples that could underly such polyphenism. Innate immune sensing of resident microbial products modulates canonical host developmental programs such as Wnt and Notch signaling to optimize host tissues for living with specific microbial consortia. A thorough understanding of how the microbiota impact host development will require discovering which microbial products are sensed by host cells, the identity and sensing mechanisms of host cells that perceive these microbial products, and the cellular responses and downstream consequences of microbial product perception.

References

Akira, S. 2003. "Toll-like receptor signaling." *Journal of Biological Chemistry* 278.40: 38105–8. doi:10.1074/jbc.R300028200.

Bates, J. M., E. Mittge, J. Kuhlman, K. N. Baden, S. E. Cheesman, and K. Guillemin. 2006. "Distinct signals from the microbiota promote different aspects of zebrafish gut differentiation." *Developmental Biology* 297(2): 374–86.

Bates, J. M, J. Akerlund, E. Mittge, and K. Guillemin. 2007. "Intestinal Alkaline phosphatase detoxifies lipopolysaccharide and prevents inflammation in Zebrafish in response to the Gut Microbiota." *Cell Host & Microbe* 2(6): 371–82. doi:10.1016/j.chom.2007.10.010.

Belkaid, Y., and O. J. Harrison. 2017. "Homeostatic immunity and the microbiota." *Immunity.* 46(4): 562–76.

Berger, M., D. W. Scheel, H. Macias et al. 2015. "Gαi/o-Coupled receptor signaling restricts pancreatic β-Cell expansion." *Proceedings of the National Academy of Sciences of the United States of America* 112(9): 2888–93. doi:10.1073/pnas.1319378112.

Burns, A. R., E. Miller, M. Agarwal et al. 2017. "Interhost dispersal alters microbiome assembly and can overwhelm host innate immunity in an experimental zebrafish model." *Proceedings of the National Academy of Sciences* 114(42): 11181–6.

Burns, A. R., and K. Guillemin. 2017. "The Scales of the Zebrafish: Host–Microbiota interactions from proteins to populations." *Current Opinion in Microbiology* 38(August): 137–41. doi:10.1016/j.mib.2017.05.011.

Cheesman, S. E., J. T. Neal, E. Mittge, B. M. Seredick, and K. Guillemin. 2011. "Epithelial cell proliferation in the developing zebrafish intestine is regulated by the Wnt pathway and microbial signaling via Myd88." *Proceedings of the National Academy of Sciences* 108(Supplement 1): 4570–7.

Crosnier, C., N. Vargesson, S. Gschmeissner et al. 2005. "Delta-notch signalling controls commitment to a secretory fate in the Zebrafish intestine." *Development* 132(5): 1093–104. doi:10.1242/dev.01644.

Dedeine, F., F. Vavre, F. Dé et al. 2001. "Removing symbiotic Wolbachia bacteria specifically inhibits oogenesis in a parasitic wasp." *Proceedings of the National Academy of Sciences* 98(11): 6247–52.

Deguine, J., and G. M. Barton. 2014. "MyD88: A central player in innate immune signaling." doi:10.12703/P6–97.

Desai, M. S., A. M. Seekatz, N. M. Koropatkin et al. 2016. "A dietary fiber-deprived gut Microbiota degrades the colonic mucus barrier and enhances pathogen susceptibility." *Cell* 167(5): 1339–1353.e21. doi:10.1016/j.cell.2016.10.043.

Falk, P., K. A. Roth, and J. I. Gordon. 1994. "Lectins are sensitive tools for defining the differentiation programs of mouse gut epithelial cell lineages." *American Journal of Physiology - Gastrointestinal and Liver Physiology* 266(6): 29–6. doi:10.1152/ajpgi.1994.266.6.g987.

Fischbach, M. A., and J. A. Segre. 2016. "Signaling in host-associated microbial communities." *Cell* 164: 1288–300. doi:10.1016/j.cell.2016.02.037.

Grunwald, D. J., and J. S. Eisen. 2002. "Headwaters of the Zebrafish—Emergence of a new model vertebrate." *Nature Reviews Genetics* 3(9): 717–24.

Hall, C., M. V. Flores, A. Chien, A. Davidson, K. Crosier, and P. Crosier. 2009. "Transgenic Zebrafish reporter lines reveal conserved toll-like receptor signaling potential in embryonic myeloid leukocytes and adult immune cell lineages." *Journal of Leukocyte Biology* 85(5): 751–65. doi:10.1189/jlb.0708405.

Hill, J. H., E. A. Franzosa, C. Huttenhower, and K. Guillemin. 2016. "A conserved bacterial protein induces pancreatic beta cell expansion during Zebrafish development." *ELife* 5: 18. doi:10.7554/eLife.20145.

Hooper, L. V., D. R. Littman, and A. J. Macpherson. 2012. "Interactions between the Microbiota and the immune system." *Science*. doi:10.1126/science.1223490.

Jault, C., L. Pichon, and J. Chluba. 2004. "Toll-like receptor gene family and TIR-domain adapters in Danio Rerio." *Molecular Immunology* 40(11): 759–71. doi:10.1016/j.molimm.2003.10.001.

Jemielita, M., M. J. Taormina, A. R. Burns et al. 2014. "Spatial and temporal features of the growth of a bacterial species colonizing the zebrafish gut." *MBio.* 5(6): e01751–14.

Johansson, M. E. V., and G. C. Hansson. 2016. "Immunological aspects of intestinal mucus and mucins." *Nature Reviews Immunology* 16(10): 639–49. doi:10.1038/nri.2016.88.

Kashyap, P. C., A. Marcobal, L. K. Ursell et al. 2013. "Genetically dictated change in host mucus carbohydrate landscape exerts a diet-dependent effect on the gut Microbiota." *Proceedings of the National Academy of Sciences of the United States of America* 110(42): 17059–64. doi:10.1073/pnas.1306070110.

Kimmel, R. A., and D. Meyer. 2010. Molecular regulation of pancreas development in Zebrafish. *Methods in Cell Biology* 100. doi:10.1016/B978-0-12-384892-5.00010-4.

Larsson, E., V. Tremaroli, Y. S. Lee et al. 2012. "Analysis of gut microbial regulation of host gene expression along the length of the gut and regulation of gut microbial ecology through MyD88." *Gut* 61(8): 1124–31. doi:10.1136/GUTJNL-2011-301104.

Ley, R. E., C. A. Lozupone, M. Hamady, R. Knight, and J. I. Gordon. 2008. "Worlds within worlds: Evolution of the vertebrate Gut Microbiota." *Nature Reviews Microbiology* 6(10): 776–88. doi:10.1038/nrmicro1978.

Lickwar, C. R., J. Gray Camp, M. Weiser et al. 2017. "Genomic dissection of conserved transcriptional regulation in intestinal epithelial cells." *PLOS Biology* 15(8). doi:https://doi.org/10.1371/journal.pbio.2002054 August.

Massaquoi, M. S., and K. Guillemin. 2018. "Evolving in a microbial soup: You are what they eat." *Developmental Cell* 47(6): 682–83. doi:10.1016/J.DEVCEL.2018.11.045.

Meijer, A. H., S. F. Gabby Krens, I. A. Medina Rodriguez et al. 2004. "Expression analysis of the toll-like receptor and TIR domain adaptor families of Zebrafish." *Molecular Immunology* 40(11): 773–83. doi:10.1016/j.molimm.2003.10.003.

Melancon, E., S. Gomez De La Torre Canny, S. Sichel et al. 2017. "Best practices for germ-free derivation and gnotobiotic zebrafish husbandry." *Methods in Cell Biology* 138: 61–100.

Murdoch, C. C., and J. F. Rawls. 2019. "Commensal Microbiota regulate vertebrate innate immunity-insights from the Zebrafish." *Frontiers in Immunology* 10: 2100. doi:10.3389/fimmu.2019.02100.

Ng, A. N. Y., T. A. de Jong-Curtain, D. J. Mawdsley et al. 2005. "Formation of the digestive system in Zebrafish: III. intestinal epithelium morphogenesis." *Developmental Biology* 286(1): 114–35. doi:10.1016/j.ydbio.2005.07.013.

Pais, R., F. M. Gribble, and F. Reimann. 2016. "Stimulation of incretin secreting cells." *Therapeutic Advances in Endocrinology and Metabolism.* 7(1): 24–42.

Park, J., D. S. Levic, K. D. Sumigray et al. 2019. "Lysosome-rich enterocytes mediate protein absorption in the vertebrate gut." *Developmental Cell. Elsevier Inc.* 1–14. doi:10.1016/j.devcel.2019.08.001.

Rawls, J. F., M. A. Mahowald, R. E. Ley, and J. I. Gordon. 2006. "Reciprocal Gut Microbiota transplants from Zebrafish and mice to germ-free recipients reveal host habitat selection." *Cell* 127(2): 423–33. doi:10.1016/j.cell.2006.08.043.

Rawls, J. F., B. S. Samuel, and J. I. Gordon. 2004. "Gnotobiotic Zebrafish Reveal Evolutionarily Conserved Responses to the Gut Microbiota." www.pnas.orgcgidoi10.1073pnas.0400706101.

Robinson, C. D., H. S. Klein, K. D. Murphy, R. Parthasarathy, K. Guillemin, and B. J. M. Bohannan. 2018. "Experimental bacterial adaptation to the Zebrafish gut reveals a primary role for immigration." *PLOS Biology* 16(12). Public Library of Science: e2006893. doi:10.1371/journal.pbio.2006893.

Rolig, A. S., E. K. Mittge, J. Ganz et al. 2017. "The enteric nervous system promotes intestinal health by constraining Microbiota composition." *PLOS Biology* 15(2): e2000689. doi:10.1371/journal.pbio.2000689.

Rolig, A. S., R. Parthasarathy, A. R. Burns, B. J. M. Bohannan, and K. Guillemin. 2015. "Individual members of the Microbiota disproportionately modulate host innate immune responses." *Cell Host and Microbe.* doi:10.1016/j.chom.2015.10.009.

Rolig, A. S., E. G. Sweeney, L. E. Kaye et al. 2018. "A bacterial immunomodulatory protein with Lipocalin-like domains facilitates host–bacteria mutualism in Larval Zebrafish." *ELife* 7(November). doi:10.7554/eLife.37172.

Sancho, R., C. A. Cremona, and A. Behrens. 2015. "Stem cell and progenitor fate in the mammalian intestine: Notch and lateral inhibition in homeostasis and disease." *EMBO Reports* 16(5): 571–81. doi:10.15252/embr.201540188.

Schlomann, B. H., T. J. Wiles, E. S. Wall, K. Guillemin, and R. Parthasarathy. 2018. "Bacterial cohesion predicts spatial distribution in the larval zebrafish intestine." *Biophysical Journal* 115(11): 2271–7.

Schroeder, B. O., and F. Bäckhed. 2016. "Signals from the Gut Microbiota to distant organs in physiology and disease." *Nature Medicine. Nature Publishing Group.* doi:10.1038/nm.4185.

Sikander, A., S. V. Rana, and K. K. Prasad. 2009. "Role of serotonin in gastrointestinal motility and irritable bowel syndrome." *Clinica Chimica Acta* 403(1–2): 47–55. doi:10.1016/j.cca.2009.01.028.

Sjögren, K., C. Engdahl, P. Henning et al. 2012. "The gut microbiota regulates bone mass in mice." *Journal of Bone and Mineral Research* 27(6): 1357–67.

Stappenbeck, T. S., L. V. Hooper, and J. I. Gordon. 2002. "Developmental regulation of intestinal angiogenesis by indigenous microbes via Paneth cells." *Proceedings of the National Academy of Sciences* 99(24): 15451–5.

Stephens, W. Z., A. R. Burns, K. Stagaman et al. 2016. "The composition of the Zebrafish intestinal microbial community varies across development." *ISME Journal* 10(3): 644–54. doi:10.1038/ismej.2015.140.

Stephens, W. Z., T. J. Wiles, E. S. Martinez et al. 2015. "Identification of population bottlenecks and colonization factors during assembly of bacterial communities within the Zebrafish intestine." *MBio* 6(6). doi:10.1128/mBio.01163-15.

Tolhurst, G., F. Reimann, and F. M. Gribble. 2012. "Intestinal sensing of nutrients." *Handbook of Experimental Pharmacology* 209: 309–35. doi:10.1007/978-3-642-24716-3_14.

Troll, J. V., M. Kristina Hamilton, M. L. Abel et al. 2018. "Microbiota promote secretory cell determination in the intestinal epithelium by modulating host notch signaling." *Development* 145(4): dev155317.

van der Sar, A. M., O. W. Stockhammer, C. van der Laan, H. P. Spaink, W. Bitter, and A. H. Meijer. 2006. "MyD88 innate immune function in a Zebrafish Embryo infection model." *Infection and Immunity* 74(4): 2436–41. doi:10.1128/IAI.74.4.2436-2441.2006.

Vincent, K. M., J. W. Sharp, and H. E. Raybould. 2011. "Intestinal glucose-induced calcium-calmodulin kinase signaling in the gut-brain axis in awake rats." *Neurogastroenterology and Motility* 23(7). doi:10.1111/j.1365-2982.2011.01673.x.

Wiles, T. J., E. S. Wall, B. H. Schlomann, E. A. Hay, R. Parthasarathy, and K. Guillemin. 2018. "Modernized tools for streamlined genetic manipulation and comparative study of wild and diverse Proteobacterial lineages." *MBio* 9(5). doi:10.1128/mBio.01877-18.

Wu, A. M., E. Lisowska, M. Duk, and Z. Yang. 2009. "Lectins as tools in glycoconjugate research." *Glycoconjugate Journal* 26(8): 899–913. doi:10.1007/s10719-008-9119-7.

15 Uncovering the history of intestinal host–microbiome interactions through vertebrate comparative genomics

Colin R. Lickwar and John F. Rawls

Contents

15.1 Introduction

In all forms of life on Earth, the genome encodes the necessary information to build and sustain that life form across generations, including its sustained interactions with other life forms (symbioses). From a scientific perspective, the genome also provides a relatively reliable record of the processes of adaptation and selection that organism has undergone over the course of its natural history. In other words, each organism's genome serves not only an *instruction manual* for that lifeform, but also as a *historical record* of that organismal lineage including its symbioses.

Natural selection acts at all levels of biological complexity—from gene content and genomic functions; to protein, organellar, cellular, tissue, and organismal functions; to symbiotic, community, and ecosystem functions (Ley et al. 2006). Natural selection at each of these levels is intrinsically interconnected with all other levels and applies to microorganisms and macroorganisms alike. Because of these interconnections, selective pressures acting at any of these levels can be expected to impact upon genomic content, structure, and function over evolutionary timescales. The tools of genome science therefore hold tremendous potential to advance our understanding of the different levels at which natural selection has shaped the natural history of organismal lineages, including their symbioses.

Indeed, advances in genome science have already facilitated dramatic advancements in our understanding of the diversity of microbial life on Earth, as well as the symbioses in which microbes engage with each other or animal and plant hosts. By coupling genomic approaches with experimental microbial manipulations (e.g., gnotobiotics, antibiotics, probiotics, prebiotics), powerful new insights have also been gained into the phenotypic consequences of host–microbe symbioses. In many cases, host–microbe symbioses have been found to contribute significantly to emergent traits that could alter fitness and thereby natural selection of the organisms involved (McFall-Ngai et al. 2013). This perspective has led to the concept that all organisms engaged in a given symbiotic relationship can be treated as a single discrete level of biological complexity, called a "holobiont" (Mindell 1992; Kutschera 2018). This concept has been further extended to the collective genomes of all organisms involved in a symbiosis, or a "hologenome" (Zilber-Rosenberg and Rosenberg 2008; Bordenstein and Theis 2015). On one hand, these holobiont/hologenome terms have been used respectively to simply describe the aggregate organisms and their genomes that are engaged in a given symbiotic relationship. On the other hand, and more contentiously, some scientists have augmented the operational definitions of the holobiont/hologenome terms to suggest that these levels of biological complexity act as distinct units of natural selection (Roughgarden et al. 2018). The debates that have ensued are important for the field, and seem to have revolved around three issues: (1) that the term "unit of selection" can be defined in different ways; (2) that host–microbe symbioses display a wide range of duration and fidelity of transmission; and (3) that natural selection acts at all levels of biological complexity, with the holobiont/hologenome representing just one of those potential levels (Moran and Sloan 2015; Douglas and Werren 2016; Roughgarden et al. 2018). This chapter will explore what genome science can contribute to this ongoing discussion and our larger understanding of host–microbe symbiosis.

Genome science offers diverse opportunities to explore the mechanisms by which organisms interact with symbiotic partners. Symbiotic relationships may be mediated by essentially any form of communication between symbiotic partners, including production of a signal by one partner, and the ability of the other partner to perceive, interpret, and respond to the signal. These symbiotic signals can take many different forms, ranging from biosynthetic, nutritional, chemical, and physical, with an equally diverse array of signal response mechanisms. If an organism has engaged in the production of, or response to, a symbiotic signal over the course of their natural history, then there should be evidence of that symbiotic signaling stored within their genome. Genomic evidence of a symbiotic signal could consist of specific gene structure or content, or aspects of genome organization or function that are specifically involved in production or reception of symbiotic signals. Genomic analysis can be used to better understand how an identified symbiotic signal is produced or perceived, but it can also be used as a discovery platform to identify new symbiosis signaling mechanisms.

15.2   A history of symbiotic interactions captured within microbial and host genomes

Ultimately, any enduring symbiotic relationship should manifest in attributable changes to the primary genomic DNA sequences of at least one organism within that symbiosis, including alterations, gain, loss, or transfer of DNA sequence. Attributable changes may also include differences in the abundance of a given microbe and its genome within a population. Important differences in the organization and evolution of multicellular eukaryotic genomes and primarily unicellular prokaryotic genomes require different strategies to identify evidence of symbiotic relationships (Koonin and Wolf 2010). This is due to significant differences in the lifecycles and properties of DNA sequence evolution in multicellular eukaryotes and prokaryotes. Whereas animals are colonized inexorably by microbiomes in each generation, the microbial lineages found in association with a given animal host may have other host-associated or environmental niches. Microbiome composition in some animals is indicative of a long-standing symbiotic relationship with a "core microbiome" consisting partly of the same microbial lineages across multiple generations. However, substantial microbiome variation can also occur between individual animal hosts and a large proportion of microbes are not necessarily shared across individuals in a particular species (Qin et al. 2010; Roeselers et al. 2011; Human Microbiome Project 2012; Hacquard et al. 2015; Adair and Douglas 2017). Furthermore, microbiome compositions can change as a function of age, developmental stage, environment, diet, genotype, and disease (Yatsunenko et al. 2012; Hacquard et al. 2015). Particularly complicating is that different host species can share or effectively inherit microbial species with no clear temporal boundaries, at times interacting directly through predation, shared environments, or coprophagia (Rosenberg and Zilber-Rosenberg 2016; Moeller et al. 2018). These behaviors substantially complicate the identification of the origin of selective pressures, and where exactly to look for evidence of symbiotic signals in microbial genomes. In this case, different host-associated microbes can contribute similar or identical functions, distributing selective pressures across multiple microorganisms (Louca et al. 2018), or transmitting function through horizontal gene transfer (HGT) (Smillie et al. 2011) or through gain/loss of plasmid or chromosomal DNA. This property of the microbiome presumably allows for relatively transient microbiome members or gene products to contribute within the symbiosis, but not be easily quantified (Booth et al. 2016). Together, these aspects of microbial ecology and DNA sequence evolution can still make it difficult to understand heredity of microbial lineages over evolutionary timescales and also to attribute symbiotic signals in microbial genomes to a particular unit of selection such as a hologenome, so much so that the very premise of a hologenome has been questioned (Moran and Sloan 2015; Douglas and Werren 2016).

Here, we will review how to detect evidence of symbiotic signals within the host genome at three levels of biological complexity—coding genome content, transcriptional responses, and cis-regulatory mechanisms. Though these concepts apply to symbioses broadly, we will focus our remarks here on the most salient form of symbiosis that occurs across bilaterian animals—that between animal hosts and the microbial communities that reside in their digestive tracts. We posit that focusing on host genomics in tissues like the intestine that are in direct contact with microbial communities will enrich for evidence of unique symbiotic signals and opportunities we can use to understand broader genomic bases of host–microbe symbioses, and the levels at which natural selection acts on upon symbiotic organisms. We will also highlight the utility of comparative genomic approaches to discern not just the symbiotic signals that exist in extant animals, but also those that have been conserved during animal evolution.

15.3 Capturing symbiotic signals within coding regions of the host genome

Substantial efforts have been directed towards interpreting conserved and evolving coding regions of transcripts in eukaryotic and prokaryotic genomes (Facco et al. 2019). The level of DNA conservation at coding regions is typically high relative to other genomic regions due to the presence of large, often syntenic protein domains with strings of amino acids specified by codons (Zheng et al. 2011). Compared to prokaryotes, rates of horizontal gene transfer in animals is very low, but have contributed to important innovations such as the evolution of placental mammals (Dupressoir et al. 2012; Boto 2014). Identification of coding regions with known functions, including facilitating symbiotic interactions, and interpreting the rate, location, or functional nature of changes in those domains across species is a critical component of understanding symbiotic relationships and phylogeny (Rosenberg and Zilber-Rosenberg 2016; Adrian et al. 2019). However, because coding regions in multicellular eukaryotic hosts are typically used by many cell types in number of contexts difficult to quantify, it is often challenging to attribute a particular change in a coding region as indicative of a host–microbe interaction, or if it is relevant in a particular tissue. However, important context can be gained from host proteins that directly interact with microbial proteins, components, or signals. For example, Toll-like receptors (TLRs), a well-described class of proteins that can directly recognize microbial products, display both substantial conservation in detecting microbial-associated molecular patterns (MAMPs) as well as expansion and evolution of additional family members across species (Roach et al. 2005; Li et al. 2017). TLR family members and components of their downstream signaling pathways are found in the genomes of basal bilaterian animals where they mediate recognition of microbial products as well as developmental processes (Tassia et al. 2017; Brennan and Gilmore 2018). Variation at TLR gene loci within a single species can also contribute to risk of infectious and inflammatory diseases (Mukherjee et al. 2019). However, TLR proteins can be expressed by different cells in different tissues at different times to serve different functions in recognizing pathogenic as well as commensal microbes (Abreu 2010). Similar challenges apply to antimicrobial proteins (AMPs), another salient class of proteins that appear to have evolved to assist animals in their encounters with microbes (Tennessen 2005).

The nature of host–microbe interactions also inherently encompasses greater conceptual complexity than simple direct protein domain functions or interactions. For example, coding regions do not inherently provide information on the amount, dynamics, or specificity of transcription of protein coding genes. Furthermore, in eukaryotes, transcript isoform usage or alternative splicing dramatically increases the number of unique transcriptional units that are utilized, but this information is not easily discerned strictly from coding region sequence. As a result, there are strengths but also significant limitations in detecting the origin or nature of the symbiotic signals that shape coding region variation in host genomes. Therefore, the following sections will highlight how additional layers of genomic information outside of coding region sequence encode a history of symbiotic interactions and associated selective pressures.

15.4 Uncovering specific symbiotic signals in host transcriptional programs

The ability of animal cells to engage in microbial symbioses is based in part on their ability to mount appropriate tissue-specific changes in gene expression. Although each cell in the animal typically harbors the same genome, each cell type may have a particular transcriptional response to a particular microbial signal or signals. Changes in gene expression, including those in response to microbes, are primarily facilitated by coordinated alterations in its chromatin organization and the activity of specific transcription factors (TFs), which we will discuss separately below. Identification of gene transcripts differentially expressed in response to microbes is particularly informative because it implicates those gene products as being potentially functionally involved in the response, and also because it indicates that the transcriptional regulatory mechanisms converging on those genes are under microbial influence (Romero et al. 2012).

The gastrointestinal tract is a common feature of all bilaterial animals that serves essential roles in dietary nutrient absorption and interaction with the complex microbial communities that assemble there. The epithelial cells that line the intestine (intestinal epithelial cells or IECs) directly interface with the intestinal microbiota and provide a physical barrier against potential microbial pathogens as well as other dietary and xenobiotic substances. IECs are highly dynamic, turning over their population every 4–5 days in mammals (Allaire et al. 2018). IECs are replenished by intestinal stem cells that differentiate into several distinct functional IEC subtypes which can vary based on developmental stage, environment, and also the particular region along the length of the intestine (Gerbe et al. 2016; Haber et al. 2017; Allaire et al. 2018). IECs are not simply passive but are instead remarkably perceptive to relevant changes in the environment and initiate changes locally and systemically, many of which influence their microbiota (Allaire et al. 2018). Much of this cellular identity and sensitivity manifest through transcriptional differences that are facilitated generally by the host genome interpreting diverse cues in a tissue- and cell-specific manner.

Increased deviation in transcript levels within orthologous tissues/organs has been shown to be positively correlated with evolutionary distance between species (Chan et al. 2009). Furthermore, the rate of evolution of transcriptional differences varies depending on the tissue/organ and evolutionary lineage of the species (Brawand et al. 2011). However, studies measuring tissue-specific transcriptional levels have shown them to be more similar in the same tissue across species than different tissues/organs within a species (Lin et al. 2014; Gilad and Mizrahi-Man 2015; Lickwar et al.

2017), suggesting maintenance of core functions during evolution. In the intestine, there is evidence of substantial conservation of gene expression programs across vertebrate lineages. Analysis of orthologous genes in zebrafish, stickleback, mouse, and human revealed almost 500 gene orthologs that showed similarly high expression in the intestinal epithelium, including evidence for regional gene expression and IEC subtype expression (Lickwar et al. 2017). Further, IECs from these four vertebrate species expressed a shared set of orthologous TFs such as HNF4A, HNF4G, FXR, GATA5, OSR2, and ELF3, which may support the larger pattern of conserved gene expression (Lickwar et al. 2017). Evidence that common gene expression programs help establish similar intestinal regionality and functions across vertebrate lineages suggest that conserved responses to microbiota may also be identifiable.

The most common approach for defining the impact of microbiota on host gene transcription has been to compare tissue-specific gene expression in host animals reared in the absence of microbiota (germ-free, GF) to those colonized with individual commensal microbes (Hooper et al. 2001; Geva-Zatorsky et al. 2017) or complex microbial communities (Rawls et al. 2006; Larsson et al. 2012; Pott et al. 2012; El Aidy et al. 2013; Camp et al. 2014; Thaiss et al. 2014). Empowered by microarray and RNA-seq technologies, this approach has been applied to diverse gnotobiotic animal models including mice (Hooper et al. 2001; Rawls et al. 2006; Chung et al. 2012; Larsson et al. 2012; Pott et al. 2012; El Aidy et al. 2013; Camp et al. 2014; Thaiss et al. 2014), zebrafish (Rawls et al. 2004, 2006; Davison et al. 2017; Koch et al. 2018; Sheng et al. 2018), stickleback (Small et al. 2017), Drosophila (Broderick et al. 2014; Elya et al. 2016), and pigs (Sun et al. 2018). Although gnotobiotic animal models are sometimes criticized because aspects of their development and physiology deviate from conventionally reared animals, those differences serve as direct evidence for the importance of microbiota on host biology, and can also help guide investigators to identify those aspects of host biology that are in communication with the microbial world. Another general approach to testing the role of microbiota on host biology is treatment of conventionally reared animals with antibiotics (Hill et al. 2010; Thaiss et al. 2016). However, this approach has its own significant caveats including retention of antibiotic-resistant bacteria and toxicity of antibiotics on the host, both of which directly and substantially affect host gene expression (Willing et al. 2011; Morgun et al. 2015). Such functional genomic studies in individual host species have proven to be very effective at identifying the host genes and represented biological processes that are under the influence of microbial stimuli.

While these methods are useful for understanding symbiotic relationships in these extant animal species, comparing results from distantly related animal host species provide opportunities to understand symbiotic relationships in their common ancestors. For example, mammals (members of Sarcopterygii) and bony fishes (members of Actinopterygii) last shared a common ancestor approximately 420 million years ago (Donoghue and Benton 2007). Direct comparison of host gene expression in the intestines of gnotobiotic zebrafish and mice revealed a number of homologous genes primarily involved in epithelial proliferation, nutrient metabolism, and immune responses that showed conserved directional responses to colonization by the microbiota (Rawls et al. 2004). Some of these identified genes have been shown to be consistently involved in the response to microbes in both species by additional studies, and implicate conservation of much larger programs of metabolism and immune response (Bäckhed et al. 2004; Rawls et al. 2006; Kanther et al. 2011; Camp et al. 2012; Murdoch et al. 2019). These patterns of conservation are striking since these animals are separated by substantial evolutionary distance and have substantial differences in diet, habitat, and microbiota composition (Rawls et al. 2006). This perspective also indicates that these symbiotic relationships existed in the common ancestor of bony fishes and mammals.

One of the major challenges emerging from the above studies is to understand the specificity of the symbiotic signals that evoke these host responses. One strategy is to test if individual members of the implicated microbiota are sufficient or necessary to evoke the host response of interest (Hooper et al. 2001; Rawls et al. 2004, 2006; Bates et al. 2006; Ivanov et al. 2009; Geva-Zatorsky et al. 2017). Once identified, strains sufficient to induce the host response can be further interrogated to identify the molecular nature of the symbiotic signal (Alegado et al. 2012; Hill et al. 2016; Matos et al. 2017). Identification of symbiosis signals remains a critically important challenge in discerning the mechanisms and ecologies that underlie symbioses.

Experiments using transplantation of intestinal microbiotas from different animals into germ-free animal hosts followed by transcriptomic surveys can also be used to test the microbial specificity of these host responses. For example, colonization of germ-free mice with intestinal microbiotas from zebrafish or mouse revealed that both microbial communities were sufficient to alter expression of transcripts involved in many metabolic pathways such as lipid and amino acid metabolism (Rawls et al. 2006). Similarly, colonization of germ-free zebrafish with the same microbial communities resulted in similar suppression of lipid metabolism genes (Rawls et al. 2006). However, other host responses typically evoked by the animal's native microbiota fail to be evoked by microbiota from other host species (Seedorf et al. 2014). For example, innate immune responses normally evoked by the zebrafish microbiota in zebrafish hosts failed to be stimulated by a mouse microbiota (Rawls et al. 2006). Similarly, gut microbiota from human or rat transplanted into germ-free mice failed to induce typical adaptive immune maturation observed in mice colonized with a normal mouse microbiota (Chung et al. 2012). These experiments also revealed that microbial communities are assembled in predictable ways. The microbial communities transplanted into germ-free hosts resembled their respective community of origin in terms of the bacterial lineages present, but the relative abundance of the bacterial lineages changed in ways to resemble the normal gut microbial community composition of the recipient host (Rawls et al. 2006; Chung et al. 2012; Seedorf et al. 2014). These studies established that differences in microbial community structure between mice, zebrafish, and other animals arise in part from selective pressures imposed within the gut habitat of each host, some of which might be mediated by host transcriptional differences. These microbiota transplant experiments may have important implications for our understanding of symbiotic signals. To the extent that nonnative microbiotas can induce the same host transcriptional responses as the native microbiota, those shared host responses are either evoked by microbial symbiotic stimuli present broadly among diverse microbiotas, or by host transcriptional regulatory programs that are able to summarize a diversity of microbial symbiotic signals into a common host transcriptional output.

To the extent that nonnative microbiotas are unable to evoke host transcriptional responses as the native microbiota, those host-specific responses may be due to differences in microbiota membership or activity.

15.5 Specific symbiotic signals within regulatory regions linked to microbiota-regulated genes

Transcription levels of a given eukaryotic gene are frequently physically linked to one or more discrete cis-regulatory regions (CRRs). There are two major classes of CRRs: (1) enhancers, which are capable of modifying transcription, are typically distal from the transcription start site, and are only a few 100 bp long; and (2) promoters, which are capable of regulating and initiating transcription often by communicating with enhancers, are immediately proximal and upstream of transcription start sites, and are typically up to a few 1000 bp long (Andersson and Sandelin 2019). Similar to transcription levels, identifying CRRs that are utilized in a particular context may allow for a clearer picture of how noncoding genomic information is involved in the response to microbes. Regulating CRR utilization is primarily possible through packaging of eukaryotic DNA around a histone protein octamer to form a repeating 147 bp complex called a nucleosome, plus a short intervening linker region of up to 80 bp that separates adjacent nucleosomes (Lai and Pugh 2017). Nucleosomes are generally thought to occlude binding of transcription factors (TFs), creating a unique opportunity to regulate access to very discrete regions of the host genome. Remodeling of nucleosomes can generate accessible regions primarily by uncovering sequence-specific transcription factor binding sites, typically to promote or regulate transcription of neighboring genes (Lai and Pugh 2017; Andersson and Sandelin 2019). Nucleosomes can be further modified with post translational modifications (PTMs) to reinforce utilization of these regions, or their continued exclusion. Importantly, these CRRs can be identified genome-wide using sequencing assays like MNase-, DNase- and ATAC-seq to identify accessible putative regulatory regions, and ChIP-seq to identify the location of TF binding or nucleosome PTMs in cell- and environment-specific contexts (Lai and Pugh 2017; Andersson and Sandelin 2019).

Though each cell within an animal contains the same genome and many CRRs are utilized in multiple tissues and contexts, these sequencing approaches can be used to identify those CRRs across the genome that are largely utilized in tissues and organs which directly interface with the microbiome (Camp et al. 2014; Davison et al. 2017). While differential CRR utilization is conceptually similar to changing transcription levels, noncoding regions account for 98% of many eukaryotic genomes, and there are likely over an order of magnitude greater noncoding regulatory elements than protein-coding regions in mammalian genomes (Bernstein et al. 2012; Shen et al. 2012; Thurman et al. 2012). This allows for many enhancers to be tissue- or environment-specific. This same freedom is not generally afforded to the approximately 25,000 protein coding regions and their promoters in mammalian genomes, many of which need to be utilized in a wide range of contexts (Long et al. 2016; Gaiti et al. 2017; Ibrahim et al. 2018). Combined with genome-wide transcription levels in multiple environments, putative direct linkages between transcriptional differences and the corresponding CRR(s) that facilitate those transcriptional differences can be mined

(Camp et al. 2014; Davison et al. 2017; Kuang et al. 2019). By identifying those linkages in the intestines of multiple related vertebrate species, for example, it may be possible to identify regulatory regions of the genome that are, or are not, conserved, how they have changed over time in their relationship with a microbiota, or as well ultimately reconstruct the regulatory information used by common ancestors to respond to a microbiota.

Genome-wide accessibility assays and ChIP-seq have identified regions of the genome narrowly utilized by a particular tissue or cell type (Bernstein et al. 2012; Shen et al. 2012; Kim et al. 2014; Yue et al. 2014). Experiments in mouse mapping differential accessible chromatin by DNase showed that microbial colonization did not significantly change the position of regions or extent of accessible chromatin in the mouse intestine (Camp et al. 2014; Davison et al. 2017). However, cultured human colonocytes have shown accessible chromatin changes by ATAC-seq following *in vitro* exposure to human microbiota with over 500 sites that were associated with transcription (Richards et al. 2019). Though these sites do not appear to explain the over 5000 genes that change significantly by mRNA levels, it does suggest that microbiotas may be able to modify host chromatin accessibility position under some circumstances (Richards et al. 2019). Complementary genome-wide assays like H3k27ac and H3k4me1 ChIP-seq identify PTMs on nucleosomes flanking CRRs and accessible chromatin regions that are associated with active transcriptional regulation (Andersson and Sandelin 2019). Recently, thousands of microbially responsive regulatory regions in the mouse intestine that generally correlate with expression levels have been mapped to accessible CRRs, dramatically expanding our awareness of regulatory regions responsible for regulating transcription in response to microbes (Davison et al. 2017; Woo and Alenghat 2017; Kuang et al. 2019). While these H3k27ac and H3k4me1 linkages are believed to be generally predictive of actual regulatory events, chromatin conformation assays can also directly detect long distance interactions by proximity ligation that can also be conserved (Smemo et al. 2014; Schoenfelder and Fraser 2019). TF motif enrichment at microbial responsive CRRs included nuclear receptor motifs associated with genes downregulated by microbes, and TFs like STAT, IRF, and ETS families associated with upregulated genes (Camp et al. 2014; Davison et al. 2017). Similar studies have also found a link between microbial responses and DNA methylation, which can also be used to influence transcription through CRRs and other means (Pan et al. 2018). A substantial effort is needed to interpret the amount of information and regulatory mechanisms uncovered by these genome-wide studies. Collectively these genomic studies can reduce by several orders of magnitude the putative search space for genes and regulatory elements that are under selective pressure to mediate responses to symbiosis signals in the intestine.

15.6   Evolutionary conservation of cis-regulatory regions

Once CRRs involved in mediating responses to microbiota are identified in one host species, understanding their evolution and conservation can identify mechanisms for how hosts have developed to adapt to microbial symbiosis. Unlike protein coding sequences which frequently have long stretches of similar DNA sequences at sites that are functionally conserved across species, regulatory regions typically show less conservation, especially across long evolutionary distances (Villar et al. 2015; Andersson and Sandelin 2019). Lower levels of conservation at CRRs presumably arise from the contribution of several factors. Sequence-specific TF binding sites in CRRs are generally required to regulate expression of neighboring genes. However, TF binding sites are generally small, usually ranging from only 8–12 bp (Stewart et al. 2012), are orientation independent (Andersson and Sandelin 2019), and degenerate, meaning that multiple similar sequences can result in functional binding for any particular factor (Zheng et al. 2011; Jolma et al. 2013; Weirauch et al. 2014). Clusters of binding sites for the same TF within a regulatory region and additional redundant enhancers harboring the same TF binding sites is also a common mechanism to regulate a gene (Hong et al. 2008; Osterwalder et al. 2018). This leads to improved robustness of transcription and transcriptional dynamics of the gene, but also facilitates the potential for reduced pressure on any individual binding site. This may support many variations of functional intermediates of enhancer organization during evolution (Hong et al. 2008; Frankel et al. 2010; Osterwalder et al. 2018). CRRs are also not strictly limited by the distance to the sites they regulate, and while cofactors and dimerization can constrain longer stretches of regulatory region, frequently interacting TF binding sites can similarly be separated in space on the DNA strand. These properties are a boon to uncoupling selective pressures on transcriptional regulation from primary DNA sequence and creating a diversity of resulting regulatory patterns. However, these properties create substantial problems when detecting conservation at nonprotein-coding regions (Villar et al. 2015). These problems are also compounded by, at times, limited or unequal whole genome sequencing for similarly and distantly related host species. Interestingly, the regulatory regions that show the highest conservation across multiple vertebrate species are linked to a subset of functional genes involved in development, RNA processing, and transcription factors. This suggests that some processes are highly conserved in their regulation, or that a certain narrow range of regulatory elements are required to faithfully transcriptionally regulate a subset of genes in a particular pattern (Bejerano et al. 2004; Hiller et al. 2013). Testing of these highly conserved noncoding elements (CNEs), using reporter expression assays, identified 13/41 (30%) had conserved transcriptional activity between zebrafish CNEs in zebrafish and their corresponding human CNE in mouse in orthologous anatomical compartments (Ritter et al. 2010). This suggests that aspects of transcriptional regulation can be highly conserved across millions of years of evolution and were present in the common ancestor of these animals. A number of studies have found the capacity for nonconserved regulatory regions surrounding the same orthologous gene to drive highly similar expression, suggesting strict conservation is not required for conserved regulatory function (Chatterjee et al. 2011; Lickwar et al. 2017). These regulatory similarities can sometimes be attributed to regulation by the same transcription factor or regulator, whose binding sites escape detection at the level of conservation, presumably due to the aforementioned small size, degeneracy, and limited spacing requirements (Fisher et al. 2006; Chatterjee et al. 2011; Villar et al. 2015; Lickwar et al. 2017). However, it is certainly not true that all similar transcription patterns seen across species are driven by conserved mechanisms, though these cases still represent one type of functional conservation (Fisher et al. 2006; Weirauch and Hughes 2010; Chatterjee et al. 2011; Stolfi et al. 2014; Villar et al. 2015). The two major classes of regulatory regions, enhancers and promoters, also do not appear to evolve at the same rate. Active promoters contain H3k4me3 and H3k27ac, and are distinguished from active enhancers which generally lack H3k4me3 (Santos-Rosa et al. 2002; Villar et al. 2015; Andersson and Sandelin 2019). An analysis of genome-wide chromatin marks for enhancers and promoters in the liver of 20 vertebrate species across 6 orders revealed higher conservation at promoters than enhancers (Villar et al. 2015). We applied a similar approach to the intestinal epithelium, focusing on accessible chromatin. Accessible chromatin data matched to RNA-seq from zebrafish, stickleback, mouse, and human adult IECs has created an initial map of regulatory elements that may be accessible and potentially specific to IECs across four species separated by 420 million years and existing in diverse environments (Lickwar et al. 2017). Very few highly conserved regions showed IEC chromatin accessibility largely limited to IECs. However, this approach did provide evidence for conserved regulators specifying regional and IEC subtype specificity, and showed that these strategies may extract additional conservation information that cannot be identified using traditional sequence conservation methods (Lickwar et al. 2017). This suggests that to fully extract conservation, additional methods need to be developed, or complementary strategies like TF ChIP-seq across multiple species in orthologous tissue will be needed to determine the sometimes elusive functional conservation signal (Schmidt et al. 2010; Ballester et al. 2014).

15.7   A case study—conserved microbial suppression of *Angptl4*

Combining transcriptional and chromatin assays in tissues across host species can help uncover conserved transcriptional responses to microbes and the underlying transcriptional regulatory networks. One of the common genes identified as microbiota-responsive across multiple species was *Angiopoetin-like 4* (*Angptl4*), which shows reduced expression in the intestinal epithelium following colonization in both mouse (Bäckhed et al. 2004) and zebrafish (Rawls et al. 2004). ANGPTL4 protein is secreted into circulation where it systemically inhibits the activity of lipoprotein lipase, and microbial suppression of *Angptl4* expression thereby promotes fat storage in adipose tissues (Bäckhed et al. 2004). We found that microbiome colonization in zebrafish similarly results in suppression of the zebrafish *angptl4* ortholog in the intestinal epithelium (Camp et al. 2012). We reasoned that by elucidating the host mechanisms through which microbes suppress transcription of *Angptl4* in the intestinal epithelium, we would gain insight into broader host transcriptional networks governed by microbial colonization. Using *in vivo* reporter assays in zebrafish, we identified multiple tissue-specific CRRs at zebrafish *angptl4* including a single intronic CRR called *in3.4* that was sufficient to recapitulate the intestinal epithelial expression and microbial suppression of endogenous *angptl4* (Camp et al. 2012). Using an unbiased screening strategy, we found that this microbiome-suppressed intestinal epithelial enhancer *in3.4* is specifically bound and activated by Hepatocyte nuclear factor 4 (HNF4) transcription factors (Davison et al. 2017). HNF4 is a family of nuclear receptor transcription factors that display conserved expression in the vertebrate intestine (Lickwar et al. 2017) and have been implicated in inflammatory bowel diseases, obesity, and diabetes in humans and mice (Fajans et al. 2001; Gerdin et al. 2006; Stegmann et al. 2006; Franke et al. 2007; Darsigny et al. 2009; Barrett et al. 2009; Jostins et al. 2012; Marcil et al. 2012; Berndt et al. 2013; Chahar et al. 2014; Baraille et al. 2015). However, a role for HNF4 members in host response to microbes was previously unknown. Our genetic analysis revealed that zebrafish Hnf4a activates nearly half of the genes that are suppressed by microbiome (Davison et al. 2017). This identified Hnf4a as a novel mediator of host–microbiome interactions and suggested that the microbiome negatively regulates Hnf4a activity. Interestingly, mutation of *hnf4a* attenuated *in3.4* enhancer activity without affecting expression of *angptl4*, which may be due to redundancy in TFs and/or CRRs regulating *angptl4* transcription (Davison et al. 2017).

To determine how microbiome colonization affects HNF4A activity and the broader chromatin landscape, we translated these zebrafish results into gnotobiotic mice, where a broader suite of functional genomic analysis tools was available. Our previous finding, that the microbiome modifies host gene transcription in mouse IECs without significantly impacting the accessible chromatin landscape (Camp et al. 2014), predicted that microbiome regulation of host gene transcription might be achieved by differential activity of specific TFs and enrichment of their binding sites in accessible CRRs near target genes. In accord, our analysis of genomic architecture in GF mouse IECs disclosed drastic genome-wide reduction of HNF4A and HNF4G occupancy compared to colonized controls (Davison et al. 2017). Interspecies meta-analysis suggested interactions between HNF4A and the microbiome promote gene expression patterns associated with the human inflammatory bowel diseases (IBD) (Davison et al. 2017). Together, these results established a novel and conserved role for HNF4A in maintaining intestinal homeostasis in response to the microbiome, and illustrated how analysis of genomic regulatory regions mediating host responses to symbiotic signals can lead to broad new insight into the upstream regulatory mechanisms that converge upon those regions.

15.8 Prospectus

Based on the research summarized above, the host genome contains abundant evidence of symbiotic signaling within coding regions, patterns of transcriptional responsiveness, and the sequence and function of regulatory regions. This strongly suggests that natural selection has acted upon the host genome to retain these elements and mechanisms, yet it remains quite difficult to determine what level(s) of biological complexity upon which natural selection acted to yield these traits, and what those selective pressures were. Most of the studies discussed here were testing the impact of complex unfractionated microbiotas, or individual microbial species, on host biology. It also remains generally unclear which of these host response mechanisms evolved to specifically perceive signals from commensals microbes, or instead evolved to provide defense against microbial pathogens but are tonically activated by commensals as well. One enduring challenge for future research is to continue to identify the specific microbes responsible for producing symbiotic signals, the molecular nature of those symbiotic signals, and the responsive host signal transduction and transcriptional pathways that converge on the host genome. Another overarching challenge for genomic analysis of symbiotic signaling is to discern transcriptional regulatory programs involved in establishing developmental identity of the tissue (i.e., the intestinal epithelium) from those that mediate that tissue's sensitivity to environmental factors like microbes. Additional experiments utilizing organoid cultures, increasingly sensitive chromatin mapping, single-cell assays, or colonization with complex or simplified microbial communities may more carefully map host genomic regions that respond to microbial symbionts, including signal-specific responses (Seedorf et al. 2014; Haber et al. 2017; Janeckova et al. 2019).

The studies summarized here also illustrate that comparative studies of symbiotic signaling across multiple host species are a powerful approach for discerning conserved and ancient modes of host–microbe symbiosis. However, systematic efforts to compare microbial responses following colonization of epithelial surfaces across multiple species have not been substantial, and we see a need for more investment in this area. We propose a number of considerations that need to be applied when designing such comparative studies. A critical early consideration is selecting the specific trait(s) to be investigated in the comparative analysis. Quantitative, physiological, or transcriptional traits will each have their own respective degree of uncertainty and conservation. For example, genomic traits like gene presence, gene expression, and TF motifs are relatively reliable, whereas traits like behavior are less so. Another important consideration is the selection of host species to be included in any comparative analysis. The evolutionary distance between any two species will affect what one can reasonably expect to find in terms of trait conservation, so inclusion of closely and distantly related host species could be helpful to calibrate conservation signals. The provenance (i.e., wild vs. domesticated), diets, life stage, and microbiome composition will also contribute to the degree of trait conservation that can be detected between host species. This is particularly relevant when searching for trait conservation within the noncoding genome, where at farther evolutionary distances there is an inability to align most regulatory sequences (Schmidt et al. 2010). This does not necessarily mean conservation doesn't exist, but instead underscores our need for more multispecies datasets and for better computational tools to detect subtle variations, turnover, and rearrangements of DNA sequence. As genomic datasets focused on microbial responsiveness arise in multiple host species, quantitative models will become feasible and more targeted studies into the rate of evolution and causal factors can be explored.

A final benefit of this comparative genomic approach is that it may assist us in understanding the natural history of animal–microbe symbioses. The ability of extant animals to engage in symbiotic relationships is a product of the evolutionary history of their ancestors and the symbiotic relationships they engaged in. Therefore, in order to fully understand symbiotic relationships in extant species, it is important to also try to understand the symbiotic relationships of their ancestors. However, such efforts are inherently challenging since the ancestors of most extant animals are now extinct, and soft tissues where symbioses often primarily occur (e.g., the digestive tract) are typically absent in the fossil record. Paleontological analysis of fossils (Shu et al. 2003), sometimes augmented by analysis of associated ancient DNA and proteins (Krause et al. 2010; Sankararaman et al. 2014; Welker et al. 2019) has helped us understand the biology and natural history of common ancestors. In cases where fossils and ancient DNA are unavailable, comparative biology in extant species coupled with evolutionary reconstruction and inference can be used to produce useful models for the evolution of the selected tissue and to discern ancestral states (Monahan-Earley et al. 2013; Poelmann and Gittenberger-de Groot 2019). Common ancestors are an important but often overlooked component of the intellectual framework guiding basic and translational research on symbiotic relationships. The genome and biology of common ancestors impose constraints for the subsequent evolution of their derivative lineages. The extent to which the biology of the common ancestor is preserved, in any two extant derivative lineages, also directly constrains the ability of discoveries in one of those extant lineages to inform the other. This is particularly relevant for the ability of animal models to be used in translational research to inform human biology and medicine. An improved comparative biological understanding of symbioses in diverse animal models should therefore not only provide greater insight into the evolutionary and mechanistic biology underlying extant symbioses, but also allow us to better define the strengths and limitations of each animal model for translational research.

References

Abreu, M. T. 2010. "Toll-Like Receptor Signalling in the Intestinal Epithelium: How Bacterial Recognition Shapes Intestinal Function." *Nat Rev Immunol* 10 (2):131–144. doi: 10.1038/nri2707.

Adair, K. L., and A. E. Douglas. 2017. "Making a Microbiome: The Many Determinants of Host-Associated Microbial Community Composition." *Curr Opin Microbiol* 35:23–29. doi: 10.1016/j.mib.2016.11.002.

Adrian, J., P. Bonsignore, S. Hammer, T. Frickey, and C. R. Hauck. 2019. "Adaptation to Host-Specific Bacterial Pathogens Drives Rapid Evolution of a Human Innate Immune Receptor." *Curr Biol* 29 (4):616–630 e5. doi: 10.1016/j.cub.2019.01.058.

Alegado, R. A., L. W. Brown, S. Cao, R. K. Dermenjian, R. Zuzow, S. R. Fairclough, J. Clardy, and N. King. 2012. "A Bacterial Sulfonolipid Triggers Multicellular Development in the Closest Living Relatives of Animals." *Elife* 1:e00013. doi: 10.7554/eLife.00013.

Allaire, J. M., S. M. Crowley, H. T. Law, S. Y. Chang, H. J. Ko, and B. A. Vallance. 2018. "The Intestinal Epithelium: Central Coordinator of Mucosal Immunity." *Trends Immunol* 39 (9):677–696. doi: 10.1016/j.it.2018.04.002.

Andersson, R., and A. Sandelin. 2019. "Determinants of Enhancer and Promoter Activities of Regulatory Elements." *Nat Rev Genet.* 21 (2):71–87. doi: 10.1038/s41576-019-0173-8. Epub 2019 Oct 11.

Bäckhed, F., H. Ding, T. Wang, L. V. Hooper, G. Y. Koh, A. Nagy, C. F. Semenkovich, and J. I. Gordon. 2004. "The Gut Microbiota as an Environmental Factor that Regulates Fat Storage." *Proc Natl Acad Sci U S A* 101 (44):15718–15723.

Ballester, B., A. Medina-Rivera, D. Schmidt, M. Gonzalez-Porta et al. 2014. "Multi-Species, Multi-Transcription Factor Binding Highlights Conserved Control of Tissue-Specific Biological Pathways." *Elife* 3:e02626. doi: 10.7554/eLife.02626.

Baraille, F., S. Ayari, V. Carriere, C. Osinski et al. 2015. "Glucose Tolerance is Improved in Mice Invalidated for the Nuclear Receptor HNF-4gamma: A Critical Role for Enteroendocrine Cell Lineage." *Diabetes* 64 (8):2744–2756. doi: 10.2337/db14-0993.

Barrett, J. C., J. C. Lee, C. W. Lees, N. J. Prescott et al. 2009. "Genome-wide Association Study of Ulcerative Colitis Identifies Three New Susceptibility Loci, Including the HNF4A Region." *Nat Genet* 41 (12):1330–1334. doi: 10.1038/ng.483.

Bates, J. M., E. Mittge, J. Kuhlman, K. N. Baden, S. E. Cheesman, and K. Guillemin. 2006. "Distinct Signals from the Microbiota Promote Different Aspects of zebrafish gut Differentiation." *Dev. Biol* 297 (2):374–386.

Bejerano, G., M. Pheasant, I. Makunin, S. Stephen, W. J. Kent, J. S. Mattick, and D. Haussler. 2004. "Ultraconserved Elements in the Human Genome." *Science* 304 (5675):1321–1325. doi: 10.1126/science.1098119.

Berndt, S. I., S. Gustafsson, R. Magi, A. Ganna et al. 2013. "Genome-Wide Meta-Analysis Identifies 11 New Loci for Anthropometric Traits and Provides Insights into Genetic Architecture." *Nat Genet* 45 (5):501–512. doi: 10.1038/ng.2606.

Bernstein, B. E., E. Birney, I. Dunham, E. D. Green, C. Gunter, and M. Snyder. 2012. "An Integrated Encyclopedia of DNA Elements in the Human Genome." *Nature* 489 (7414):57–74. doi: nature11247 [pii] 10.1038/nature11247.

Booth, A., C. Mariscal, and W. F. Doolittle. 2016. "The Modern Synthesis in the Light of Microbial Genomics." *Annu Rev Microbiol* 70:279–297. doi: 10.1146/annurev-micro-102215-095456.

Bordenstein, S. R., and K. R. Theis. 2015. "Host Biology in Light of the Microbiome: Ten Principles of Holobionts and Hologenomes." *PLoS Biol* 13 (8):e1002226. doi: 10.1371/journal.pbio.1002226.

Boto, L. 2014. "Horizontal Gene Transfer in the Acquisition of Novel Traits by Metazoans." *Proc Biol Sci* 281 (1777):20132450. doi: 10.1098/rspb.2013.2450.

Brawand, D., M. Soumillon, A. Necsulea, P. Julien et al. 2011. "The Evolution of Gene Expression Levels in Mammalian Organs." *Nature* 478 (7369):343–348. doi: 10.1038/nature10532.

Brennan, J. J., and T. D. Gilmore. 2018. "Evolutionary Origins of Toll-like Receptor Signaling." *Mol Biol Evol* 35 (7):1576–1587. doi: 10.1093/molbev/msy050.

Broderick, N. A., N. Buchon, and B. Lemaitre. 2014. "Microbiota-Induced Changes in Drosophila Melanogaster Host Gene Expression and Gut Morphology." *MBio* 5 (3):e01117–14. doi: 10.1128/mBio.01117–14.

Camp, J. G., C. L. Frank, C. R. Lickwar, H. Guturu, T. Rube, A. M. Wenger, J. Chen, G. Bejerano, G. E. Crawford, and J. F. Rawls. 2014. "Microbiota Modulate Transcription in the Intestinal Epithelium Without Remodeling the Accessible Chromatin Landscape." *Genome Res* 24 (9):1504–1516. doi: 10.1101/gr.165845.113.

Camp, J. G., A. L. Jazwa, C. M. Trent, and J. F. Rawls. 2012. "Intronic cis-Regulatory Modules Mediate Tissue-Specific and Microbial Control of angptl4/fiaf Transcription." *PLoS Genet* 8 (3):e1002585. doi: 10.1371/journal.pgen.1002585 PGENETICS-D-11-01855 [pii].

Chahar, S., V. Gandhi, S. Yu, K. Desai et al. 2014. "Chromatin Profiling Reveals Regulatory Network Shifts and a Protective Role for Hepatocyte Nuclear Factor 4alpha during Colitis." *Mol Cell Biol* 34 (17):3291–3304. doi: 10.1128/MCB.00349-14.

Chan, E. T., G. T. Quon, G. Chua, T. Babak et al. 2009. "Conservation of Core Gene Expression in Vertebrate Tissues." *J Biol* 8 (3):33. doi: 10.1186/jbiol130.

Chatterjee, S., G. Bourque, and T. Lufkin. 2011. "Conserved and non-Conserved Enhancers Direct Tissue Specific Transcription in Ancient Germ Layer Specific Developmental Control Genes." *BMC Dev Biol* 11:63. doi: 10.1186/1471-213X-11-63.

Chung, H., S. J. Pamp, J. A. Hill, N. K. Surana et al. 2012. "Gut Immune Maturation Depends on Colonization with a Host-Specific Microbiota." *Cell* 149 (7):1578–1593. doi: 10.1016/j.cell.2012.04.037.

Darsigny, M., J. P. Babeu, A. A. Dupuis, E. E. Furth, E. G. Seidman, E. Levy, E. F. Verdu, F. P. Gendron, and F. Boudreau. 2009. "Loss of Hepatocyte-Nuclear-Factor-4alpha Affects Colonic Ion Transport and Causes Chronic Inflammation Resembling Inflammatory Bowel Disease in Mice." *PLoS One* 4 (10):e7609. doi: 10.1371/journal.pone.0007609.

Davison, J. M., C. R. Lickwar, L. Song, G. Breton, G. E. Crawford, and J. F. Rawls. 2017. "Microbiota Regulate Intestinal Epithelial Gene Expression by Suppressing the Transcription Factor Hepatocyte Nuclear Factor 4 Alpha." *Genome Res* 27 (7):1195–1206. doi: 10.1101/gr.220111.116.

Donoghue, P. C., and M. J. Benton. 2007. "Rocks and Clocks: Calibrating the Tree of Life Using Fossils and Molecules." *Trends Ecol Evol* 22 (8):424–431. doi: 10.1016/j.tree.2007.05.005.

Douglas, A. E., and J. H. Werren. 2016. "Holes in the Hologenome: Why Host-Microbe Symbioses Are Not Holobionts." *MBio* 7 (2):e02099. doi: 10.1128/mBio.02099-15.

Dupressoir, A., C. Lavialle, and T. Heidmann. 2012. "From Ancestral Infectious Retroviruses to Bona Fide Cellular Genes: Role of the Captured Syncytins in Placentation." *Placenta* 33 (9):663–671. doi: 10.1016/j.placenta.2012.05.005.

El Aidy, S., C. A. Merrifield, M. Derrien, P. Kleerebezem et al. 2013. "The Gut Microbiota Elicits a Profound Metabolic Reorientation in the Mouse Jejunal Mucosa During Conventionalisation." *Gut* 62 (9):1306–1314. doi: gutjnl-2011-301955 [pii] 10.1136/gutjnl-2011-301955.

Elya, C., V. Zhang, W. B. Ludington, and M. B. Eisen. 2016. "Stable Host Gene Expression in the Gut of Adult Drosophila Melanogaster with Different Bacterial Mono-Associations." *PLoS One* 11 (11):e0167357. doi: 10.1371/journal.pone.0167357.

Facco, E., A. Pagnani, E. T. Russo, and A. Laio. 2019. "The Intrinsic Dimension of Protein Sequence Evolution." *PLoS Comput Biol* 15 (4):e1006767. doi: 10.1371/journal.pcbi.1006767.

Fajans, S. S., G. I. Bell, and K. S. Polonsky. 2001. "Molecular Mechanisms and Clinical Pathophysiology of Maturity-Onset Diabetes of the Young." *N Engl J Med* 345 (13):971–980. doi: 10.1056/NEJMra002168.

Fisher, S., E. A. Grice, R. M. Vinton, S. L. Bessling, and A. S. McCallion. 2006. "Conservation of *RET* Regulatory Function from Human to Zebrafish Without Sequence Similarity." *Science* 312 (5771):276–279.

Franke, A., J. Hampe, P. Rosenstiel, C. Becker et al. 2007. "Systematic Association Mapping Identifies NELL1 as a Novel IBD Disease Gene." *PLoS One* 2 (8):e691. doi: 10.1371/journal.pone.0000691.

Frankel, N., G. K. Davis, D. Vargas, S. Wang, F. Payre, and D. L. Stern. 2010. "Phenotypic Robustness Conferred by Apparently Redundant Transcriptional Enhancers." *Nature* 466 (7305):490–493. doi: 10.1038/nature09158.

Gaiti, F., A. D. Calcino, M. Tanurdzic, and B. M. Degnan. 2017. "Origin and Evolution of the Metazoan Non-Coding Regulatory Genome." *Dev Biol* 427 (2):193–202. doi: 10.1016/j.ydbio.2016.11.013.

Gerbe, F., E. Sidot, D. J. Smyth, M. Ohmoto et al. 2016. "Intestinal Epithelial Tuft Cells Initiate Type 2 Mucosal Immunity to Helminth Parasites." *Nature* 529 (7585):226–230. doi: 10.1038/nature16527.

Gerdin, A. K., V. V. Surve, M. Jonsson, M. Bjursell et al. 2006. "Phenotypic Screening of Hepatocyte Nuclear Factor (HNF) 4-Gamma Receptor Knockout Mice." *Biochem Biophys Res Commun* 349 (2):825–832. doi: S0006-291X(06)01929-2 [pii] 10.1016/j.bbrc.2006.08.103.

Geva-Zatorsky, N., E. Sefik, L. Kua, L. Pasman et al. 2017. "Mining the Human Gut Microbiota for Immunomodulatory Organisms." *Cell* 168 (5):928–943 e11. doi: 10.1016/j.cell.2017.01.022.

Gilad, Y., and O. Mizrahi-Man. 2015. "A reanalysis of Mouse ENCODE Comparative Gene Expression Data." *F1000Res* 4:121. doi: 10.12688/f1000research.6536.1.

Haber, A. L., M. Biton, N. Rogel, R. H. Herbst et al. 2017. "A Single-Cell Survey of the Small Intestinal Epithelium." *Nature* 551 (7680):333–339. doi: 10.1038/nature24489.

Hacquard, S., R. Garrido-Oter, A. Gonzalez, S. Spaepen et al. 2015. "Microbiota and Host Nutrition across Plant and Animal Kingdoms." *Cell Host Microbe* 17 (5):603–616. doi: 10.1016/j.chom.2015.04.009.

Hill, D. A., C. Hoffmann, M. C. Abt, Y. Du, D. Kobuley, T. J. Kirn, F. D. Bushman, and D. Artis. 2010. "Metagenomic Analyses Reveal Antibiotic-Induced Temporal and Spatial Changes in Intestinal Microbiota with Associated Alterations in Immune Cell Homeostasis." *Mucosal Immunol* 3 (2):148–158. doi: mi2009132 [pii] 10.1038/mi.2009.132.

Hill, J. H., E. A. Franzosa, C. Huttenhower, and K. Guillemin. 2016. "A Conserved Bacterial Protein Induces Pancreatic Beta Cell Expansion during Zebrafish Development." *Elife* 5. doi: 10.7554/eLife.20145.

Hiller, M., S. Agarwal, J. H. Notwell, R. Parikh, H. Guturu, A. M. Wenger, and G. Bejerano. 2013. "Computational Methods to Detect Conserved Non-Genic Elements in Phylogenetically Isolated Genomes: Application to Zebrafish." *Nucleic Acids Res* 41 (15):e151. doi: 10.1093/nar/gkt557.

Hong, J. W., D. A. Hendrix, and M. S. Levine. 2008. "Shadow Enhancers as a Source of Evolutionary Novelty." *Science* 321 (5894):1314. doi: 10.1126/science.1160631.

Hooper, L. V., M. H. Wong, A. Thelin, L. Hansson, P. G. Falk, and J. I. Gordon. 2001. "Molecular Analysis of Commensal Host-Microbial Relationships in the Intestine." *Science* 291 (5505):881–884.

Human Microbiome Project, Consortium. 2012. "Structure, Function and Diversity of the Healthy Human Microbiome." *Nature* 486 (7402):207–214. doi: 10.1038/nature11234.

Ibrahim, M. M., A. Karabacak, A. Glahs, E. Kolundzic, A. Hirsekorn, A. Carda, B. Tursun, R. P. Zinzen, S. A. Lacadie, and U. Ohler. 2018. "Determinants of Promoter and Enhancer Transcription Directionality in Metazoans." *Nat Commun* 9 (1):4472. doi: 10.1038/s41467-018-06962-z.

Ivanov, II, K. Atarashi, N. Manel, E. L. Brodie et al. 2009. "Induction of Intestinal Th17 Cells by Segmented Filamentous Bacteria." *Cell* 139 (3):485–498. doi: S0092-8674(09)01248-3 [pii] 10.1016/j.cell.2009.09.033.

Janeckova, L., K. Kostovcikova, J. Svec, M. Stastna et al. 2019. "Unique Gene Expression Signatures in the Intestinal Mucosa and Organoids Derived from Germ-Free and Monoassociated Mice." *Int J Mol Sci* 20 (7). doi: 10.3390/ijms20071581.

Jolma, A., J. Yan, T. Whitington, J. Toivonen et al. 2013. "DNA-Binding Specificities of Human Transcription Factors." *Cell* 152 (1–2):327–339. doi: 10.1016/j.cell.2012.12.009.

Jostins, L., S. Ripke, R. K. Weersma, R. H. Duerr et al. Genetics Consortium International. 2012. "Host-Microbe Interactions Have Shaped the Genetic Architecture of Inflammatory Bowel Disease." *Nature* 491 (7422):119–124. doi: 10.1038/nature11582.

Kanther, M., X. Sun, M. Muhlbauer, L. C. Mackey, E. J. Flynn, 3rd, M. Bagnat, C. Jobin, and J. F. Rawls. 2011. "Microbial Colonization Induces Dynamic Temporal and Spatial Patterns of NF-KappaB Activation in the Zebrafish Digestive Tract." *Gastroenterology* 141 (1):197–207. doi: 10.1053/j.gastro.2011.03.042.

Kim, T. H., F. Li, I. Ferreiro-Neira, L. L. Ho, A. Luyten, K. Nalapareddy, H. Long, M. Verzi, and R. A. Shivdasani. 2014. "Broadly Permissive Intestinal Chromatin Underlies Lateral Inhibition and Cell Plasticity." *Nature* 506 (7489):511–515. doi: 10.1038/nature12903.

Koch, B. E. V., S. Yang, G. Lamers, J. Stougaard, and H. P. Spaink. 2018. "Intestinal Microbiome Adjusts the Innate Immune Setpoint during Colonization through Negative Regulation of MyD88." *Nat Commun* 9 (1):4099. doi: 10.1038/s41467-018-06658-4.

Koonin, E. V., and Y. I. Wolf. 2010. "Constraints and Plasticity in Genome and Molecular-Phenome Evolution." *Nat Rev Genet* 11 (7):487–498. doi: 10.1038/nrg2810.

Krause, J., Q. Fu, J. M. Good, B. Viola, M. V. Shunkov, A. P. Derevianko, and S. Paabo. 2010. "The Complete Mitochondrial DNA Genome of an Unknown Hominin from Southern Siberia." *Nature* 464 (7290):894–897. doi: 10.1038/nature08976.

Kuang, Z., Y. Wang, Y. Li, C. Ye, K. A. Ruhn, C. L. Behrendt, E. N. Olson, and L. V. Hooper. 2019. "The Intestinal Microbiota Programs Diurnal Rhythms in Host Metabolism through Histone Deacetylase 3." *Science* 365 (6460):1428–1434. doi: 10.1126/science.aaw3134.

Kutschera, U. 2018. "Systems Biology of Eukaryotic Superorganisms and The Holobiont Concept." *Theory Biosci* 137 (2):117–131. doi: 10.1007/s12064-018-0265-6.

Lai, W. K. M., and B. F. Pugh. 2017. "Understanding Nucleosome Dynamics and Their Links to gene Expression and DNA Replication." *Nat Rev Mol Cell Biol* 18 (9):548–562. doi: 10.1038/nrm.2017.47.

Larsson, E., V. Tremaroli, Y. S. Lee, O. Koren, I. Nookaew, A. Fricker, J. Nielsen, R. E. Ley, and F. Backhed. 2012. "Analysis of Gut Microbial Regulation of Host Gene Expression Along the Length of the Gut and Regulation of Gut Microbial Ecology through MyD88." *Gut* 61 (8):1124–1131. doi: gutjnl-2011-301104 [pii] 10.1136/gutjnl-2011-301104.

Ley, R. E., D. A. Peterson, and J. I. Gordon. 2006. "Ecological and Evolutionary Forces Shaping Microbial Diversity in the Human Intestine." *Cell* 124 (4):837–848.

Li, Y., Y. Li, X. Cao, X. Jin, and T. Jin. 2017. "Pattern Recognition Receptors in Zebrafish Provide Functional and Evolutionary Insight into Innate Immune Signaling Pathways." *Cell Mol Immunol* 14 (1):80–89. doi: 10.1038/cmi.2016.50.

Lickwar, C. R., J. G. Camp, M. Weiser, J. L. Cocchiaro, D. M. Kingsley, T. S. Furey, S. Z. Sheikh, and J. F. Rawls. 2017. "Genomic Dissection of Conserved Transcriptional Regulation in Intestinal Epithelial Cells." *PLoS Biol* 15 (8):e2002054. doi: 10.1371/journal.pbio.2002054.

Lin, S., Y. Lin, J. R. Nery, M. A. Urich et al. 2014. "Comparison of the Transcriptional Landscapes between Human and Mouse Tissues." *Proc Natl Acad Sci U S A* 111 (48):17224–9. doi: 10.1073/pnas.1413624111.

Long, H. K., S. L. Prescott, and J. Wysocka. 2016. "Ever-Changing Landscapes: Transcriptional Enhancers in Development and Evolution." *Cell* 167 (5):1170–1187. doi: 10.1016/j.cell.2016.09.018.

Louca, S., M. F. Polz, F. Mazel, M. B. N. Albright et al. 2018. "Function and Functional Redundancy in Microbial Systems." *Nat Ecol Evol* 2 (6):936–943. doi: 10.1038/s41559-018-0519-1.

Marcil, V., D. Sinnett, E. Seidman, F. Boudreau et al. 2012. "Association between Genetic Variants in the HNF4A Gene and Childhood-Onset Crohn's Disease." *Genes Immun* 13 (7):556–565. doi: 10.1038/gene.2012.37.

Matos, R. C., M. Schwarzer, H. Gervais, P. Courtin et al. 2017. "D-Alanylation of Teichoic Acids Contributes to Lactobacillus Plantarum-Mediated Drosophila Growth during Chronic Undernutrition." *Nat Microbiol* 2 (12):1635–1647. doi: 10.1038/s41564-017-0038-x.

McFall-Ngai, M., M. G. Hadfield, T. C. Bosch, H. V. Carey et al. 2013. "Animals in a Bacterial World, A New Imperative for The Life Sciences." *Proc Natl Acad Sci U S A.* 26;110 (9):3229–3236. doi: 10.1073/pnas.1218525110. Epub 2013 Feb 7.

Mindell, D. P. 1992. "Phylogenetic Consequences of Symbioses: Eukarya and Eubacteria are not Monophyletic Taxa." *Biosystems* 27 (1):53–62. doi: 10.1016/0303-2647(92)90046-2.

Moeller, A. H., T. A. Suzuki, M. Phifer-Rixey, and M. W. Nachman. 2018. "Transmission Modes of the Mammalian Gut Microbiota." *Science* 362 (6413):453–457. doi: 10.1126/science.aat7164.

Monahan-Earley, R., A. M. Dvorak, and W. C. Aird. 2013. "Evolutionary Origins of the Blood Vascular System and Endothelium." *J Thromb Haemost* 11 (Suppl 1):46–66. doi: 10.1111/jth.12253.

Moran, N. A., and D. B. Sloan. 2015. "The Hologenome Concept: Helpful or Hollow?" *PLoS Biol* 13 (12):e1002311. doi: 10.1371/journal.pbio.1002311.

Morgun, A., A. Dzutsev, X. Dong, R. L. Greer, D. J. Sexton, J. Ravel, M. Schuster, W. Hsiao, P. Matzinger, and N. Shulzhenko. 2015. "Uncovering Effects of Antibiotics on the Host and Microbiota Using Transkingdom Gene Networks." *Gut* 64 (11):1732–1743. doi: 10.1136/gutjnl-2014-308820.

Mukherjee, S., S. Huda, and S. P. Sinha Babu. 2019. "Toll-Like Receptor Polymorphism in Host Immune Response to Infectious Diseases: A Review." *Scand J Immunol* 90 (1):e12771. doi: 10.1111/sji.12771.

Murdoch, C. C., S. T. Espenschied, M. A. Matty, O. Mueller, D. M. Tobin, and J. F. Rawls. 2019. "Intestinal Serum Amyloid A Suppresses Systemic Neutrophil Activation and Bactericidal Activity In Response to Microbiota Colonization." *PLoS Pathog* 15 (3):e1007381. doi: 10.1371/journal.ppat.1007381.

Osterwalder, M., I. Barozzi, V. Tissieres, Y. Fukuda-Yuzawa et al. 2018. "Enhancer Redundancy Provides Phenotypic Robustness in Mammalian Development." *Nature* 554 (7691):239–243. doi: 10.1038/nature25461.

Pan, W. H., F. Sommer, M. Falk-Paulsen, T. Ulas et al. 2018. "Exposure to the Gut Microbiota Drives Distinct Methylome and Transcriptome Changes in Intestinal Epithelial Cells During Postnatal Development." *Genome Med* 10 (1):27. doi: 10.1186/s13073-018-0534-5.

Poelmann, R. E., and A. C. Gittenberger-de Groot. 2019. "Development and Evolution of the Metazoan Heart." *Dev Dyn* 248 (8):634–656. doi: 10.1002/dvdy.45.

Pott, J., S. Stockinger, N. Torow, A. Smoczek et al. 2012. "Age-Dependent TLR3 Expression of the Intestinal Epithelium Contributes To Rotavirus Susceptibility." *PLoS Pathog* 8 (5):e1002670. doi: 10.1371/journal.ppat.1002670 PPATHOGENS-D-11-02432 [pii].

Qin, J., R. Li, J. Raes, M. Arumugam et al. 2010. "A human Gut Microbial Gene Catalogue Established by Metagenomic Sequencing." *Nature* 464 (7285):59–65. doi: nature08821 [pii] 10.1038/nature08821.

Rawls, J. F., M. A. Mahowald, R. E. Ley, and J. I. Gordon. 2006. "Reciprocal Gut Microbiota Transplants from Zebrafish and Mice to Germ-Free Recipients Reveal Host Habitat Selection." *Cell* 127 (2):423–433.

Rawls, J. F., B. S. Samuel, and J. I. Gordon. 2004. "Gnotobiotic Zebrafish Reveal Evolutionarily Conserved Responses to the Gut Microbiota." *Proc. Natl. Acad. Sci. U. S. A.* 101 (13):4596–4601.

Richards, A. L., A. L. Muehlbauer, A. Alazizi, M. B. Burns et al. 2019. "Gut Microbiota Has a Widespread and Modifiable Effect on Host Gene Regulation." *mSystems* 4 (5). doi: 10.1128/mSystems.00323-18.

Ritter, D. I., Q. Li, D. Kostka, K. S. Pollard, S. Guo, and J. H. Chuang. 2010. "The Importance of Being Cis: Evolution of Orthologous Fish and Mammalian Enhancer Activity." *Mol Biol Evol* 27 (10):2322–2332. doi: 10.1093/molbev/msq128.

Roach, J. C., G. Glusman, L. Rowen, A. Kaur, M. K. Purcell, K. D. Smith, L. E. Hood, and A. Aderem. 2005. "The Evolution of Vertebrate Toll-Like Receptors." *Proc Natl Acad Sci U S A* 102 (27):9577–9582. doi: 10.1073/pnas.0502272102.

Roeselers, G., E. K. Mittge, W. Z. Stephens, D. M. Parichy, C. M. Cavanaugh, K. Guillemin, and J. F. Rawls. 2011. "Evidence for a Core Gut Microbiota in the Zebrafish." *ISME J* 5 (10):1595–1608. doi: ismej201138 [pii] 10.1038/ismej.2011.38.

Romero, I. G., I. Ruvinsky, and Y. Gilad. 2012. "Comparative Studies of Gene Expression and the Evolution of Gene Regulation." *Nat Rev Genet* 13 (7):505–516. doi: 10.1038/nrg3229.

Rosenberg, E., and I. Zilber-Rosenberg. 2016. "Microbes Drive Evolution of Animals and Plants: The Hologenome Concept." *MBio* 7 (2):e01395. doi: 10.1128/mBio.01395-15.

Roughgarden, J., S. F. Gilbert, E. Rosenberg, I. Zilber-Rosenberg, and E. A. Lloyd. 2018. "Holobionts as Units of Selection and a Model of Their Population Dynamics and Evolution." *Biological Theory* 13 (1):44–65. doi: 10.1007/s13752-017-0287-1.

Sankararaman, S., S. Mallick, M. Dannemann, K. Prufer, J. Kelso, S. Paabo, N. Patterson, and D. Reich. 2014. "The Genomic Landscape of Neanderthal Ancestry in Present-Day Humans." *Nature* 507 (7492):354–357. doi: 10.1038/nature12961.

Santos-Rosa, H., R. Schneider, A. J. Bannister, J. Sherriff, B. E. Bernstein, N. C. Emre, S. L. Schreiber, J. Mellor, and T. Kouzarides. 2002. "Active Genes are Tri-Methylated at K4 of Histone H3." *Nature* 419 (6905):407–411. doi: 10.1038/nature01080.

Schmidt, D., M. D. Wilson, B. Ballester, P. C. Schwalie et al. 2010. "Five-Vertebrate ChIP-seq Reveals the Evolutionary Dynamics of Transcription Factor Binding." *Science* 328 (5981):1036–1040. doi: science.1186176 [pii] 10.1126/science.1186176.

Schoenfelder, S., and P. Fraser. 2019. "Long-Range Enhancer-Promoter Contacts in Gene Expression Control." *Nat Rev Genet* 20 (8):437–455. doi: 10.1038/s41576-019-0128-0.

Seedorf, H., N. W. Griffin, V. K. Ridaura, A. Reyes et al. 2014. "Bacteria from Diverse Habitats Colonize and Compete in the Mouse Gut." *Cell* 159 (2):253–266. doi: 10.1016/j.cell.2014.09.008.

Shen, Y., F. Yue, D. F. McCleary, Z. Ye et al. 2012. "A Map of the Cis-Regulatory Sequences in the Mouse Genome." *Nature* 488 (7409):116–120. doi: 10.1038/nature11243.

Sheng, Y., H. Ren, S. M. Limbu, Y. Sun, F. Qiao, W. Zhai, Z. Y. Du, and M. Zhang. 2018. "The Presence or Absence of Intestinal Microbiota Affects Lipid Deposition and Related Genes Expression in Zebrafish (Danio rerio)." *Front Microbiol* 9:1124. doi: 10.3389/fmicb.2018.01124.

Shu, D. G., S. C. Morris, J. Han, Z. F. Zhang et al. 2003. "Head and Backbone of the Early Cambrian Vertebrate Haikouichthys." *Nature* 421 (6922):526–529. doi: 10.1038/nature01264.

Small, C. M., K. Milligan-Myhre, S. Bassham, K. Guillemin, and W. A. Cresko. 2017. "Host Genotype and Microbiota Contribute Asymmetrically to Transcriptional Variation in the Threespine Stickleback Gut." *Genome Biol Evol* 9 (3):504–520. doi: 10.1093/gbe/evx014.

Smemo, S., J. J. Tena, K. H. Kim, E. R. Gamazon et al. 2014. "Obesity-Associated Variants within FTO form Long-Range Functional Connections with IRX3." *Nature* 507 (7492):371–375. doi: 10.1038/nature13138.

Smillie, C. S., M. B. Smith, J. Friedman, O. X. Cordero, L. A. David, and E. J. Alm. 2011. "Ecology Drives a Global Network of Gene Exchange Connecting the Human Microbiome." *Nature* 480 (7376):241–244. doi: 10.1038/nature10571.

Stegmann, A., M. Hansen, Y. Wang, J. B. Larsen et al. 2006. "Metabolome, Transcriptome, and Bioinformatic Cis-Element Analyses Point to HNF-4 as a Central Regulator of Gene Expression During Enterocyte Differentiation." *Physiol Genomics* 27 (2):141–155. doi: 10.1152/physiolgenomics.00314.2005.

Stewart, A. J., S. Hannenhalli, and J. B. Plotkin. 2012. "Why Transcription Factor Binding Sites are Ten Nucleotides Long." *Genetics* 192 (3):973–985. doi: 10.1534/genetics.112.143370.

Stolfi, A., E. K. Lowe, C. Racioppi, F. Ristoratore, C. T. Brown, B. J. Swalla, and L. Christiaen. 2014. "Divergent Mechanisms Regulate Conserved Cardiopharyngeal Development and Gene Expression in Distantly Related Ascidians." *Elife* 3:e03728. doi: 10.7554/eLife.03728.

Sun, J., H. Zhong, L. Du, X. Li, Y. Ding, H. Cao, Z. Liu, and L. Ge. 2018. "Gene Expression Profiles of Germ-Free and Conventional Piglets from the Same Litter." *Sci Rep* 8 (1):10745. doi: 10.1038/s41598-018-29093-3.

Tassia, M. G., N. V. Whelan, and K. M. Halanych. 2017. "Toll-Like Receptor Pathway Evolution in Deuterostomes." *Proc Natl Acad Sci U S A* 114 (27):7055–7060. doi: 10.1073/pnas.1617722114.

Tennessen, J. A. 2005. "Molecular Evolution of animal Antimicrobial Peptides: Widespread Moderate Positive Selection." *J Evol Biol* 18 (6):1387–1394. doi: 10.1111/j.1420-9101.2005.00925.x.

Thaiss, C. A., M. Levy, T. Korem, L. Dohnalová et al. 2016. "Microbiota Diurnal Rhythmicity Programs Host Transcriptome Oscillations." *Cell* 167 (6):1495–1510.e12. doi: 10.1016/j.cell.2016.11.003.

Thaiss, C. A., D. Zeevi, M. Levy, G. Zilberman-Schapira et al. 2014. "Transkingdom Control of Microbiota Diurnal Oscillations Promotes Metabolic Homeostasis." *Cell* 159 (3):514–529. doi: 10.1016/j.cell.2014.09.048.

Thurman, R. E., E. Rynes, R. Humbert, J. Vierstra et al. 2012. "The Accessible Chromatin Landscape of the Human Genome." *Nature* 489 (7414):75–82. doi: nature11232 [pii] 10.1038/nature11232.

Villar, D., C. Berthelot, S. Aldridge, T. F. Rayner et al. 2015. "Enhancer Evolution Across 20 Mammalian Species." *Cell* 160 (3):554–566. doi: 10.1016/j.cell.2015.01.006.

Weirauch, M. T., and T. R. Hughes. 2010. "Conserved Expression without Conserved Regulatory Sequence: The More Things Change, The More They Stay The Same." *Trends Genet* 26 (2):66–74. doi: 10.1016/j.tig.2009.12.002.

Weirauch, M. T., A. Yang, M. Albu, A. G. Cote et al. 2014. "Determination and Inference of Eukaryotic Transcription Factor Sequence Specificity." *Cell* 158 (6):1431–1443. doi: 10.1016/j.cell.2014.08.009.

Welker, F., J. Ramos-Madrigal, M. Kuhlwilm, W. Liao et al. 2019. "Enamel Proteome Shows that Gigantopithecus was an Early Diverging Pongine." *Nature*. 576 (7786):262–265. doi: 10.1038/s41586-019-1728-8. Epub 2019 Nov 13.

Willing, B. P., S. L. Russell, and B. B. Finlay. 2011. "Shifting the Balance: Antibiotic Effects on Host-Microbiota Mutualism." *Nat Rev Microbiol* 9 (4):233–243. doi: 10.1038/nrmicro2536.

Woo, V., and T. Alenghat. 2017. "Host-Microbiota Interactions: Epigenomic Regulation." *Curr Opin Immunol* 44:52–60. doi: 10.1016/j.coi.2016.12.001.

Yatsunenko, T., F. E. Rey, M. J. Manary, I. Trehan et al. 2012. "Human Gut Microbiome Viewed Across Age and Geography." *Nature* 486 (7402):222–227. doi: nature11053 [pii] 10.1038/nature11053.

Yue, F., Y. Cheng, A. Breschi, J. Vierstra et al. and Encode Consortium Mouse. 2014. "A comparative Encyclopedia of DNA Elements in the Mouse Genome." *Nature* 515 (7527):355–364. doi: 10.1038/nature13992.

Zheng, W., T. A. Gianoulis, K. J. Karczewski, H. Zhao, and M. Snyder. 2011. "Regulatory Variation Within and Between Species." *Annu Rev Genomics Hum Genet* 12:327–346. doi: 10.1146/annurev-genom-082908-150139.

Zilber-Rosenberg, I., and E. Rosenberg. 2008. "Role of Microorganisms in the Evolution of Animals and Plants: The Hologenome Theory of Evolution." *FEMS Microbiol Rev* 32 (5):723–735. doi: 10.1111/j.1574-6976.2008.00123.x.

16 Molecular interactions of microbes and the plant phyllosphere

The phyllosphere-microbiome is shaped by the interplay of secreted microbial molecules and the plant immune system

*Janine Haueisen, Cecile Lorrain,
and Eva H. Stukenbrock*

Contents

16.1 Introduction: Multi-partite microbial interactions in the plant phyllosphere

Humans are inhabited by a large number of microbes – recent calculations estimated an approximate 1:1 ratio between bacterial and human cells in a body (Sender, Fuchs, and Milo 2016) – and microbiota hold a key role for human and animal health and function (Cho and Blaser 2012; Ezenwa et al. 2012). Likewise, plants host a broad diversity of microbes that colonize interior spaces (endophytic colonization) and surfaces (epiphytic colonization) of all below- and above-ground organs. On the global scale the plant phyllosphere, defined as the space in and around all aerial plant parts (Figure 16.1), is estimated to exceed 10^8 km^2, and thereby comprises the largest environmental surface for microbial colonization on the planet (Peñuelas and Terradas 2014; Vorholt 2012). The phyllosphere represents a structurally hyper diverse ecosystem and is densely populated (Vacher et al. 2016). On average, every square centimetre of leaf surface is colonized by 5.4×10^8 bacterial cells, and epiphytic bacteria alone outnumber plant cells (Peñuelas and Terradas 2014; Remus-Emsermann et al. 2014). In terms of abundance, phyllosphere microbiota are dominated by bacteria and most of them belong to a few phyla (Bulgarelli et al. 2013; Delmotte et al. 2009). Fungi are less abundant, but leaf-colonizing taxa are highly diverse (Jumpponen and Jones 2009). Together with oomycetes, archaea and viruses, bacteria and fungi in the phyllosphere assemble complex epi- and endophytic multi-kingdom microbial communities that impact plant physiology, health, nutrition, and the ability of plants to adapt to environmental changes (Hassani et al. 2019; Vacher et al. 2016; Vandenkoornhuyse et al. 2015; Vannier, Agler, and Hacquard 2019).

Despitethe gigantic dimension of the phyllosphere, research on plant-associated microbiota initially focused on rhizosphere communities (Bulgarelli et al. 2012; Lundberg et al. 2012). The beneficial effects of root-associated mutualistic symbionts like mycorrhiza and nitrogen-fixing rhizobia on plant health and nutrition have long been known (Martin, Uroz, and Barker 2017). Moreover,

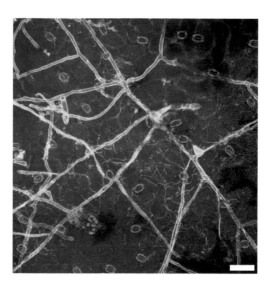

FIGURE 16.1   Phyllosphere densely populated by a diversity of microbial colonizers.
Micrograph shows epiphytic colonization of the *Plantago lanceolata* phyllosphere. Plant cells, including stomatal guard cells and nuclei, including prokaryotic DNA, are visualized in purple, filamentous microbial structures are displayed in green. Scale bar = 50 μm.

rhizosphere-colonizing microbes can play important roles for plant health by increasing nutrient availability, production of plant hormones or suppression of disease (Haas and Défago 2005).

Research on plant–microbe symbioses in the phyllosphere has mainly focused on plant–pathogen interactions because leaf pathogens greatly impact crop production (Savary et al. 2019). Nevertheless, the phyllosphere also hosts a diverse spectrum of symbionts including mutualists, commensals, and parasites. A well-studied example of a phyllosphere fungus, that can be both a mutualist and parasite, is the ascomycete *Epichloë festucae*. Fungi belonging to the genus *Epichloë* colonize the aerial tissues of several Pooideae grasses. *Epichloë* endophytes synthesize toxic alkaloids that protect their host grasses against herbivores (Kauppinen et al. 2018; Schardl 1996; Schardl et al. 2013); they simultaneously induce dramatic reprogramming of the hosts' metabolism (Dupont et al. 2015) that can lead to 'choke disease' in which the fungal infection leads to abortion of host inflorescences (Schardl, Leuchtmann, and Spiering 2004). This interaction thereby constitutes a dynamic balance between mutualism and parasitism (Scott, Green, and Berry 2018). The phyllosphere also hosts beneficial bacteria. For example, nitrogen-fixing cyano- and proteobacteria have been identified in phyllospheres of tropical rainforest plants. These bacteria associate with other leaf-colonizing epiphytes and likely play a role in nitrogen cycling in tropical ecosystems (Fürnkranz et al. 2008). A multitude of phyllosphere-inhabiting pathogenic microbes have also been described. Fungal leaf pathogens, including the wheat-infecting *Zymoseptoria tritici* and the *Plantago lanceolata*-infecting powdery mildew *Podosphaera plantaginis,* harm their host plants by causing disease. Pathogen infections in the phyllosphere can have economic impacts such as severe yield losses caused by *Z. tritici* (Fones and Gurr 2015). Negative fitness consequences can extend beyond the infected host plant to insect populations and other organisms that associate with plants (Laine 2004).

A whole field of research is dedicated to unravelling the molecular mechanisms underlying the establishment and maintenance of bipartite plant–microbe interactions. The model plant species *Arabidopsis thaliana* (hereafter only referred to as *Arabidopsis*) has been a primary model organism in the study of plant–microbe interactions and plant-associated microbiota (Müller et al. 2016; Nishimura and Dangl 2010). More recent studies have addressed the diversity and role of microbial communities associated with a variety of plant species, including *Arabidopsis* and diverse crop species (reviewed in Müller et al. 2016). These studies underline the diversity of interactions between plants and their associated microbes. One plant simultaneously engages in different symbioses with an enormous number of microbes that compete, co-exist or cooperate with each other across taxa and kingdoms (Abdullah et al. 2017; Hassani, Durán, and Hacquard 2018). These interactions create dynamic networks consisting of direct plant–microbe and microbe–microbe interactions, as well as indirect interactions mediated by microbes or by the host itself. Microbes that hold central positions in these networks become disproportionally important and have been termed 'hubs' or keystone species (Agler et al. 2016). One of the hub species identified, among natural *Arabidopsis* leaf microbiota, is the obligate biotrophic pathogen *Albugo laibachii*. Although *Arabidopsis* endo- and epiphytic communities were less diverse in the presence of *Albugo*, the microbiota were more stable (Agler et al. 2016). This emphasizes the role of microbial hubs in shaping host microbiota and also demonstrates that the influence of plant pathogens extends beyond manipulation and exploitation of host physiology (Seybold et al. 2020).

As in animals, plants acquire microbes from their environment, and community assembly is not random. Rhizosphere microbiome community compositions are mainly driven by the availability of microbes in surrounding soil (Bulgarelli et al. 2012). Soil also represents a main reservoir for phyllosphere microbiota (Zarraonaindia et al. 2015) because root and phyllosphere taxa overlap (Bai et al. 2015). Additional microbes can be added by biotic and abiotic vectors, including neighbouring plants, insects, wind or precipitation (Hassani et al. 2019). In higher plants, a fraction of the microbiota is inherited from the mother plants. Seed-associated microbes represent an important source of inoculum that is vertically transmitted and may represent prolonged co-existence and co-evolution with the plant (Özkurt et al. 2019).

Due to environmental fluctuations, phyllosphere microbiota are less stable compared to rhizosphere communities, and host genotype and seasonal environmental changes can influence leaf microbiota (Copeland et al. 2015; Grady et al. 2019). Endophytic microbes have to deal with host plant physiology and have adapted strategies to infect and explore host tissues (Haueisen and Stukenbrock 2016). In contrast, epiphytic microbes use the host mainly as a surface for colonization but still need strategies to cope with diverse and rapidly changing environmental influences.

Our current understanding of how plants manage root- and leaf-associated microbes is mainly based on studies focusing on specific one-to-one interactions between model plants – or a few crop species – and individual symbionts. We are only beginning to understand how the fundamental molecular principles underlying plant–microbe interactions apply in natural, multipartite interactions. How plants simultaneously defend against pathogens and recruit and maintain symbioses with mutualistic symbionts is largely unknown and are top questions in plant–microbe research. Similarly, how can plants discriminate between beneficial and pathogenic microbes that often use highly similar colonization strategies? This chapter aims to outline what is known about how plants shape the composition and structure of their phyllosphere microbiota. We propose that insights from molecular phytopathology, especially research on the plant immune system and how it is manipulated by microbial molecules, are instrumental to decipher how plants assemble and maintain their associated microbiota. Furthermore, we summarize recent findings that support a key role of plant pathogens in shaping host microbiota through indirect microbe–microbe interactions mediated by modulation of host immunity.

16.2   The plant immune system as a microbial management system

While the environment, and in particular the soil, is the main source of plant-associated microbes, the actual assembly of the phyllo-microbiome is not random. Instead, plant-derived factors impact its structure and composition (Hassani et al. 2019). A study on rhizosphere bacterial consortia associated with several *Arabidopsis thaliana* relatives demonstrated that microbial community compositions overlap between genotypes, but host-specific microbiota increased with increasing phylogenetic distance (Schlaeppi et al. 2014). Moreover, research on plant pathogens provides examples of highly specific interactions between plant and microbial species that are sometimes limited to individual host genotypes because of host adaptation (Haueisen and Stukenbrock 2016; van der Does and Rep 2017). To identify the host genetic traits that underlie differences in bacterial and fungal phyllosphere microbiota compositions, Horton and colleagues performed a genome-wide association study (GWAS) including 196 *Arabidopsis thaliana* genotypes grown together at the same field site. Differences among the most abundant taxa (comprising more than 99% of all sequenced reads) could be explained by host genotype. In particular, genetic variation in genes involved in defence responses, signal transduction/kinase-related activities and cell wall integrity all correlated with microbial composition (Horton et al. 2014). These gene categories are highly relevant for abiotic and biotic interactions of plants. It is plausible that genetic variation in these candidate sequences can translate into phenotypic differences affecting plant microbiota.

Genes involved in defence and immune signalling may contribute to the assembly of plant microbiota. It has even been proposed that the plant immune system should be considered as a 'microbial management system' in the context of microbiota assembly (Hacquard et al. 2017). In contrast to the adaptive immune system of animals, the plant immune system is innate and every cell is involved in recognition, signalling and response. Plant immunity is conferred by a multi-layered surveillance system consisting of inter- and intra-cellular receptors able to detect a wide range of microbial and cell-wall damage-associated molecules (Dodds and Rathjen 2010; Jones and Dangl 2006). Recognition signals become integrated and 'tuned' by a complex signal-integration layer comprising plant mitogen-activated protein kinase (MAPK) cascades, transcription factors, plant hormones (including salicylic acid, jasmonic acid and ethylene), reactive oxygen and nitrogen species and small RNAs (Wang, Tyler, and Wang 2019). Adjusted immune responses result in a range of plant defences from local cell wall enforcements, synthesis and secretion of antimicrobial and toxic compounds to programmed host cell death and systemic effects (Wang, Tyler, and Wang 2019). Consequently, plant endospheres are only colonized by microbes able to avoid recognition or withstand immune responses – either by host specialization or by synergistic interactions with host-adapted microbes (McMullan et al. 2015; Seybold et al. 2020).

16.2.1   Do pattern recognition receptors (PRRs) direct microbiota assembly?

The first line of non-self recognition in plants consists of cell surface-localized pattern recognition receptors (PRRs) that serve to detect broadly distributed, conserved microbial molecules such as bacterial flagellin, lipopolysaccharides, peptidoglycans and fungal chitin (Dodds and Rathjen 2010). These essential molecules are summarized as microbe-associated molecular patterns (MAMPs) and can elicit MAMP-triggered immunity (MTI) leading to a cascade of antimicrobial responses. MTI plays an important role for non-host resistance – the reason why not all pathogens attempt infection of all plants (Jones and Dangl 2006) – and is proposed to play a pivotal role in shaping and managing the plant microbiome

(Hacquard et al. 2017). Plant PRRs contain (1) extracellular domains for ligand binding, (2) transmembrane domains attaching the receptor to the cell membrane and, (3) in the case of receptor kinases (RKs), intracellular kinase domains involved in downstream signalling. PRRs without a kinase domain are called receptor-like proteins (RLPs) and function in complexes with one or several RKs that facilitate intracellular transduction of ligand binding (Macho and Zipfel 2014). Well-studied examples of PRRs are the RLKs CERK1 (chitin elicitor receptor kinase 1) and FLS2 (flagellin sensing 2) (Gómez-Gómez and Boller 2000; Miya et al. 2007). Both have been identified and characterized in *Arabidopsis*. CERK1 possesses three extracellular LysM domains that bind fungal chitin. Binding of long oligomers leads to CERK1 homodimerization and the generation of an active receptor complex (Liu et al. 2012). Further research suggests that the structurally similar lysin motif receptor kinase 5 (LYK5) is the major chitin receptor in *Arabidopsis*. This receptor forms complexes with CERK1 to initiate chitin-triggered immunity (Cao et al. 2014). The 22-amino acid flagellin epitope flg22 is bound by the leucin-rich repeat ectodomain of FLS2, thereby stabilizing heterodimerization between FLS2 and another RLK called BAK1, which will initiate signalling (Chinchilla et al. 2007; Sun et al. 2013). Signalling upon PRR activation leads to MTI whereby defence-related genes are up-regulated and induce, for example, deposition of callose to enhance cell walls and the secretion of antimicrobial compounds (Wang, Tyler, and Wang 2019). These defence responses confer local resistance against most of the microbes attempting to invade the plant. While several MAMPS and the respective PRRs are well characterized, there is still a large number of potential MAMPS including bacterial quorum sensing molecules, siderophores and microbial proteins that have not yet been characterized (Boller and Felix 2009).

If plants are able to detect microbial elicitors in a generalist manner and induce defence responses, how is microbiota establishment even possible? Due to their essential role in fundamental biological processes such as basic cell wall building blocks, MAMPs have been considered to be conserved and broadly distributed. However, more and more evidence has accumulated showing that MAMPs are also subject to diversification. Indeed, it was shown in six plant pathogenic bacteria that the genes encoding putative elicitors are part of the core genome, and exhibit patterns of strong purifying selection required to maintain function. Still, distinct regions of these genes, likely encoding structurally and functionally dispensable residues, accumulate mutations and show elevated rates of non-synonymous substitutions indicative of positive selection to avoid host immune recognition (Cai et al. 2011; McCann et al. 2012). Further, microbial populations can harbour different variants of the same MAMP and these variants can cause stronger or weaker defence responses in a given host genotype (M. Vetter, Karasov, and Bergelson 2016; W. Sun et al. 2006; Cai et al. 2011). Established microbiota presumably possess MAMP variants that are either non-detectable or cause weak defence responses upon recognition in the local host genotypes.

Considerable variation in MAMP recognition by plants has also been demonstrated, and this variation may allow plants to tailor defence response intensities (Hacquard et al. 2017). Studies in *Arabidopsis* showed that the strength of plant defence responses upon recognition of flg22 is different from responses to elongation factor TU (EF-Tu), another bacterial MAMP (M. Vetter, Karasov, and Bergelson 2016). This demonstrates that plant responses to different MAMPs are not redundant, and this non-redundancy may enable selection for specific microbes while defending against others. Another possibility is that the plant is 'blind' to certain MAMPs. PRR abundance on the cell surface exhibit variation in spatial distribution (Beck et al. 2012). Expression of receptor-encoding genes varies between different plant tissues and with plant age (Beck et al. 2014). As a result, temporal, spatial or permanent lack of specific PRRs might indirectly select for colonization of certain microbes. Except for the flg22-binding FS2, which has been identified in a large number of plants across different families (Robatzek et al. 2007; Takai et al. 2008; Zipfel et al. 2004), different plant species are equipped with different PRR repertoires. Even at the genus or species level, perception of the identical microbial epitope by the same PRR and subsequent response induction can vary. Analyses of flg22 binding, FLS2 receptor abundance and flg22-elicited responses in 45 accessions of *Arabidopsis* and its relatives in the Brassicaceae revealed high levels of quantitative variation in flg22 perception and severity of the elicited defence response (M. M. Vetter et al. 2012). These results indicate that MAMP perception evolves quantitatively, possibly indicating adaptation to recruit and defend against local microbial communities.

16.2.2   Do intra-cellular NLR proteins contribute to assembly of microbial consortia?

While the first layer of plant defence detects and responds to general non-self patterns, the second layer involves the recognition and protection against specific microbes. Every plant cell has intracellular sensors, constituted of nucleotide-binding leucine-rich repeat-containing proteins (NLRs), also called resistance (R) proteins (Wang, Tyler, and Wang 2019). These receptors are directly or indirectly involved in the detection of cytoplasmic elicitors; these elicitors are mostly secreted microbial molecules called effectors (see below). Some R proteins recognize effectors through direct binding and others sense their actions e.g. by monitoring effector targets ('guardee') or target mimics ('decoy') (Cui, Tsuda, and Parker 2015). Recognition leads to effector-triggered immunity (ETI) (Jones and Dangl 2006), the molecular manifestation of the classic gene-for-gene concept (Flor 1971). ETI can be triggered by all microbial effectors, irrespective whether they were secreted by a pathogen or a mutualist and is usually associated with a hypersensitive response and programmed cell death (Wang, Tyler, and Wang 2019). The interplay of microbial effectors and plant NLRs is highly specific and determines the outcome of the interactions between the plant and microbe (i.e. plant susceptibility or resistance). As a result, effector- and plant NLR-encoding genes are usually under strong selection pressure. The co-evolution of plant and pathogen genes can follow an 'Arms-race scenario' (directional selection of specific alleles) or a 'Trench warfare scenario' (diversifying selection favouring different alleles in the population) (Möller and Stukenbrock 2017). A small number of non-synonymous nucleotide changes in the plant or microbe can occur rapidly and alter the efficiency of effector recognition and downstream signalling (Daverdin et al. 2012; Yang et al. 2010). At the same time, exposure to strong selective forces might also contribute indirectly to selection and recruitment of specific strains for the assemblage, namely by non-recognition of particular species or strains. In the context of beneficial host–microbe interactions in the rhizosphere, it was demonstrated that distinct alleles of the soybean genes *Rj2* and *Rfg1* determine the outcome of nodulation with Rhizobium (Yang et al. 2010). Both encode typical NLR receptors that serve as nodulation-restrictive R proteins. Recognition of yet unknown effectors secreted by specific strains of *Bradyrhizobium japonicum* and *Sinorhizobium fredii* activates defence responses that terminate colonization and nodulation. This variation in symbiotic specificity may serve to recruit rhizobial strains with high nitrogen-fixing efficiency and may block establishment of less efficient or cheating genotypes (Yang et al. 2010). In addition to their central function in plant pathogen recognition, NLRs may also play an important role in 'protecting' symbiosis pathways against exploitation by non-beneficial microbes (Hacquard et al. 2017). As a result, NLR proteins may be essential for the establishment and maintenance of beneficial microbes in the phyllosphere.

16.2.3   Interactions of plant pathogens and the plant microbiota: Systemic effects in susceptible and resistant plants

Plant–microbe interactions may activate the plant immune system, or microbes may suppress immunity by secreting effector molecules (see below). Activation of the plant immune system by MTI or ETI triggers antimicrobial defences that can comprise local or systemic responses that alter the microbial host environment (Jones and Dangl 2006; Wang, Tyler, and Wang 2019). Incompatible pathogen infections on resistant hosts can lead to systemic acquired resistance (SAR) that causes the production of defence compounds in non-infected tissues (Durrant and Dong 2004). Stimulation of the plant immune system by beneficial microbes can also cause an alert state with similar positive effects. This immune stimulation is referred to as 'priming' or 'induced systemic resistance' (ISR) (Pieterse et al. 2014). Accumulation and signalling of the plant hormone salicylic acid (SA) are essential for the establishment of SAR. Several metabolites including methyl ester of SA, azaleic acid and pipecolic acid are involved in the spread of SAR signals from local to systemic tissues. A consequence of SAR is the coordinated up-regulation of Pathogenesis-Related (PR) genes that predominantly encode proteins with anti-microbial properties (van Loon, Rep, and Pieterse 2006). Signalling by jasmonic acid (JA) and ethylene are central for ISR that is mostly independent of SA (Pieterse et al. 2014). These hormones contribute to priming of the plant and allow faster defence against pathogen attack, stronger defence against pathogen attack, or both (Conrath et al. 2015). Because pathogen-induced SAR and mutalist-induced ISR are regulated by different pathways, they also protect against different microbes (Ton et al. 2002). As a result, previous mutualistic interactions and pathogen infections indirectly affect future plant–microbe interactions and thereby shape plant-associated microbiota. Moreover, pathogen infections in the *Arabidopsis* phyllosphere can alter root exudate profiles that in turn attract different sets of microbes, which may be beneficial for the plant (Berendsen et al. 2018; Stringlis et al. 2018; Yuan et al. 2018). Endophytic bacteria in sugar beet roots that produce specific disease-suppressive compounds, including chitin degrading enzymes and antifungal metabolites, become enriched when the roots are attacked by the fungal pathogen *Rhizoctonia solani* (Carrión et al. 2019). Although the underlying molecular interactions are still unknown, this study reveals a mechanism by which microbes provide another level of immunity to their host plants, and underlines the functional importance of specific microbiota for plant health (Carrión et al. 2019).

In contrast, infections by virulent pathogens that successfully manipulate host immunity can enhance susceptibility to the extent that they promote the growth of otherwise incompatible microbes (McMullan et al. 2015; Menardo et al. 2016; Seybold et al. 2020). A study that compared compatible and incompatible wheat infections by the fungal pathogen *Zymoseptoria tritici* demonstrated that infection of the susceptible wheat cultivar Obelisk leads to enhanced microbial colonization, both locally at the fungal infection site and also systemically in other leaves (Figure 16.2) (Seybold et al. 2020). During compatible infections, *Z. tritici* colonizes leaf tissue for a prolonged phase of biotrophic, symptomless growth whereby the host immune system is actively suppressed by the fungus. *Pseudomonas syringae* pathovars inoculated (1) locally with the fungal pathogen, (2) at an adjacent leaf area, and (3) on another leaf benefit from this immune suppression as their growth was significantly increased in plants successfully infected by *Z. tritici*. Manipulation of wheat immunity by *Z. tritici* also facilitated significantly increased proliferation of the non-adapted *P. syringae* pathovar *tomato* which is otherwise unable to colonize wheat leaves (Seybold et al. 2020). Seybold and co-authors propose calling this phenomenon 'systemic induced susceptibility (SIS)' because active suppression of the immune system renders the plant more vulnerable to microbial colonization, even in non-infected tissues. Furthermore, incompatible infections of the resistant wheat cultivar Chinese Spring have the opposite effect on bacterial growth, possibly because of SAR (Figure 16.2). Immune-related biosynthetic pathways became activated and antimicrobial metabolites accumulated locally as well as systemically in the resistant wheat when infected with *Z. tritici* (Seybold et al. 2020). These compounds likely help to block fungal infections, but they have the potential to exert negative effects on other phyllosphere microbiota because they are non-specific. While both SIS and SAR induced by *Z. tritici* influenced phyllosphere bacterial community structure, only incompatible *Z. tritici* infections of the resistant wheat Chinese Spring affected community composition and led to significantly reduced microbial alpha diversity (Seybold et al. 2020). While the biological relevance of SIS for *Z. tritici* still needs to be clarified, one may speculate that it plays a role in facilitating secondary infections of the same host, or that it contributes to phyllosphere microbiota recruitment beneficial for the pathogen. On the other hand, SIS may merely be an unavoidable consequence of compatible infections that can be exploited by non-adapted microbes. It could even provide a route for host jumps and the emergence of new pathogens (Feurtey et al. 2019; McMullan et al. 2015; Menardo et al. 2016). By activation or silencing of the host immune system, the compatibility of individual host–microbe interactions has cascading consequences for other microorganisms, and can thereby indirectly shape microbiota composition by induction of SAR or SIS.

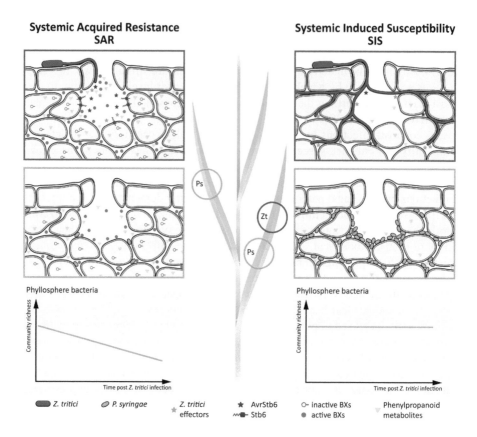

FIGURE 16.2 *Zymoseptoria tritici* **infections impact bacterial phyllosphere colonization by systemic acquired resistance (SAR) and systemic induced susceptibility (SIS) of wheat cultivars**. SAR: Resistance against *Z. tritici* in the wheat cultivar Chinese Spring is mediated by Stb6, a wall-associated receptor kinase-like protein involved in detection of the secreted fungal effector AvrStb6 (Kema et al. 2018; Saintenac et al. 2018; Zhong et al. 2017). Antimicrobial benzoxazinoids (BXs) and metabolites derived from the phenylpropanoid biosynthetic pathway (green triangles) accumulate locally as well as systemically, both adjacent to the fungal infection site and in another leaf. Invading fungal hyphae are blocked early during wheat infection. Growth of *Pseudomonas syringae* pathovars is impaired both locally and systemically, and alpha diversity of phyllosphere bacterial communities decreases. **SIS**: Immunity of the susceptible wheat cultivar Obelisk is actively suppressed by *Z. tritici*, most likely by secreted effectors including AvrStb6. Fungal hyphae can then colonize the intercellular space of the wheat mesophyll. Very small quantities of antimicrobial compounds can be detected and *P. syringae* pathovars *oryzae* and *tomato* proliferate in the mesophyll tissue of Obelisk adjacent to the fungal infection as well as in another leaf. Richness of bacterial communities in the phyllosphere is stable. (Seybold et al. 2020) Zt: Fungal infection site (red frame). Ps: Bacterial infection site adjacent to fungal infection or on another leaf (orange frames).

16.3   Microbial effectors mediate plant-microbe interactions

An intriguing question in the field of plant microbiota is how microbes that are tightly associated with plant tissues cope with plant immune responses, and how the plant distinguishes foe from friend. Little is known about the molecular interactions of plants with commensal microorganism. However, years of intensive research on the molecular biology of pathogenic and beneficial microbes have shed light on microbial molecules that are produced and secreted to manipulate host defences and avoid immune recognition. A diversity of small proteinaceous and non-proteinaceous molecules such as toxins, metabolites and small RNAs have been identified to play a central role in the interaction of microbes with plants (Collemare, O'Connell, and Lebrun 2019; Franceschetti et al. 2017). Collectively, these secreted molecules are termed effectors according to their effect on the immune system of the plant, and these are produced by plant pathogens as well as mutualistic symbionts to facilitate plant colonization (Rovenich, Boshoven, and Thomma 2014). Some effectors function in the cytoplasm of the plant while others are translocated into the plant cell where they interfere with processes like immune signalling and plant metabolism.

Studies aimed to identify effectors in microbial genomes have provided hundreds of candidate effectors in plant-associated species from parasites to symbionts. However, functional research involving gene deletions has mainly addressed the effect of effector candidates on the plant interaction. Using reverse genetics, a pathogenicity related function has only been documented for a small subset of the multitude of effector candidates that are predicted in the genomes of plant symbionts. So, what is the role of these secreted small molecules? Likely, many of these molecules play a role in microbe–microbe interactions (Table 16.1), for example, in competition or cooperation between different microbes (Rovenich, Boshoven, and Thomma 2014). Snelders and colleagues proposed a classification of effector proteins that takes into account the potential role of effectors in multipartite interactions (Snelders et al. 2018). Hereby, effector molecules produced by plant-associated microbiota can be classified

---

TABLE 16.1
Examples of Effectors Involved in Microbe–Microbe Interactions

Effector	Species	Type	Function(s)	Reference
Zt6	*Zymoseptoria tritici*	Plant pathogen	Ribonuclease activity and toxicity against bacteria and fungi	Kettles et al. (2018)
Tse2	*Pseudomonas aeruginosa*	Non-pathogenic saprophyte	Cytotoxic activity against prokaryote cells	Hood et al. (2010)
Tfe1; Tfe2	*Serratia marcescens* Db10	Insect pathogen	Antifungal and antibacterial toxin	Fritsch et al. (2013)
EI pairs	*Agrobacterium tumefaciens*	Plant pathogen	Antimicrobial activities	Wu et al. (2019)
X-Tfes	*Xanthomonas citri*	Plant pathogen	Lysis of Gram-positive cells	Souza et al. (2015)

as (i) effectors that specifically target plant molecules, (ii) effectors that have more than one target and engage in both plant–microbe and microbe–microbe interaction, and (iii) effectors that are involved in interactions with other microbial species.

So far effectors with host targets are the best described among these. Interestingly, there are examples of plant pathways which are targets of different effectors produced by the same bacteria. This is the case for a plant immune kinase pathway that can be silenced by different effectors produced by the bacterium *P. syringae*. The effectors AvrPto, AvrPtoB, HopF2, and HopB1 all target the receptor kinase BAK1 in *Arabidopsis* (Li et al. 2016; Shan et al. 2008; Zhou et al. 2014). This pattern underlines the frequent occurrence of convergent evolution in species that co-exist, although in different ways, with plants.

There are so far few examples of effectors that have more than one target and engage in both host–microbe and microbe–microbe interaction. One example is the effector Zt6 in the wheat pathogen *Z. tritici* (Kettles et al. 2018). Zt6 is a secreted ribonuclease that cleaves both plant and animal rRNA species, and is toxic to wheat, tobacco, bacterial and yeast cells, but not to the fungus itself.

Effectors that play a role in the interaction with other microbial species often comprise non-proteinaceous effectors, such as secondary metabolites with antibacterial or antifungal properties. These effectors have been described in the broadest range of microbial lifestyles including parasite, symbionts as well as saprotrophs (Snelders et al. 2018). The non-proteinaceous effector phenazine-1-carboxamide secreted by the bacterium *Pseudomonas piscium* directly inhibits the growth of *Fusarium graminearum* through deregulation of histone acetylation (Chen et al. 2018). Also, bacterial T6SS-effectors are widespread in both plant- and animal-associated bacteria. These effectors appear to be particularly relevant in microbe–microbe interactions as several have been shown to hold antimicrobial properties (Fritsch et al. 2013; Trunk, Coulthurst, and Quinn 2019). Clearly, as we increase culture collections of plant-associated microbes, we will be able to gain more insight into the diversity and function of effectors involved in microbe–microbe interactions.

The distribution of effectors in plant-associated microorganisms has been shown to be highly variable. Some effectors are highly conserved across taxa, while others can show highly diverse patterns with respect to distribution among microbial taxa. Thereby, effectors can be distinguished as 'core' and 'non-core' effectors depending on their distribution among microbial taxa (Irieda et al. 2019). Core effectors are conserved and present across a diversity of microbial species, while non-core or accessory effectors are present at varying frequencies, even between individuals of the same species. The fungal effector NIS1 represents a core effector that is encoded in the genome of a variety of fungal plant pathogen species, including species of Ascomycetes and Basidiomycetes (Irieda et al. 2019). This protein targets the innate immunity of plants by suppressing the PRR-associated kinases BAK1 and BIK1. Effector proteins otherwise often show little sequence conservation because they co-evolve with their host targets. However, a number of specific motifs have been identified in effectors, and shown to be crucial for their interaction and function in host cells. These include motifs that are important for protein translocation or binding of conserved plant motifs, for example RXLR, CRN, LysM, RGD, DELD, EAR, RYWT, Y/F/WXC or CFEM motifs (Dong and Wang 2016).

Comparative genome studies that have focused on the secretomes of plant-associated fungi, have identified common features among parasites and symbionts (Lo Presti et al. 2015). Both beneficial and parasitic fungi comprise large repertoires of secreted protein-encoding genes (Lo Presti et al. 2015). It is interesting to note that genomes of mutualistic symbionts and biotrophic pathogens encode reduced sets of secreted plant cell wall-degrading enzymes and secondary metabolites reflecting their shared biotrophic lifestyle. The same genomes harbour large sets of genes encoding secreted proteins predicted to be effectors with yet unknown functions. This suggests that effectors are ancestral traits of plant-associated microbes. If microbiota-manipulating effectors are crucial in microbial community assembly, some effectors involved in microbe–microbe interactions could be shared among species that co-exist in the same niche. Metagenome sequencing and comparative genome analyses of microbes inhabiting the same plant niche will, in the future, allow us to assess the distribution of effectors in microbial networks, notably effectors encoded by hub species.

16.4   Perspective

A key question to be addressed in the field of plant–microbiota research is how plants distinguish beneficial and antagonistic microbes (among the 'Top 10 Question of the 2019 IS-MPMI Congress'). Many years of research have shed light on the complex interactions of mutualistic symbionts like mycorrhizal fungi and nitrogen-fixing rhizobia, as well as antagonistic pathogens including biotrophs and necrotrophs. Common for all these interactions are microbial produced effector molecules that interfere with the plant immune system and defence signalling. Genes encoding effector proteins can be predicted in microbial genomes according to the presence of a signal peptide, conserved domains, expression patterns and gene size (Sperschneider et al. 2018). Interestingly, most of these predicted effectors have no known function, and many functional studies using reverse genetics approaches have failed to identify infection-related phenotypes. Accumulating evidence suggests that effector molecules also are instrumental in microbe–microbe interactions. A future challenge will be to study these interactions and identify effector targets in complex microbial communities. To this end we rely on culture collections from different model plant species.

So far, plant–microbe interactions have been studied in highly reduced systems with mainly two components, the plant and one microbial species. In nature plants engage simultaneously with a huge diversity of microorganisms, some being beneficial other pathogenic. Future research on plant microbiota will address how plants integrate diverse signalling pathways to optimize the composition of their associated microbiota. An ultimate goal in this field of research is to understand and use the complexity of plant–microbial associations to improve plant health.

Acknowledgments

Microbiota research in the lab of E.H. Stukenbrock is funded by the DFG supported Collaborative Research Center SFB1182 and CIFAR. Research by C. Lorrain is funded by a post-doctoral fellowship from the French "Institut national de la recherche agronomique" (INRA).

References

Abdullah, A. S., C. S. Moffat, F. J. Lopez-Ruiz, M. R. Gibberd, J. Hamblin, and A. Zerihun. 2017. "Host–Multi-Pathogen Warfare: Pathogen Interactions in Co-Infected Plants." *Frontiers in Plant Science* 8 (October): 1–12. doi:10.3389/fpls.2017.01806.

Agler, M. T., J. Ruhe, S. Kroll, C. Morhenn, S.-T. Kim, D. Weigel et al. 2016. "Microbial Hub Taxa Link Host and Abiotic Factors to Plant Microbiome Variation." *PLOS Biology* 14 (1): e1002352. doi:10.1371/journal.pbio.1002352.

Bai, Y., D. B. Müller, G. Srinivas, R. Garrido-Oter, E. Potthoff, M. Rott et al. 2015. "Functional Overlap of the Arabidopsis Leaf and Root Microbiota." *Nature* 528 (7582): 364–9. doi:10.1038/nature16192.

Beck, M., I. Wyrsch, J. Strutt, R. Wimalasekera, A. Webb, T. Boller et al. 2014. "Expression Patterns of Flagellin Sensing 2 Map to Bacterial Entry Sites in Plant Shoots and Roots." *Journal of Experimental Botany* 65 (22): 6487–98. doi:10.1093/jxb/eru366.

Beck, M., J. Zhou, C. Faulkner, D. L. Mac, and S. Robatzek. 2012. "Spatio-Temporal Cellular Dynamics of the Arabidopsis Flagellin Receptor Reveal Activation Status-Dependent Endosomal Sorting." *Plant Cell* 24 (10): 4205–19. doi:10.1105/tpc.112.100263.

Berendsen, R. L., G. Vismans, K. Yu, Y. Song, R. De Jonge, W. P. Burgman et al. 2018. "Disease-Induced Assemblage of a Plant-Beneficial Bacterial Consortium." *ISME Journal* 12 (6): 1496–1507. doi:10.1038/s41396-018-0093-1.

Boller, T., and G. Felix. 2009. "A Renaissance of Elicitors: Perception of Microbe-Associated Molecular Patterns and Danger Signals by Pattern-Recognition Receptors." *Annual Review of Plant Biology* 60 (1): 379–406. doi:10.1146/annurev.arplant.57.032905.105346.

Bulgarelli, D., M. Rott, K. Schlaeppi, E. Ver Loren van Themaat, N. Ahmadinejad, F. Assenza et al. 2012. "Revealing Structure and Assembly Cues for Arabidopsis Root-Inhabiting Bacterial Microbiota." *Nature* 488 (7409): 91–95. doi:10.1038/nature11336.

Bulgarelli, D., K. Schlaeppi, S. Spaepen, E. Ver Loren van Themaat, and P. Schulze-Lefert. 2013. "Structure and Functions of the Bacterial Microbiota of Plants." *Annual Review of Plant Biology* 64 (1): 807–38. doi:10.1146/annurev-arplant-050312-120106.

Cai, R., J. Lewis, S. Yan, H. Liu, C. R. Clarke, F. Campanile et al. 2011. "The Plant Pathogen Pseudomonas Syringae Pv. Tomato Is Genetically Monomorphic and under Strong Selection to Evade Tomato Immunity." *PLoS Pathogens* 7 (8). doi:10.1371/journal.ppat.1002130.

Cao, Y., Y. Liang, K. Tanaka, C. T. Nguyen, R. P. Jedrzejczak, A. Joachimiak et al. 2014. "The Kinase LYK5 Is a Major Chitin Receptor in Arabidopsis and Forms a Chitin-Induced Complex with Related Kinase CERK1." *ELife* 3: 1–19. doi:10.7554/eLife.03766.

Carrión, V. J., J. Perez-Jaramillo, V. Cordovez, V. Tracanna, M. de Hollander, D. Ruiz-Buck et al. 2019. "Pathogen-Induced Activation of Disease-Suppressive Functions in the Endophytic Root Microbiome." *Science (New York, N.Y.)* 366 (6465): 606–12. doi:10.1126/science.aaw9285.

Chen, Y., J. Wang, N. Yang, Z. Wen, X. Sun, Y. Chai et al. 2018. "Wheat Microbiome Bacteria Can Reduce Virulence of a Plant Pathogenic Fungus by Altering Histone Acetylation." *Nature Communications* 9 (1): 1–14. doi:10.1038/s41467-018-05683-7.

Chinchilla, D., C. Zipfel, S. Robatzek, B. Kemmerling, T. Nürnberger, J. D. G. Jones et al. 2007. "A Flagellin-Induced Complex of the Receptor FLS2 and BAK1 Initiates Plant Defence." *Nature* 448 (7152): 497–500. doi:10.1038/nature05999.

Cho, I., and M. J. Blaser. 2012. "The Human Microbiome: At the Interface of Health and Disease." *Nature Reviews Genetics* 13 (4): 260–70. doi:10.1038/nrg3182.

Collemare, J., R. O'Connell, and M-H. Lebrun. 2019. "Nonproteinaceous Effectors: The Terra Incognita of Plant–Fungal Interactions." *New Phytologist* 223 (2): 590–96. doi:10.1111/nph.15785.

Conrath, U., G. J. M. Beckers, C. J. G. Langenbach, and M. R. Jaskiewicz. 2015. "Priming for Enhanced Defense." *Annual Review of Phytopathology* 53 (1): 97–119. doi:10.1146/annurev-phyto-080614-120132.

Copeland, J. K., L. Yuan, M. Layeghifard, P. W. Wang, and D. S. Guttman. 2015. "Seasonal Community Succession of the Phyllosphere Microbiome." *Molecular Plant-Microbe Interactions* 28 (3): 274–85. doi:10.1094/MPMI-10-14-0331-FI.

Cui, H., K. Tsuda, and J. E. Parker. 2015. "Effector-Triggered Immunity: From Pathogen Perception to Robust Defense." *Annual Review of Plant Biology* 66 (1): 487–511. doi:10.1146/annurev-arplant-050213-040012.

Daverdin, G., T. Rouxel, L. Gout, J. N. Aubertot, I. Fudal, M. Meyer et al. 2012. "Genome Structure and Reproductive Behaviour Influence the Evolutionary Potential of a Fungal Phytopathogen." *PLoS Pathogens* 8 (11). doi:10.1371/journal.ppat.1003020.

Delmotte, N., C. Knief, S. Chaffron, G. Innerebner, B. Roschitzki, R. Schlapbach et al. 2009. "Community Proteogenomics Reveals Insights into the Physiology of Phyllosphere Bacteria." *Proceedings of the National Academy of Sciences of the United States of America* 106 (38): 16428–33. doi:10.1073/pnas.0905240106.

Dodds, P. N., and J. P. Rathjen. 2010. "Plant Immunity: Towards an Integrated View of Plant–Pathogen Interactions." *Nature Reviews Genetics* 11 (8): 539–48. doi:10.1038/nrg2812.

Does, H. C. van der, and M. Rep. 2017. "Adaptation to the Host Environment by Plant-Pathogenic Fungi." *Annual Review of Phytopathology* 55 (1): 427–50. doi:10.1146/annurev-phyto-080516-035551.

Dong, S., and Y. Wang. 2016. "Nudix Effectors: A Common Weapon in the Arsenal of Plant Pathogens." *PLOS Pathogens* 12 (8): e1005704.

Dupont, P. Y., C. J. Eaton, J. J. Wargent, S. Fechtner, P. Solomon, J. Schmid et al. 2015. "Fungal Endophyte Infection of Ryegrass Reprograms Host Metabolism and Alters Development." *New Phytologist* 208 (4): 1227–40. doi:10.1111/nph.13614.

Durrant, W. E., and X. Dong. 2004. "Systemic Acquired Resistance." *Annual Review of Phytopathology* 42 (1): 185–209. doi:10.1146/annurev.phyto.42.040803.140421.

Ezenwa, V. O., N. M. Gerardo, D. W. Inouye, M. Medina, and J. B. Xavier. 2012. "Animal Behavior and the Microbiome." *Science* 338 (6104): 198–99. doi:10.1126/science.1227412.

Feurtey, A., D. M. Stevens, W. Stephan, and E. H. Stukenbrock. 2019. "Interspecific Gene Exchange Introduces High Genetic Variability in Crop Pathogen." *Genome Biology and Evolution.* 11 (11): 3095–3105. doi:10.1093/gbe/evz224.

Flor, H. H. 1971. "Current Status of the Gene-for-Gene Concept." *Annual Review of Phytopathology* 9: 275–96.

Fones, H., and S. Gurr. 2015. "The Impact of Septoria Tritici Blotch Disease on Wheat: An EU Perspective." *Fungal Genetics and Biology* 79: 3–7. doi:10.1016/j.fgb.2015.04.004.

Franceschetti, M., A. Maqbool, M. J Jimenez-Dalmaroni, H. G. Pennington, S. Kamoun, and M. J Banfield. 2017. "Effectors of Filamentous Plant Pathogens: Commonalities amid Diversity." *Microbiology and Molecular Biology Reviews : MMBR* 81 (2). doi:10.1128/MMBR.00066-16.

Fritsch, M. J., K. Trunk, J. A. Diniz, M. Guo, M. Trost, and S. J. Coulthurst. 2013. "Proteomic Identification of Novel Secreted Antibacterial Toxins of the Serratia Marcescens Type VI Secretion System." *Molecular & Cellular Proteomics* 12 (10): 2735 LP – 2749. doi:10.1074/mcp.M113.030502.

Fürnkranz, M., W. Wanek, A. Richter, G. Abell, F. Rasche, and A. Sessitsch. 2008. "Nitrogen Fixation by Phyllosphere Bacteria Associated with Higher Plants and Their Colonizing Epiphytes of a Tropical Lowland Rainforest of Costa Rica." *ISME Journal* 2 (5): 561–70. doi:10.1038/ismej.2008.14.

Gómez-Gómez, L., and T. Boller. 2000. "FLS2: An LRR Receptor-like Kinase Involved in the Perception of the Bacterial Elicitor Flagellin in Arabidopsis." *Molecular Cell* 5 (6): 1003–11. doi:10.1016/S1097-2765(00)80265-8.

Grady, K. L., J. W. Sorensen, N. Stopnisek, J. Guittar, and A. Shade. 2019. "Assembly and Seasonality of Core Phyllosphere Microbiota on Perennial Biofuel Crops." *Nature Communications* 10 (1): 1–10. doi:10.1038/s41467-019-11974-4.

Haas, D., and G. Défago. 2005. "Biological Control of Soil-Borne Pathogens by Fluorescent Pseudomonads." *Nature Reviews Microbiology* 3 (4): 307–19. doi:10.1038/nrmicro1129.

Hacquard, S., S. Spaepen, R. Garrido-Oter, and P. Schulze-Lefert. 2017. "Interplay Between Innate Immunity and the Plant Microbiota." *Annual Review of Phytopathology* 55 (1): 565–89. doi:10.1146/annurev-phyto-080516-035623.

Hassani, M. A., P. Durán, and S. Hacquard. 2018. "Microbial Interactions within the Plant Holobiont." *Microbiome* 6 (1): 58. doi:10.1186/s40168-018-0445-0.

Hassani, M. A., E. Özkurt, H. Seybold, T. Dagan, and E. H. Stukenbrock. 2019. "Interactions and Coadaptation in Plant Metaorganisms." *Annual Review of Phytopathology* 57 (1): 483–503. doi:10.1146/annurev-phyto-082718-100008.

Haueisen, J., and E. H. Stukenbrock. 2016. "Life Cycle Specialization of Filamentous Pathogens - Colonization and Reproduction in Plant Tissues." *Current Opinion in Microbiology.* 32:31–7. doi:10.1016/j.mib.2016.04.015.

Hood, R. D., P. Singh, F. Hsu, T. Guvener, M. A. Carl, R. R. S. Trinidad et al. 2010. "A Type VI Secretion System of Pseudomonas Aeruginosa Targets a Toxin to Bacteria." *Cell Host & Microbe* 7 (1): 25–37. doi:10.1016/j.chom.2009.12.007.

Horton, M. W., N. Bodenhausen, K. Beilsmith, D. Meng, B. D. Muegge, S. Subramanian et al. 2014. "Genome-Wide Association Study of Arabidopsis Thaliana Leaf Microbial Community." *Nature Communications* 5 (May): 1–7. doi:10.1038/ncomms6320.

Irieda, H., Y. Inoue, M. Mori, K. Yamada, Y. Oshikawa, H. Saitoh et al. 2019. "Conserved Fungal Effector Suppresses PAMP-Triggered Immunity by Targeting Plant Immune Kinases." *Proceedings of the National Academy of Sciences* 116 (2): 496–505. doi:10.1073/pnas.1807297116.

Jones, J. D. G., and J. L. Dangl. 2006. "The Plant Immune System." *Nature* 444 (7117): 323–29. doi:10.1038/nature05286.

Jumpponen, A., and K. L. Jones. 2009. "Massively Parallel 454 Sequencing Indicates Hyperdiverse Fungal Communities in Temperate Quercus Macrocarpa Phyllosphere." *New Phytologist* 184 (2): 438–48. doi:10.1111/j.1469-8137.2009.02990.x.

Kauppinen, M., M. Helander, N. Anttila, I. Saloniemi, and K. Saikkonen. 2018. "Epichloë Endophyte Effects on Leaf Blotch Pathogen (Rhynchosporium Sp.) of Tall Fescue (Schedonorus Phoenix) Vary among Grass Origin and Environmental Conditions." *Plant Ecology and Diversity* 11 (5–6): 625–35. doi:10.1080/17550874.2019.1613451.

Kema, G. H. J., A. M. Gohari, L. Aouini, H. A. Y. Gibriel, S. B. Ware, F. Van Den Bosch et al. 2018. "Stress and Sexual Reproduction Affect the Dynamics of the Wheat Pathogen Effector AvrStb6 and Strobilurin Resistance." *Nature Genetics* 50 (3): 375–80. doi:10.1038/s41588-018-0052-9.

Kettles, G. J., C. Bayon, C. A. Sparks, G. Canning, K. Kanyuka, and J. J. Rudd. 2018. "Characterization of an Antimicrobial and Phytotoxic Ribonuclease Secreted by the Fungal Wheat Pathogen Zymoseptoria Tritici." *New Phytologist* 217 (1): 320–31. doi:10.1111/nph.14786.

Laine, A. L. 2004. "A Powdery Mildew Infection on a Shared Host Plant Affects the Dynamics of the Glanville Fritillary Butterfly Populations." *Oikos* 107 (2): 329–37. doi:10.1111/j.0030-1299.2004.12990.x.

Li, L., P. Kim, L. Yu, G. Cai, S. Chen, J. R. Alfano et al. 2016. "Activation-Dependent Destruction of a Co-Receptor by a Pseudomonas Syringae Effector Dampens Plant Immunity." *Cell Host & Microbe* 20 (4): 504–14. doi:10.1016/j.chom.2016.09.007.

Liu, T., Z. Liu, C. Song, Y. Hu, Z. Han, Ji She et al. 2012. "Chitin-Induced Dimerization Activates a Plant Immune Receptor." *Science* 336 (6085): 1160–64. doi:10.1126/science.1218867.

van Loon, L. C., M. Rep, and C. M. J. Pieterse. 2006. "Significance of Inducible Defense-Related Proteins in Infected Plants." *Annual Review of Phytopathology* 44 (1): 135–62. doi:10.1146/annurev.phyto.44.070505.143425.

Lundberg, D., S. Lebeis, S. H. Paredes, S. Yourstone, J. Gehring, S. Malfatti et al. 2012. "Defining the Core Arabidopsis Thaliana Root Microbiome." *Nature.* doi:10.1038/nature11237.

Macho, A. P., and C. Zipfel. 2014. "Plant PRRs and the Activation of Innate Immune Signaling." *Molecular Cell* 54 (2): 263–72. doi:10.1016/j.molcel.2014.03.028.

Martin, F. M., S. Uroz, and D. G. Barker. 2017. "Ancestral Alliances: Plant Mutualistic Symbioses with Fungi and Bacteria." *Science* 356 (6340). doi:10.1126/science.aad4501.

McCann, H. C., H. Nahal, S. Thakur, and D. S. Guttman. 2012. "Identification of Innate Immunity Elicitors Using Molecular Signatures of Natural Selection." *Proceedings of the National Academy of Sciences of the United States of America* 109 (11): 4215–20. doi:10.1073/pnas.1113893109.

McMullan, M., A. Gardiner, K. Bailey, E. Kemen, B. J. Ward, V. Cevik et al. 2015. "Evidence for Suppression of Immunity as a Driver for Genomic Introgressions and Host Range Expansion in Races of Albugo Candida, a Generalist Parasite." *ELife* 4: 1–24. doi:10.7554/eLife.04550.

Menardo, F., C. Praz, S. Wyder, S. A. Bourras, K. E. McNally, F. Parlange et al. 2016. "Hybridization of Powdery Mildew Strains Gives Raise to Pathogens on Novel Agricultural Crop Species." *Nature Genetics* 48 (January): 1–24. doi:10.1038/ng.3485.

Miya, A., P. Albert, T. Shinya, Y. Desaki, K. Ichimura, K. Shirasu et al. 2007. "CERK1, a LysM Receptor Kinase, Is Essential for Chitin Elicitor Signaling in Arabidopsis." *Proceedings of the National Academy of Sciences of the United States of America* 104 (49): 19613–8. doi:10.1073/pnas.0705147104.

Möller, M., and E. H. Stukenbrock. 2017. "Evolution and Genome Architecture in Fungal Plant Pathogens." *Nature Reviews Microbiology* 15 (12):756–71. doi:10.1038/nrmicro.2017.76.

Müller, D. B., C. Vogel, Y. Bai, and J. A. Vorholt. 2016. "The Plant Microbiota: Systems-Level Insights and Perspectives." *Annual Review of Genetics* 50 (1): 211–34. doi:10.1146/annurev-genet-120215-034952.

Nishimura, M. T., and J. L. Dangl. 2010. "Arabidopsis and the Plant Immune System." *Plant Journal* 61 (6): 1053–66. doi:10.1111/j.1365-313X.2010.04131.x.

Özkurt, E., M. A. Hassani, U. Sesiz, S. Künzel, T. Dagan, H. Özkan et al. 2019. "Higher Stochasticity of Microbiota Composition in Seedlings of Domesticated Wheat Compared to Wild Wheat." *BioRxiv* 54 (1): 685164. doi:10.1101/685164.

Peñuelas, J., and J. Terradas. 2014. "The Foliar Microbiome." *Trends in Plant Science* 19 (5): 278–80. doi:10.1016/j.tplants.2013.12.007.

Pieterse, C. M. J., C. Zamioudis, R. L. Berendsen, D. M. Weller, S. C. M. Van Wees, and P. A. H. M. Bakker. 2014. "Induced Systemic Resistance by Beneficial Microbes." *Annual Review of Phytopathology* 52 (1): 347–75. doi:10.1146/annurev-phyto-082712-102340.

Presti, L. Lo, D. Lanver, G. Schweizer, S. Tanaka, L. Liang, M. Tollot et al. 2015. "Fungal Effectors and Plant Susceptibility." *Annual Review of Plant Biology* 66: 513–45. doi:10.1146/annurev-arplant-043014-114623.

Remus-Emsermann, M. N. P., S. Lücker, D. B. Müller, E. Potthoff, H. Daims, and J. A. Vorholt. 2014. "Spatial Distribution Analyses of Natural Phyllosphere-Colonizing Bacteria on Arabidopsis Thaliana Revealed by Fluorescence in Situ Hybridization." *Environmental Microbiology* 16 (7): 2329–40. doi:10.1111/1462-2920.12482.

Robatzek, S., P. Bittel, D. Chinchilla, P. Köchner, G. Felix, S. H. Shiu et al. 2007. "Molecular Identification and Characterization of the Tomato Flagellin Receptor LeFLS2, an Orthologue of Arabidopsis FLS2 Exhibiting Characteristically Different Perception Specificities." *Plant Molecular Biology* 64 (5): 539–47. doi:10.1007/s11103-007-9173-8.

Rovenich, H., J. C. Boshoven, and B. P. H. J. Thomma. 2014. "Filamentous Pathogen Effector Functions: Of Pathogens, Hosts and Microbiomes." *Current Opinion in Plant Biology* 20. Elsevier Ltd: 96–103. doi:10.1016/j.pbi.2014.05.001.

Saintenac, C., W.-S. Lee, F. Cambon, J. J. Rudd, R. C. King, W. Marande et al. 2018. "Wheat Receptor-Kinase-like Protein Stb6 Controls Gene-for-Gene Resistance to Fungal Pathogen Zymoseptoria Tritici." *Nature Genetics*. 50 (3): 368–74. doi:10.1038/s41588-018-0051-x.

Savary, S., L. Willocquet, S. J. Pethybridge, P. Esker, N. McRoberts, and A. Nelson. 2019. "The Global Burden of Pathogens and Pests on Major Food Crops." *Nature Ecology and Evolution* 3 (3): 430–9. doi:10.1038/s41559-018-0793-y.

Schardl, C. L. 1996. "EPICHLOE SPECIES: Fungal Symbionts of Grasses." *Annu. Rev. Phytopathol* 34 (67): 109–30. doi:10.1146/annurev.phyto.34.1.109.

Schardl, C. L., A. Leuchtmann, and M. J. Spiering. 2004. "Symbioses of Grasses with Seedborne Fungal Endophytes." *Annual Review of Plant Biology* 55 (1): 315–40. doi:10.1146/annurev.arplant.55.031903.141735.

Schardl, C. L., C. A. Young, U. Hesse, S. G. Amyotte, K. Andreeva, P. J. Calie et al. 2013. "Plant-Symbiotic Fungi as Chemical Engineers: Multi-Genome Analysis of the Clavicipitaceae Reveals Dynamics of Alkaloid Loci." *PLoS Genetics* 9 (2): e1003323. doi:10.1371/journal.pgen.1003323.

Schlaeppi, K., N. Dombrowski, R. Garrido Oter, E. Ver Loren Van Themaat, and P. Schulze-Lefert. 2014. "Quantitative Divergence of the Bacterial Root Microbiota in Arabidopsis Thaliana Relatives." *Proceedings of the National Academy of Sciences of the United States of America* 111 (2): 585–92. doi:10.1073/pnas.1321597111.

Scott, B., K. Green, and D. Berry. 2018. "The Fine Balance between Mutualism and Antagonism in the Epichloë Festucae–Grass Symbiotic Interaction." *Current Opinion in Plant Biology* 44: 32–8. doi:10.1016/j.pbi.2018.01.010.

Sender, R., S. Fuchs, and R. Milo. 2016. "Revised Estimates for the Number of Human and Bacteria Cells in the Body." *PLOS Biology* 14 (8): 1–14. doi:10.1371/journal.pbio.1002533.

Seybold, H., T. Demetrowitsch, M. Amine Hassani, S. Szymczak, E. Reim, J. Haueisen et al. 2020. "A Fungal Pathogen Induces Systemic Susceptibility and Systemic Shifts in Wheat Metabolome and Microbiome Composition." *Nature Communications* 11 (1): 1910. doi:10.1038/s41467-020-15633-x.

Shan, L., P. He, J. Li, A. Heese, S. C. Peck, T. Nurnberger et al. 2008. "Bacterial Effectors Target the Common Signaling Partner BAK1 to Disrupt Multiple MAMP Receptor-Signaling Complexes and Impede Plant Immunity." *Cell Host & Microbe* 4 (1). United States: 17–27. doi:10.1016/j.chom.2008.05.017.

Snelders, N. C., G. J. Kettles, J. J. Rudd, and B. P. H. J. Thomma. 2018. "Plant Pathogen Effector Proteins as Manipulators of Host Microbiomes?" *Molecular Plant Pathology* 19 (2): 257–9. doi:10.1111/MPP.12628.

Souza, D. P., G. U. Oka, C. E. Alvarez-Martinez, A. W. Bisson-Filho, G. Dunger, L. Hobeika et al. 2015. "Bacterial Killing via a Type IV Secretion System." *Nature Communications* 6 (March). England: 6453. doi:10.1038/ncomms7453.

Sperschneider, J., P. N. Dodds, D. M. Gardiner, K. B. Singh, and J. M. Taylor. 2018. "Improved Prediction of Fungal Effector Proteins from Secretomes with EffectorP 2.0." *Molecular Plant Pathology* 19 (9): 2094–110. doi:10.1111/mpp.12682.

Stringlis, I. A., K. Yu, K. Feussner, R. De Jonge, S. Van Bentum, M. C. Van Verk et al. 2018. "MYB72-Dependent Coumarin Exudation Shapes Root Microbiome Assembly to Promote Plant Health." *Proceedings of the National Academy of Sciences of the United States of America* 115 (22): E5213–22. doi:10.1073/pnas.1722335115.

Sun, W., F. M. Dunning, C. Pfund, R. Weingarten, and A. F. Bent. 2006. "Within-Species Flagellin Polymorphism in Xanthomonas Campestris Pv Campestris and Its Impact on Elicitation of Arabidopsis Flagellin Sensing2-Dependent Defenses." *Plant Cell* 18 (3): 764–79. doi:10.1105/tpc.105.037648.

Sun, Y., L. Li, A. P. Macho, Z. Han, Z. Hu, C. Zipfel et al. 2013. "Structural Basis for Flg22-Induced Activation of the Arabidopsis FLS2-BAK1 Immune Complex." *Science* 342 (6158): 624–8. doi:10.1126/science.1243825.

Takai, R., A. Isogai, S. Takayama, and F. S. Che. 2008. "Analysis of Flagellin Perception Mediated by Flg22 Receptor OsFLS2 in Rice." *Molecular Plant-Microbe Interactions* 21 (12): 1635–42. doi:10.1094/MPMI-21-12-1635.

Ton, J., J. A. Van Pelt, L. C. Van Loon, and C. M. J. Pieterse. 2002. "Differential Effectiveness of Salicylate-Dependent and Jasmonate/Ethylene-Dependent Induced Resistance in Arabidopsis." *Molecular Plant-Microbe Interactions* 15 (1): 27–34. doi:10.1094/MPMI.2002.15.1.27.

Trunk, K., J. S. Coulthurst, and J. Quinn. 2019. "A New Front in Microbial Warfare—Delivery of Antifungal Effectors by the Type VI Secretion System." *Journal of Fungi* 5 (2). doi:10.3390/jof5020050.

Vacher, C., A. Hampe, A. J. Porté, U. Sauer, S. Compant, and C. E. Morris. 2016. "The Phyllosphere: Microbial Jungle at the Plant–Climate Interface." *Annual Review of Ecology, Evolution, and Systematics* 47 (1): 1–24. doi:10.1146/annurev-ecolsys-121415-032238.

Vandenkoornhuyse, P., A. Quaiser, M. Duhamel, A. Le Van, and A. Dufresne. 2015. "The Importance of the Microbiome of the Plant Holobiont." *New Phytologist* 206 (4): 1196–1206. doi:10.1111/nph.13312.

Vannier, N., M. Agler, and S. Hacquard. 2019. "Microbiota-Mediated Disease Resistance in Plants." *PLoS Pathogens* 15 (6): e1007740. doi:10.1371/journal.ppat.1007740.

Vetter, M., T. L. Karasov, and J. Bergelson. 2016. "Differentiation between MAMP Triggered Defenses in Arabidopsis Thaliana." *PLoS Genetics* 12 (6): 1–18. doi:10.1371/journal.pgen.1006068.

Vetter, M., I. Kronholm, F. He, H. Häweker, M. Reymond, J. Bergelson et al. 2012. "Flagellin Perception Varies Quantitatively in Arabidopsis Thaliana and Its Relatives." *Molecular Biology and Evolution* 29 (6): 1655–67. doi:10.1093/molbev/mss011.

Vorholt, J. A. 2012. "Microbial Life in the Phyllosphere." *Nature Reviews. Microbiology* 10 (12). Nature Publishing Group: 828–40. doi:10.1038/nrmicro2910.

Wang, Y., B. M. Tyler, and Y. Wang. 2019. "Defense and Counterdefense During Plant-Pathogenic Oomycete Infection." *Annual Review of Microbiology* 73 (1): annurev-micro-020518-120022. doi:10.1146/annurev-micro-020518-120022.

Wu, C-F., M. N. M. Santos, S-T. Cho, H-H. Chang, Y-M. Tsai, D. A. Smith et al. 2019. "Plant-Pathogenic Agrobacterium Tumefaciens Strains Have Diverse Type VI Effector-Immunity Pairs and Vary in In-Planta Competitiveness." *Molecular Plant-Microbe Interactions* 32 (8). Scientific Societies: 961–71. doi:10.1094/MPMI-01-19-0021-R.

Yang, S., F. Tang, M. Gao, H. B. Krishnan, and H. Zhu. 2010. "R Gene-Controlled Host Specificity in the Legume-Rhizobia Symbiosis." *Proceedings of the National Academy of Sciences of the United States of America* 107 (43): 18735–40. doi:10.1073/pnas.1011957107.

Yuan, J., J. Zhao, T. Wen, M. Zhao, R. Li, P. Goossens et al. 2018. "Root Exudates Drive the Soil-Borne Legacy of Aboveground Pathogen Infection." *Microbiome* 6 (1). Microbiome: 1–12. doi:10.1186/s40168-018-0537-x.

Zarraonaindia, I., S. M. Owens, P. Weisenhorn, K. West, J. Hampton-Marcell, S. Lax et al. 2015. "The Soil Microbiome Influences Grapevine-Associated Microbiota." *MBio* 6 (2): 1–10. doi:10.1128/mBio.02527-14.

Zhong, Z., T. C. Marcel, F. E. Hartmann, X. Ma, C. Plissonneau, M. Zala et al. 2017. "A Small Secreted Protein in Zymoseptoria Tritici Is Responsible for Avirulence on Wheat Cultivars Carrying the Stb6 Resistance Gene." *New Phytologist* 214 (2): 619–31. doi:10.1111/nph.14434.

Zhou, J., S. Wu, X. Chen, C. Liu, J. Sheen, L. Shan et al. 2014. "The Pseudomonas Syringae Effector HopF2 Suppresses Arabidopsis Immunity by Targeting BAK1." *The Plant Journal: For Cell and Molecular Biology* 77 (2). England: 235–45. doi:10.1111/tpj.12381.

Zipfel, C., S. Robatzek, L. Navarro, E. J. Oakeley, J. D. G. Jones, G. Felix et al. 2004. "Bacterial Disease Resistance in Arabidopsis through Flagellin Perception." *Nature* 428 (6984): 764–7. doi:10.1038/nature02485.

17 Cellular dialogues between hosts and microbial symbionts
Generalities emerging

Michael G. Hadfield and Thomas C. G. Bosch

Contents

Microbes are the oldest and, at the same time, most abundant living organisms, having existed for almost four billion years. Although eukaryotic protists, the pioneer metazoans, and the earliest plants were surely around much earlier, it was not until about 550–600 million years ago that the massive evolution of multicellular animal and plant species occurred. It is perhaps unsurprising that microbes remain the dominant life form on our planet, since the explosion of multicellular life provided a huge and abundant diversity of environments for bacteria to colonize. Microbes display a wide diversity of specialized interactions with eukaryotic organisms. As documented in the chapters of this book, research on microbial communities inhabiting protists, animals, and plants has progressed at a spectacular rate over the past few decades. This progress is due in large part to the application of modern technologies, including "metagenomic" methods, that allow studies of the biology of symbiosis using reductionistic approaches. This research has yielded catalogues of thousands of microbial species, many previously unknown and belonging to all three domains of life, as well as lists of millions of microbial genes. Research on host–microbe interactions has also begun to uncover the biochemical architecture of the language used between the different members of the holobiont. Studying them in the dual contexts of complex free-living microbial communities and microbial interactions with protist, plant, animal, and human hosts using metagenomic tools is changing the way we view all of biological function, diversity, and evolution.

Although we are accustomed to thinking about symbioses as associations that tie microbes intimately to the surfaces or insides of protists and multicellular organisms, in many marine invertebrates completion of the life cycle and progression from pelagic larva to bottom-dwelling juvenile depends on cues from specific biofilm-residing bacteria, including lipopolysaccharides and flagellar proteins (Chapter 1). The Chapter 1 authors address the important question, is this symbiosis? At what point in macroorganism–microorganism dependence can we consider the relationship obligate, and might a dependence of an animal on cues from specific bacteria for completing its developmental program be an evolutionary "first step" toward a more complex relationship? We anticipate that application of metagenomic approaches to the study of marine and freshwater biofilms and, especially, identification of their resident bacteria will significantly expand such studies.

Well established symbioses take many forms, not all of which could be reviewed in this volume. The modern emphasis focuses on mutualisms, distinguishing them from parasitism, although the nature of chemical interactions is not necessarily different between the two. These relationships may well be portrayed as "power struggles in time and space" (Chapter 2). The participants in the battle are manifold and include not only host–bacteria interactions, but also more complicated associations such as the tripartite interactions of sponge cells, bacteria, and bacteriophages (Chapter 4); or between Hydra cells, bacteria, and an algae photobiont (Chapter 5). Multipartite microbial interactions take place as well, such as those in corals (Chapter 7), vertebrates (Chapter 15), and in the plant phyllosphere, including symbiotic epiphytic bacteria and ascomycete fungi (Chapter 16). The basic search for commonalities across this diversity is frustrating, perhaps because they don't exist. However, what we do see is both enlightening and hugely complex (e.g., Chapter 15, on a genomics analysis to understand the complex, multi-microbe symbioses in vertebrate guts), making an in-depth, comparative analysis difficult.

One of the most crucial questions in all symbioses systems described in this volume is asked by Hadfield and McFall-Ngai (Chapter 3) when reflecting on Trichoplax–symbiont interactions: *"What maintains the balance? What keeps the host from killing its symbionts and the symbionts from overwhelming the cells and tissues they reside in?"* Surely the answer lies in a precise communication between the partners, regardless of how many are involved. As noted in the Preface, discovering the "language of symbiosis," if there is one, should expose the common evolutionary roots of symbiosis, if they exist. Judging from many of the contributions in this volume, if anything can be derived from the search across symbiotic life for a "common language," it is the involvement of elements of an innate immune system, mentioned in eleven of the chapters and covering plants and most animal groups. Eukaryotic organisms arose via symbiotic interactions in seas teeming with viruses, bacteria and archaea, many already obligately living on or within each other. A major concern of the first eukaryotes in a sea of nutrient-seeking bacteria was how to keep themselves from becoming living nutrient sources. The development of defensive mechanisms encoded in their genomes must have been an early evolutionary event. These mechanisms, in the form of innate immune systems, surely underlaid the evolution of many symbiosis and are still present today. A role for innate immunity is described for most of the organisms discussed in this volume.

Perhaps best characterized here in vertebrates (Chapter 14), and previously thought to exist only in cnidarians and bilaterians, genetic elements of innate immunity have been described in the last two years in choanoflagellate protists (Richter et al. 2018), placozoans (Kamm et al. 2019), sponges (see Chapter 4), and ctenophores (Taylor-Knowles et al. 2019). Today, there is abundant experimental evidence that the innate immune system in invertebrates, as well as in vertebrates, is crucial for establishing and maintaining host–symbiont homeostasis. Similarly, in plants the immune system plays a crucial role in controlling host–microbe interactions (Chapter 16). A major function of the innate immune system is the detection of specific bacterial surface molecules and secretion of antimicrobial peptides. Toll-like receptors are the uniting element here, and they are known now from every major animal phylum. Pattern-recognition receptors such as the intracellular NLR proteins are considered major pillars of the plant "microbial management system" (Chapter 16). Haueisen et al. (Chapter 16) also suggest here that to answer an important question in any symbiosis research — e.g., how do plants distinguish between beneficial and antagonistic microbes?— a closer look at microbe-produced effector molecules will be useful.

Among many invertebrates, including green Hydra and corals, microbes serve as important sources of metabolites and supplements towards upgrading their host´s nutritional ecology (Chapter 5, 6, and 8). While comparable in their roles of indispensable nutrient supply to their hosts, mainly indispensable amino acids and vitamins, microbial symbionts in many insects are remarkable for having lost almost all elements of their genomes except for those involved in genome inheritance and the production of indispensable amino acids and vitamins indispensable for host nutrition. As noted by Bennett (Chapter 11), in sap-sucking insects the intracellular bacterial symbionts are little more than self-replicating organelles; they do not and cannot exist outside their hosts. These indentured slaves appear to be unique to insects in an evolutionary sense. Nothing described in other chapters of this volume suggest evolutionary kinship with these intracellular bacterial remnants. It is apparent that a dialogue between these insects and their bacterial slaves has been reduced to little more than nutrient exchange. The bacteria are bound to their hosts and passed down strictly by vertical transmission mechanisms. We might conclude that they are at an evolutionary dead end.

Things are different in the guts of many insects, where symbiosis with specific extracellular bacteria, horizontally obtained from the environment, are common. As an intriguing example, Berasategui and Salem (Chapter 13) show that leaf-feeding behavior in insects is intimately dependent on a microbiome which is upgrading the nutritional ecology. One even more extreme form of nutritional symbiosis is found in the *Nardonella*–weevil endosymbiotic system, where the symbionts drastically affect the host phenotype by providing tyrosine which is critical for the beetle's cuticle formation (Chapter 12). Significantly, Bennett reports that recognition factors involving innate immunity are involved in the relationships between at least some insects and their intestinal bacteria (Chapter 11).

How do hosts communicate with their symbionts and symbionts with their host cells? The model where this language allowing partner recognition, specificity determination, colonization, and persistence is best characterized is the binary symbiosis of the bobtail squid and its bioluminescent symbiont *Vibrio fischeri* (Chapter 10). Establishing the symbiosis by each new generation involves major participation of innate-immune elements. In the freshwater polyp *Hydra,* we see an animal host that can directly interfere with the bacterial communication system by modifying bacterial acyl-homoserine lactones (AHL). Thus, quorum-sensing molecules seem to be involved in the interaction between the *Hydra* host and its bacterial symbiont *Curvibacter* (Chapter 6). Symbiont-produced molecules such as bryostatin targeting eukaryotic protein kinase C (PKC) (Chapter 9) may be another crucial mechanism involved in maintaining the homeostasis between the bacterial symbiont and the host. However, since not all bryozoan hosts are colonized by bryostatin producing symbionts, the interaction seems to be context dependent and only effective under certain environmental conditions. Examination of more bryozoans with regard to potential symbioses is definitely warranted. Using the nematode *C. elegans*, Choi et al. (Chapter 8) show that TGF-beta signaling can regulate the host–symbiont homeostasis. Comparative genomic analyses of host–microbe interactions in related nematodes promise to reveal much more about the natural history of animal–microbe symbioses (Chapter 15).

It is clear from the discussion presented by Lickwar and Rawls (Chapter 15) that the relationships between gut-dwelling bacteria in vertebrates have either evolved to uniquely complex gene-level interactions, or that the symbioses in the guts of nonvertebrates await far greater exploration at the level of the genome. In the guts of mice and zebrafish, products from each partner impact the other at the level of genes, transcription, cis-regulatory regions, and translational products. Yet, Lickwar and Rawls note, "Identification of symbiosis signals remains a critically important challenge in discerning the mechanisms and ecologies that underlie symbioses."

Although the chapters presented here provide ample evidence that many "host" protists, plants, and animals employ elements of innate immunity to recognize surface molecules on their symbiont bacteria, none address the question, do the microbes recognize very specific hosts? Surely, a dialogue should work in both directions. While it is well known in human medicine that some pathogenic bacteria target host cell-type specific surface glycoproteins for attachment, and that when legumes experience low nitrogen, they synthesize flavonoids that induce rhizobia to secrete nod factors leading to root infection, solid data are lacking for the vast majority of symbioses discussed here. That is, half the dialogue is missing.

Symbiosis is a widespread phenomenon that has obviously driven the evolution of important developmental impulses. To fully understand symbiotic interactions, and to uncover the role of the microbiome in evolution, it will now be important to integrate experimental results from different symbiosis systems across the tree of life. Indeed, it is becoming clear that many of the topics currently debated raise problems that call for tight collaboration between macro- and microbiologists, as well as other members of the natural sciences, to understand the fundamental drivers of symbiosis initiation and maintenance. New knowledge of the diversity of host–microbe interactions will provide insights into how animals communicate with their symbionts and how this affects development, evolution, and absence of disease. This CRC book brought together molecular and biochemical researchers, zoologists and botanists, ecologists, and bioinformaticians on multiple topics related to the microbiome. By presenting findings on the dialogue between hosts and their symbionts across species, we consider the volume per se an important step in concretizing this essential interdisciplinary dialogue.

References

Kamm, K., B. Schierwater, and R. DeSalle. 2019. Innate immunity in the simplest animals – placozoans. *BMC Genomics* 20:5. Doi: org/10.1186/s12864-018-5337-3

Richter, D. J., P. Fozouni, M. B. Eisen, and N. King. 2018. Gene family innovation, conservation and loss on the animal stem lineage. *eLIFE* 2018;7e34226. Doi: org/10.7554/eLife.34226

Taylor-Knowles, N., L. E. Vandepas, and W. E. Browne. 2019. Still enigmatic: Innate immunity in the ctenophore *Mnemiopsis leidyi*. *Integrative and Comparative Biology*, 59: 811–818. doi:10.1093/icb/icz116

Index

A

Acanthamoeba, 20
   A. polyphaga, 20, 21
Accessory nidamental gland (ANG), 153–154
Acromyrmex echinatior, 221
Acropora, 103
   A. humilis, 93
Acyl-homoserine lactones (AHL), 84, 289
Acyrthosiphon pisum, 2, 176
Adomerus spp., 210
Aerobic respiration, 222
Aeromonas immune modulator A (AimA), 236
AHL, *see* Acyl-homoserine lactones
AHLs, *see* N-acyl homoserine lactones
AI, *see* Autoinducers
AimA, *see* *Aeromonas* immune modulator A
Akkermansia muciniphila, 122
Albugo laibachii, 269
Algoriphagus machipongonensis, 3
Aliivibrio fischeri, 154
Alkaline phosphatase (AP), 238
Alphaproteobacteria, 56
Alteromonadaceae, 99
Alteromonas-affiliated sequences, 101
Alteromonas, 101
Ammonia-oxidizing archaea (AOA), 104
Amoeba–Legionella symbiosis, 19
Amoeba proteus, 18
Amphibalanus amphitrite, 5
Amphimedon queenslandica, 52–53
AMPs, *see* Anti-microbial peptides
Anaerobiosis, 26
ANG, *see* Accessory nidamental gland
Angiopoetin-like 4 (*Angptl4*), 255
   microbial suppression of, 255–256
Animal–microbe interactions, 49
Ankyphages aid symbionts in immune evasion, 58–59
Anti-microbial peptides (AMPs), 71–73, 123, 248
Anticarsia gemmatalis, 224
AOA, *see* Ammonia-oxidizing archaea
AP, *see* Alkaline phosphatase
ApGLNT1, 189
Aphidoidea, 176
Aplysina aerophoba, 52
Arabidopsis
   A. thaliana, 269–271
   pathogen infections, 274
Archaea associated with coral holobiont, 103–104
Archaeocytes, 50

Archeocytes, 53
Arcobacter, 27
Aster Yellows (AY), 224
Atta cephalotes, 221
Autoinducers (AI), 84
AY, *see* Aster Yellows

B

Bacillus aquimaris, 8
Bacteria–bacteria dialogue, 55
   QQ, 57
   QS, 56–57
Bacteria(l), 3, 84, 98
   cell densities, 50
   effect on host physiology, 74
   light organs, 153
   resident, 235
   species, 201
Bacterially induced metamorphosis
   in *Hydroides elegans*, 5–6
   of marine invertebrate animals, 4–5
Bacterial symbionts, 174
   of placozoans transmitted between
     generations, 44–45
Balanus amphitrite, see *Amphibalanus
     amphitrite*
Banana aphid (*Pentalonia nigronervosa*), 187
Baumannia, 187
BCAT, *see* Branched-chain amino acid
     aminotransferase gene
BefA, *see* Beta Cell Expansion Factor A
Bemisia tabaci, see Whitefly
Bemisia tabaci, 177
Beta Cell Expansion Factor A (BefA), 239–240
Black hard weevil (*Pachyrhynchus infernalis*), 203
   Nardonella contribution to adult cuticle
     formation in, 207
   Nardonella endosymbiotic system in, 204
Bradyrhizobia, 101
Bradyrhizobium japonicum, 274
Branched-chain amino acid aminotransferase
     gene (BCAT), 189
Bryopsis, 105
Bryostatins, 139–140
   bryostatin 1, 139
   defensive role of, 141–143
   production by bacterial symbiont of
     B. neritina, 140–141
   and symbionts in closely related genera,
     144–145

T - #0119 - 111024 - C314 - 234/156/15 - PB - 9780367513757 - Gloss Lamination